ENERGY
and Human Needs

ENERGY and Human Needs

SAMUEL C. CURRAN
M.A., B.Sc., Ph.D., D.Sc. (Glas.), Ph.D. (Cantab.), F.R.S.
University of Strathclyde

and

JOHN S. CURRAN
B.Sc. (Glas.), D.Phil. (Oxford)
EURATOM Centre, Ispra, Italy

A HALSTED PRESS BOOK

JOHN WILEY & SONS
NEW YORK TORONTO

Published in the U.S.A. and
Canada by Halsted Press, a
Division of John Wiley & Sons Inc., New York

Library of Congress Cataloging in Publication Data
Curran, Samuel Crowe, Sir.
Energy and Human needs.
"A Halsted Press Book."
Includes index.
1. Power resources. 2. Power (Mechanics).
3. Human ecology. I. Curran, John S., joint author.
II. Title.
TJ163.2.C86 1979 333.7 79-16728
ISBN 0-470-26818-2

Printed in Great Britain by
R. & R. Clark Ltd., Edinburgh

Contents

PART TWO
FOSSIL FUELS AND THEIR DEPLETION

PART THREE
THE NUCLEAR OPTION

Part Four
The Provision of Energy

PART FIVE
SOCIAL, ENVIRONMENTAL, AND INTERNATIONAL ASPECTS

Abbreviated Names of Journals used in the Text

A. Rev. Energy	Annual Review of Energy
Appl. Energy	Applied Energy
Appl. Ergonom.	Applied Ergonomics
B.S.E.	British Services Engineer
Chem. Econ. Eng. Rev.	Chemical Economy and Engineering Review
Electron. & Power	Electronics and Power
Energy Res.	Energy Research Reports
Exp. Agric.	Experimental Agriculture
Geol. Soc. Misc. Pap.	Geological Society [of London] Miscellaneous Papers
Heat. Piping Air Cond.	Heating, Piping and Air-Conditioning
Heat. Vent. Eng.	Heating and Ventilation Engineer
J. Br. Nucl. Energy Soc.	Journal of the British Nuclear Energy Society
New Sci.	New Scientist
Nucl. Eng. Int.	Nuclear Engineering International
Phil. Trans. R. Soc.	Philosophical Transactions of the Royal Society
Phys. Bull.	Physics Bulletin
Phys. Technol.	Physics in Technology
Proc. Inst. Mech. Eng.	Proceedings of the Institution of Mechanical Engineers
Sci. Am.	Scientific American
Trans. NE. Cst Instn Engrs Shipbldrs	Transactions of the North-East Coast Institution of Engineers and Shipbuilders
US Geol. Surv. Bull.	United States Geological Survey Bulletin

List of Abbreviations, Other than of Journal Names, used in the Text

ADP	Adenosine Diphosphate
AEI	Associated Electrical Industries
AERE	Atomic Energy Research Establishment
AGR	Advanced Gas-cooled Reactor
ATP	Adenosine Triphosphate
BEEP	Building Energy Estimating Programme
BRE	Building Research Establishment
BRS	Building Research Station
BWR	Boiling Water Reactor
CANDU	Canadian Deuterium Uranium Reactor
CEGB	Central Electricity Generating Board
CERL	Central Electricity Research Laboratories
CERN	Centre Européen pour Recherche Nucléaire
CFR	Commercial Fast Reactor
CIB	Congrès International de Bâtiment
DFR	Dounreay Fast Reactor
ECRC	Electricity Council Research Centre (Capenhurst)
EPRI	Electric Power Research Institute
ERDA	Energy Research and Development Agency
ERW	Enhanced Radiation Weapon
EUR	Euratom (European Atomic Energy Community)
FBR	Fast Breeder Reactor
HTR	High Temperature Reactor
IAEA	International Atomic Energy Agency
IBM	International Business Machines
ICI	Imperial Chemical Industries
IEA	International Energy Agency
IIASA	International Institute for Applied Systems Analysis

JET	Joint European Torus
LNG	Liquid Natural Gas
LWR	Light Water Reactor
NASA	National Aeronautics and Space Agency
NEA	Nuclear Energy Agency
NRPB	Natural Radiological Protection Board
NRX	National Research Experimental (Canadian Reactor)
NSF	National Science Foundation (Washington DC)
OECD	Organization for Economic Co-operation and Development
OPEC	Organization of Petroleum Exporting Countries
PFR	Prototype Fast Reactor
PWR	Pressurized Water Reactor
RCA	Radio Corporation of America
SGHWR	Steam Generating Heavy Water Reactor
SNG	Synthetic Natural Gas
TNT	Trinitrotoluene
UKAEA	United Kingdom Atomic Energy Authority
UNESCO	United Nations Educational Scientific and Cultural Organisation

Scientific Units used in the Text

The Système International d'Unités (SI) has been increasingly adopted in recent times in an effort to standardise scientific units, which, over the previous two centuries, had proliferated to an almost unmanageable degree. Although most research workers now employ the SI system, many of the older units, particularly those of the Centigrade-Gram-Second (CGS) and Foot-Pound-Second (FPS) systems, remain in use to some extent. Below, we list the units normally used in the course of this book. The reader should also refer to fig. 1, which illustrates diagrammatically the relationships between various systems of measurement. As a system, the SI one is more comprehensive than its predecessors and it relies on six defined basic units from which all the others are derived. The units are:

length	metre	(m)
mass	kilogramme	(kg)
time	second	(s)
electric current	ampere	(A)
thermodynamic temperature	kelvin	(K)
luminous intensity	candela	(cd)

Ampere (A)	The basic SI unit of current, approximately equal to a flow of 6×10^{18} electrons per second, or the transfer of one coulomb of charge per second.
Ångström (Å)	A unit of length. One Ångstrom equals 10^{-10} metre.
Bar	The international unit of pressure, 10^6 dyne per cm^2 (=0·987 atmospheres).
British Thermal Unit (btu)	One British Thermal Unit is the quantity of heat necessary to raise the temperature of 1 lb of water through 1 °F (=1055·06 joules or 252 calories).
Calorie (cal)	The basic CGS unit of heat. One calorie is the amount of heat required to raise 1 gram of water from 3·5 °C to 4·5 °C (=4·184 joules). The large calorie, or kilocalorie, is 1,000 calories.
Centimetre (cm)	The basic CGS unit of length.
Curie (Ci)	A measure of radioactivity. One curie is the radiation corresponding to $3{\cdot}7 \times 10^{10}$ disintegrations per second, which is approximately the activity of 1 g of radium.

Electron-Volt (eV)	A unit of energy widely used in atomic and nuclear physics. One electron-volt is the energy gained by an electron falling freely through a potential difference of 1 volt and equals $1{\cdot}60 \times 10^{-12}$ erg.
Erg	The derived CGS unit of work. One erg is the work done by a force of 1 dyne acting over a distance of 1 cm ($=10^{-7}$ joules).
Gramme or Gram (g)	The basic CGS unit of mass.
Hertz (Hz)	The derived SI unit of frequency. One hertz is a rate of 1 cycle per second.
Horse Power (hp)	The British Unit of Power. One horsepower is a rate of working and equals 550 foot-pounds per second, or 745·7 watts.
Joule (J)	The derived SI unit for all forms of energy. One joule is the work done when the point of application of a force of 1 newton moves through 1 metre in the direction of the force; and the work done per second by a current of 1 ampere flowing against a resistance of 1 ohm; also the unit of heat energy.
Mole (mol)	The basic SI unit of substance. One mole is the amount of substance which contains 6×10^{23} elementary units; that is, as many atoms as there are in 0·012 kilogramme of carbon 12.
Newton (N)	The derived SI unit of force. One newton is the force required to give a mass of 1 kilogramme an acceleration of 1 metre per second per second.
Ohm (Ω)	The SI unit of resistance. One ohm is the resistance which allows a current of 1 ampere to flow through a wire when the potential difference between its ends is 1 volt.
Rad	A unit of absorbed dose of radiation, equivalent to ·01 joule per kilogramme, or 100 ergs per gramme, of material. (Contraction of 'Radiation Absorbed Dose'.)
Rem	A unit of absorbed dose which has the same effect as that due to one roentgen of X-rays. (Contraction of 'Roentgen Equivalent Man'.)
Roentgen	A unit of radiation. One roentgen is the amount of X- or gamma-radiation that will produce ions carrying 1 electrostatic unit of electricity in 1 cc of dry air. (One coulomb $= 3 \times 10^{9}$ e.s. units.)

Siemens (S)	The SI unit of electrical conductance. One siemens is the conductance of material that has a resistance of 1 ohm.
Ton	2,240 lb.
Tonne (t)	The metric tonne of 1,000 kilogrammes, equivalent to ·9842 tons.
Torr	A unit of pressure used in defining low pressures. One torr is equivalent to 1 mm of mercury, or 133 newtons per sq. metre.
Volt (V)	The derived SI unit of electric potential, potential difference, and electromotive force. One volt is defined as the potential difference required to make a current of 1 ampere flow through a resistance of 1 ohm. The work expended per second in the resistance is 1 joule.
Watt (W)	The derived SI unit of power. One watt equals 1 joule per second or, in electrical terms, the product of 1 ampere and 1 volt.

Prefixes

nano	n	10^{-9}
micro	μ	10^{-6}
milli	m	10^{-3}
centi	c	10^{-2}
kilo	k	10^{3}
mega	M	10^{6}
giga	G	10^{9}
tera	T	10^{12}

Area and Volume

For the benefit of the non-scientific reader, area and volume are denoted in the text in the old manner, that is by 'sq.' for 'square' and 'cu.' for 'cubic'. In the diagrams, where space is short, we follow scientific practice, using a superscript 2 after the unit of length to denote area; and a superscript 3 after the unit of length to denote volume; for example, m^2 and ft^3.

	—	kg	t	long ton	short ton	lb	—
MASS	1 kilogramme (kg)	1	0·001	0·000984	0·001102	2·20462	—
	1 tonne (t)	1,000	1	0·9841	1·10231	2,204·6	—
	1 long ton (UK)	1,016	1·016	1	1·12	2,240	—
	1 short ton (US)	907·19	0·90719	0·89286	1	2,000	—
	1 pound (lb)	0·45359	$4·5359 \times 10^{-4}$	$4·4643 \times 10^{-4}$	0·0005	1	—
REFERENCE FUELS AND EQUIVALENTS	—	10^6 Btu gross	10^6 kcal gross	boe	ton oe	tce	GJ
	1 barrel oil equivalent (boe)	5·8	1·462	1	0·1349	0·2122	6·12
	1 tonne coal equivalent (tce)	27·337	6·888	4·713	0·6357	1	28·84
	1 ton oil equivalent (ton oe)	43	10·836	7·414	1	1·5730	45·37
	1 ton fuel oil equivalent	41·4	10·433	7·138	0·9628	1·5144	43·68
POWER AND HEAT FLOW	—	W	hp	metric horsepower	kcal/h	Btu/h	—
	1 watt (W)=1 J/s	1	$1·341 \times 10^{-3}$	$1·360 \times 10^{-3}$	0·860	3·412	—
	1 horsepower (hp)	745·7	1	1·014	641·2	2,544	—
	1 metric horsepower	735·5	0·986	1	632·4	2,510	—
	1 kilocalorie per hour (kcal/h)	1·163	$1·560 \times 10^{-3}$	$1·581 \times 10^{-3}$	1	3·968	—
	1 British thermal unit per hour (Btu/h)	0·293	$0·393 \times 10^{-3}$	$0·398 \times 10^{-3}$	0·252	1	—

List of Elements in Order of Atomic Number, together with Atomic Weights

(The Atomic Weights are based on the Isotope C^{12})

Atomic number	Element	Atomic weight	Atomic number	Element	Atomic weight
1	Hydrogen	1·00797	54	Xenon	131·30
2	Helium	4·0026	55	Cesium	132·905
3	Lithium	6·939	56	Barium	137·34
4	Beryllium	9·0122	57	Lanthanum	138·91
5	Boron	10·811	58	Cerium	140·12
6	Carbon	12·01115	59	Praseodymium	140·91
7	Nitrogen	14·0067	60	Neodymium	144·24
8	Oxygen	15·9994	61	Promethium	(145)
9	Fluorine	18·9984	62	Samarium	150·35
10	Neon	20·183	63	Europium	151·96
11	Sodium	22·9898	64	Gadolinium	157·25
12	Magnesium	24·312	65	Terbium	158·92
13	Aluminium	26·9815	66	Dysprosium	162·50
14	Silicon	28·086	67	Holmium	164·93
15	Phosphorus	30·9738	68	Erbium	167·26
16	Sulphur	32·064	69	Thulium	168·93
17	Chlorine	35·453	70	Ytterbium	173·04
18	Argon	39·948	71	Lutetium	174·97
19	Potassium	39·102	72	Hafnium	178·49
20	Calcium	40·08	73	Tantalum	180·95
21	Scandium	44·956	74	Tungsten	183·85
22	Titanium	47·90	75	Rhenium	186·2
23	Vanadium	50·942	76	Osmium	190·2
24	Chromium	51·996	77	Iridium	192·2
25	Manganese	54·938	78	Platinum	195·09
26	Iron	55·847	79	Gold	196·967
27	Cobalt	58·933	80	Mercury	200·59
28	Nickel	58·71	81	Thallium	204·37
29	Copper	63·54	82	Lead	207·19
30	Zinc	65·37	83	Bismuth	208·98
31	Gallium	69·72	84	Polonium	(210)
32	Germanium	72·59	85	Astatine	(210)
33	Arsenic	74·922	86	Radon	(222)
34	Selenium	78·96	87	Francium	(223)
35	Bromine	79·909	88	Radium	(226)
36	Krypton	83·80	89	Actinium	(227)
37	Rubidium	85·47	90	Thorium	232·038
38	Strontium	87·62	91	Protactinium	(231)
39	Yttrium	88·905	92	Uranium	238·03
40	Zirconium	91·22	93	Neptunium	(237)
41	Niobium	92·906	94	Plutonium	(242)
42	Molybdenum	95·94	95	Americium	(243)
43	Technetium	(99)	96	Curium	(247)
44	Ruthenium	101·07	97	Berkelium	(249)
45	Rhodium	102·905	98	Californium	(251)
46	Palladium	106·4	99	Einsteinium	(254)
47	Silver	107·87	100	Fermium	(253)
48	Cadmium	112·40	101	Mendelevium	(256)
49	Indium	114·82	102	Nobelium	(254)
50	Tin	118·69	103	Lawrencium	—
51	Antimony	121·75		+other man-made	
52	Tellurium	127·60		unstable actinides	
53	Iodine	126·90			

The values in brackets are the mass numbers of the most stable isotopes of particular elements. T atomic weights are not whole numbers because of the mixture of isotopes. Between numbers 57 a 71 are the rare earth elements and between numbers 90 and 103 are the actinides. Hundreds unstable forms of the various elements can be made by neutron irradiation and other means.

Foreword

'Getting and spending we lay waste our powers'
WORDSWORTH,
Miscellaneous Sonnets pt. i, xxxiii

We hope that this book gives the reader with little or no scientific knowledge a reasonably comprehensive treatment of the very topical and important subject of energy. Energy studies have by no means been neglected over the last two or three centuries, but they are presently a matter of supreme importance to all intelligent men and women, and one calling for the constant attention of politicians and economists as well as scientists and engineers. The twentieth century has been one of unprecedented increase in the power of machines and the consumption of fossil fuels. Man has been 'getting' and 'spending' in order to generate power in a way unimagined in any previous century. How long can our energy supplies last? Can our way of life endure?

This book examines most of the resources available to us, the means of production, questions of conversion and storage, and the implications of energy use on the environment and our way of life. As the subject has so many ramifications, certain parts of the book may seem a matter of common sense while others are somewhat technical. We trust that, in spite of this difficulty, the level of information is tolerably balanced and that the book will serve as an introduction to energy studies and to more specialized literature. Scientific education is an indispensable part of democratic government and, in itself, a powerful support to political and individual freedom.

The figures were drawn in the Department of Engineering Design and Drawing at the University of Strathclyde. We thank the artists for their patient interpretation of our original sketches.

We are greatly indebted to Miss Carole Wilson, Miss Janice Horn, and Mrs. Margaret Reddish, for the care with which they handled the preparation of the volume, and we are happy to acknowledge their most valuable assistance.

We are indebted to Dr. James Curran for his helpful discussions on the material of parts of the volume and for his work on the index. Lady Curran contributed a great deal in very many ways.

Finally, we owe much to the thought and care of Mrs. Jean Jones, the scientific editor.

Part One

ENERGY: SOURCES AND CONVERTED FORMS

Chapter 1

Understanding Energy

The idea of energy has a long and shadowy history. Many religions, past and present, invoke ideas about life-force—for example, power from within, or power from without. In a vague, but nevertheless real, sense the white-coated scientist and the devil-masked priest both refer to the same entity, the power or capability to do things. Despite the complex trappings of mathematical treatments, in practice this power is essentially what the scientist means by 'energy'. Though knowledge of energy has been important to man throughout the ages, there has been a remarkable increase in our understanding during the last hundred years.

Many ancient religions involved Sun-worship; modern scientists realize the central importance of radiation from the Sun, and the heat thus provided to support life in all its forms. This is a point to which we shall return frequently. In his work Copernicus overthrew some of the religious beliefs of his time in proving that the Sun was at the centre of our whole planetary system. But neither Copernicus nor Galileo (who was more explicit in establishing that our planetary system was helio-centric) could have realized how our oceans, climate, and the whole gamut of living things depend utterly on the Sun's energy. This is no longer a matter of devoted faith, but an easily demonstrable scientific truth.

In the earlier days of modern science—between the time of Newton and Darwin—it was tacitly assumed that a few satisfactory experiments were sufficient to establish a theory valid for the whole of space and time. Nowadays most scientists would not be so extravagant in any claim. Science is an on-going process of hypothesis, discovery, confirmation, and, not infrequently, refutation: every theory is open to scrutiny, and belief in the 'truth' of a theory is always tentative and conditional. Sometimes new work is able to extend old laws; for example, Einstein did not disprove Newton's theories of motion—rather, he extended them and showed that Newton's theory was only part, albeit a critically important part, of a more complete whole.

It is worth stressing these points because in the following pages the reader may find that the present state of knowledge on many points is not complete enough for us to reach definitely correct conclusions. We must appreciate that scientific knowledge is always expanding, and that present limitations can give rise to uncertainties which irritate policy-makers. Nevertheless, decisions which can be made with such knowledge as modern science provides are much to be preferred to those made in ignorance. In some important fields it is confidently expected that future research will help to resolve present uncertainties, and, when science advances as rapidly as it does at present, scientists and engineers will sometimes ask that decisions be taken on the assumption that their experiments will later vindicate their arguments. Too much reliance on the probable outcome of future work should not, however, blind one to the fact that the arguments might well prove wrong in the light of later studies.

There is today a reason for appreciating the science and technology that affects the way of life of all of us. Rapid exploitation of energy resources has made us dependent on fuels to an extent unknown throughout human history and the supply of the most important fuels has become a matter of grave concern.

When the representatives of the Organization of Petroleum Exporting Countries (OPEC) met in 1973, they decided that oil was such an important commodity the price could be raised very substantially. After that it increased by a factor of five in less than four years. Many highly industrialized countries, together with a large number of developing countries, were faced with a critical situation. Their economies, and even their way of life, had come to depend heavily on the importation of substantial amounts of comparatively cheap oil and, in some cases, natural gas also. In several instances the imported oil and gas provided the energy to support 60 per cent or more of the industrial activity of the nations. It could be said that oil is the very life-blood of some countries. Even the US uses a large amount of imported oil as well as the enormous output of its own wells. The huge investment in generating plant and machinery of all kinds, e.g. diesel-powered locomotives and automobiles, means that an alternative energy policy, although theoretically feasible, cannot be implemented quickly. This lesson is important for its own sake. It can take between a decade and several decades to change to a marked degree the pattern of energy consumption of a country.

After 1973 oil-importing countries realized that although the OPEC countries could supply all the oil required for several decades the price was going to continue increasing steeply and that their energy policies would have to be modified in a very substantial way. The existing rate of use, and the anticipated increase in the rate of use of oil, were seriously studied. It was realized also that oil was a non-renewable fuel of the highest value, and that the rate of discovery of completely new wells was not matching the rise in the rate of consumption so that, by the year 2000 or so, supply would begin to fail to meet demand. This is a fundamental fact that has to be faced whatever the price at which oil may be made available. Almost all experts have come to the conclusion that an energy gap will emerge as a crucially important feature of the world scene sometime between the late 1980s and the year 2010, the time depending on the nature of the economy of the various countries, on the steps taken to conserve energy, on the rate of expansion or diminution in the use of fuels other than oil itself, and on the capacity to meet new prices and new investments. The total amount of energy potentially available is not the decisive factor. In many countries, for example, vast coal reserves still remain to be tapped; the question is, can they be mined, and can coal, in practice, replace oil as a fuel? Nuclear energy supplies in the form of electricity can be increased, but is electricity a feasible alternative to oil in many of its applications, and is it possible (or desirable) to construct generating plants at a rate sufficient to provide an answer? Statesmen everywhere speak of the gravity of the situation that has been brought sharply to the fore by these expert studies.

Discussion of the 'energy gap' recurs throughout this book and is dealt with in some detail in chapter 13. This book, however, is not intended to be primarily concerned with the problem of the gap; rather it is an effort to bring together the main ideas about energy, the resources and their exploitation, the environmental effects, the conservation questions and the questions of health and safety. For readers who wish to be further informed about the predicted gap a Report of the Workshop of Alternative Energy Studies, sponsored by the Massachusetts Institute of Technology and entitled *Energy: Global Prospects 1985–2000* (McGraw-Hill, 1977), demonstrates most thoroughly that an extremely serious supply problem will arise unless decisive policies are pursued from now on. A body of thirty internationally recognised experts from fifteen industrialized countries reached agreed conclusions on a world basis. The extent of the agreement, in views attained with Carroll L. Wilson as Director of the Project, is impressive and at the same time alarming. Detail on the data supporting the Project is given in associated technical volumes. It is a world study for countries in the non-Communist world, and we believe that the major conclusion is unquestionably sound. While accepting the findings of the Workshop team, and while we are concerned to find all possible ways of minimizing the repercussions of serious temporary deficiencies of supply, we are optimistic about the longer-term future.

It is unfortunate that there is not a single word for what is variously described under headings such as 'energy studies'. Energy is so important it seems most desirable that the science of energy should be described in a single word. We suggest that the Greek word *ergon*, meaning work, provides an answer. 'Ergonology' could be a suitable word to indicate succinctly the large body of material that constitutes this whole field of knowledge.

Chapter 2

Textbook Concepts of Energy and Work

Most of us are familiar with the idea of work. We know that we do a good deal of work and expend a lot of energy when we climb a hill or mountain; in the same fashion, we know we are certainly working when we dig in the garden or labour in the fields. From the earliest recorded time man sought ways of making work easier, and he made extremely rapid advances when he realized his capacity to make tools. Another advance, perhaps one of the most important, was made when he brought the horse to his aid—indeed it is not surprising that the physicists of later days studied the rate at which a horse could do work and that this rate came to be expressed in horsepower (hp).

No other animal has achieved man's rate of progress but some are aware, in however primitive a fashion, of the advantage of doing work by the exploitation of energy forms. For example, a bird stores gravitational energy as it takes a seashell aloft, and it converts this stored energy instinctively and to good account when it drops the shell from a height so that it strikes the earth forcefully and breaks. In this simple act some important basic ideas are introduced: the bird does work, and work is one form of energy. More generally, and extending beyond the idea of gravity, we say that the energy possessed by a body due to its position in a field of force is its potential energy. All of us are well aware of the existence of potential energy: for example, in running downhill we achieve more speed than in going uphill. In this case we are converting some of the potential energy into kinetic energy, or energy of motion. When water falls from a height, the kinetic energy obtained by the conversion of its potential energy can be used to drive a water wheel, one of the important landmarks in the history of man's development of energy resources.

We have mentioned both energy and work. The three classical laws of Newton are adequate to explain most everyday phenomena involving these concepts. These state (1) that every body continues in its state of rest, or of motion in a straight line, except in so far as a force acting on it compels it to change that state; (2) the rate of change is proportional to the force applied, and takes place in the direction of that force; (3) action and reaction are always equal and opposite, i.e. the mutual actions of two bodies act in opposite directions and are equal in force. Work is done in moving an object against a force, for example, the gravitational force. The work is equal to the product of the force F and the distance moved against the force denoted by S—that is, the work (or energy) expended is FS. In the same way, when a force F on an object causes it to move through a distance S, the product FS is the work done on the object.

Though the illustrations of work and energy that we have given are simple, it must be admitted that an exact scientific formulation of the energy concept is fraught with difficulties, even for scientists and philosophers. The idea is basically mathematical and linked with invariance; that is, the sum of certain quantities (heat, work, potential energy, etc.) always remains constant. The reader should also understand that when the correct energy units are used these different forms can be regarded as equivalent, so that addition, subtraction, and comparison can be made one with the other. Since the multiplicity of units in use often causes confusion, the interrelationships between the most important of them are displayed diagrammatically in fig. 1.

Different forms of energy may be partially or entirely interconvertible. For example, the work involved in rubbing two sticks together is entirely convertible to heat. The reverse process, however, is not possible. The science of thermodynamics is concerned with reversible and irreversible processes and enables us to predict such matters. The fossil fuels are stores of chemical energy and in burning them the chemical energy is converted to heat energy. The processes are irreversible.

Man as a Machine. Man himself is not impressive if regarded merely as a working machine. He requires a relatively high food intake to keep himself fit for work and, in energy terms, a reasonable average intake value is 12,000 kilojoules per day. One kilojoule equals 0·0004 horsepower-hour (hp-hr), so 12,000 kJ corresponds to $12{,}000 \times 0{\cdot}0004$ hp-hr or $12{,}000 \times 0{\cdot}0004/24$ hp-day, which is 0·20 hp-day. Hence a working man can, at most, work at an average rate of 0·2 hp. In practice, he could not sustain a high working rate for more than half a day, which means his average daily rate would be around 0·1 hp at best. By using a horse, man managed to multiply his effectiveness by a factor of about ten. There was of course the snag that the horse also had to eat, and indeed the average horse eats a lot more than the average man; the horse's food, however, could often be provided more readily.

Much of the energy consumed as food is found to be emitted as heat; the rate for a human is about 100 watts. The 12,000 kJ contained in man's daily food corresponds to $1 \times 12{,}000 \times 1{,}000/24 \times 60 \times 60$, or approximately 140 watt-days. If the 100 watts emitted as heat is compared with 140 watts (the maximum rate of release of energy), we realize that just over two-thirds of man's total available energy is emitted as heat. This somewhat surprising result arises from the fact that we have to keep our bodies warm in order to stay alive and the heat produced is always leaking away.

The total machine energy available to each inhabitant of the more advanced countries (such as the US, West Germany, the UK, and Sweden) corresponds to the effort which would be available from one hundred, or even more, human slaves. This drives home the point that modern living standards depend, so far as material possessions and personal mobility are concerned, on the availability of large amounts of energy per person.

Conservation of Energy. All of us are familiar with the fact that when motion of various kinds occurs and friction between moving bodies is present some of the kinetic energy of motion is converted to heat. For example, we see smoke when aircraft tyres encounter the ground. The conversion of

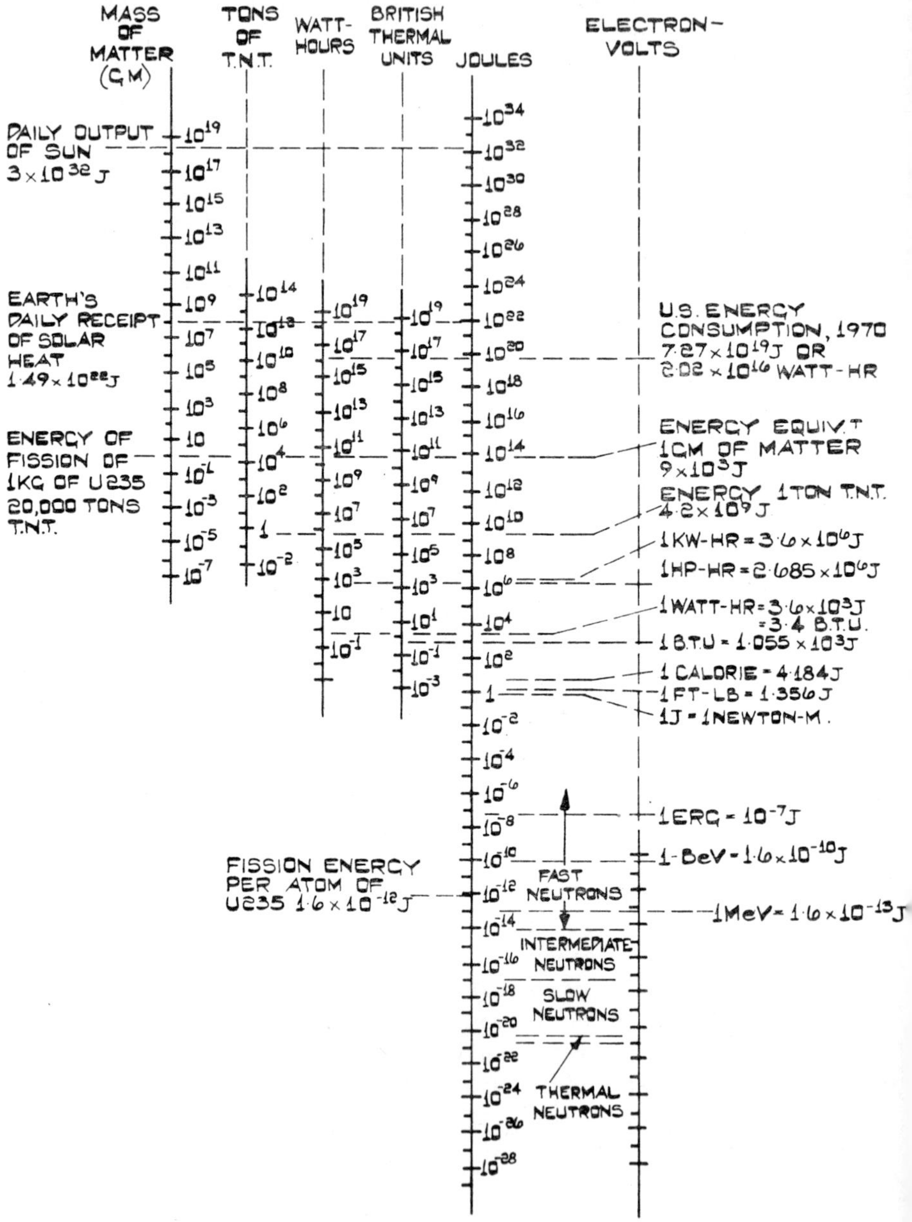

FIG. 1. Energy units and conversion factors. [From Chauncey Starr, *Scient. Am.* Sept. 1971, p. 48]

mechanical energy to heat has been long appreciated, although no measurements were properly done till the phenomenon was examined in detail by Joule in the nineteenth century. The great importance of his measurements resides in the fact that they established that energy as such is invariant, a principle enshrined in the Law of Conservation of Energy. This law states unequivocally that energy cannot be created or destroyed, or, in other words, that energy is conserved in all processes in physics. Indeed the law holds good in all phenomena in all fields of science—it is fundamental in thermodynamic processes, in mechanics it underpins most of our calculations, and likewise in chemistry and chemical engineering it is assumed to hold good. It is considered to be the first of the Laws of Thermodynamics—the science which attracted so much of the attention of outstanding scientists of the eighteenth and nineteenth centuries, such as Lavoisier, Clausius, and Kelvin. They realized that though energy considerations were not always simple, and many forms of energy had to be included in the analysis of systems undergoing physical, chemical, or biological change, nevertheless if a closed system were considered in sufficient detail then energy could be shown, without exception, to be conserved. A stimulating modern appraisal of the central position of energy in the universe has been given by Dyson in his notable article in *Scientific American* (Sept. 1971, p. 51).

The Second Law of Thermodynamics is concerned with the rather elusive concept of entropy, the degree of disorder within a system. The Second Law can be expressed in a variety of ways and possibly the briefest form states that the entropy of a system always tends to increase. The law may be illustrated by dropping a grain or pellet of soluble material into a glass of clear water. The pellet soon begins to dissolve, becoming so completely intermingled with the solvent that it is distributed throughout the entire vessel. The order obtaining when all the molecules of the solute are tidily together in the pellet has been replaced with the disorder obtaining when the pellet molecules move about in the liquid within the vessel.

The reasons for the profound interest in the principle expressed in the Second Law are many, and some are certainly philosophical in nature. Possibly the Law seemed to present a disturbing element in the ordered, systematic world of nineteenth-century science when it was first discovered, and some scientists did not like to be faced with the fact that order tends to increasing disorder. However, to see our universe as entirely tending to a state of increasing disorder can be remarkably wrong in some ways. One has to remember that, from some points of view, the order of our universe is one of the outstanding characteristics of the entire system. In the next chapter we consider the extraordinary stability of molecules, atoms, and nuclei of matter. We know that in these ordered forms of material there are immense stores of energy. Indeed, a very large fraction of the total energy of the universe has been wonderfully well stored for an extremely long time and no tendency whatsoever to disorder is observed in the nuclei of atoms, save in the most extreme conditions.

If a system, ordered or not, is able to store energy, then there must be (*a*) a lower total energy state available to it, and (*b*) an energy barrier to be overcome before reaching this lower state and releasing energy. As an example, let us consider a mixture of hydrogen and oxygen gases. There is a lower energy state available to the system through the reaction which produces water. In chemical symbols this is usually expressed:

$$2H_2 + O_2 \rightarrow 2H_2O + \text{heat}$$

Here H_2 denotes a hydrogen molecule, O_2 an oxygen molecule, H_2O a water molecule, and the equation indicates that two hydrogen molecules can react with one oxygen molecule to yield two molecules of water and energy as heat. The reaction, however, will not normally occur because an energy barrier exists and this must be overcome to enable the reaction to proceed. In practice, this energy barrier is removed by triggering the reaction with an electric spark. It is important the reader understands that this kind of condition frequently exists in nature, and that energy barriers prevent many reactions from proceeding even though these reactions would release more energy than is required to overcome the barrier. In an energy-producing device some method is employed which lowers or circumvents this barrier and enables the process to occur at a useful controllable rate. For example, an electric spark in a mixture of hydrogen and oxygen gases overcomes the energy barriers of molecules near the spark, and the heat of reaction released by these molecules allows the reaction to spread through the whole volume.

The same arguments that we have advanced in the case of chemical reactions hold true in the case of reactions between nuclei of atoms. The nucleus of an atom contains the main part of the mass of the atom and is usually very dense. Nuclei may often be able to react with each other releasing huge amounts of energy, but they do not usually do so unless a high energy barrier is overcome. One way of overcoming this high energy barrier is to provide other nuclei which collide violently with them (pp 17–21 below).

Returning to the order/disorder puzzle, the science of thermodynamics shows in a precise way that the tendency to disorder must always be considered in conjunction with the tendency for a decrease in energy. For example, if in a given process the energy of part of a system would increase a great deal while the order would decrease a little, the process would not occur. A general theory of change can develop from such considerations, but the theory does not tell us how fast these changes may occur. Here then we have a matter of great practical importance, particularly when we wish to store or release energy. We must have some means of controlling the *rate* of the energy-releasing process.

Electromagnetic Radiation. All of us are familiar with the fact that heat can be radiated. We feel it when we hold our hands before a fire and heat is radiated to us through the intervening air. This transfer of heat is almost as effective through a vacuum, for example that through which the Sun's rays pass. Current theories suggest that some of the heat energy in flames is transformed into tiny packets of electromagnetic energy which are called photons. Photons can exhibit properties typical of particles and of waves. All electromagnetic radiation consists of a flux of photons and in vacuum they are able to transfer energy at the speed of light from one place to another, and with 100 per cent efficiency. Depending on the frequency of oscillation of the electromagnetic waves, we refer to the principal forms as radio waves, heat, light, ultra-violet, X-rays, and gamma rays (see fig. 2). All are considered to be fundamentally the same, though we may have to detect their presence in quite different ways. The work of Clerk Maxwell resulted in elucidation of the laws of electromagnetic radiation, laws unifying in one theory the wide spectrum of its different forms.

Energy travels around the universe in these forms of radiation, crossing huge tracts of empty space. Most of the energy that we receive from the Sun reaches us on Earth in the form of light and heat; as we shall see in chapter 6, we are utterly dependent on this primary energy input to our ecosystem.

Fig. 3 illustrates our planetary system, but we have to remember that it is extremely difficult to represent it since the distances are very great indeed compared with the size of the Sun and its planets. The Earth itself is an average distance of 93 million miles from the Sun, and the Earth is only a little over 4,000 miles in radius. The Sun is but one of millions of millions of similar stars in the whole universe. The stars are themselves usually part of galaxies, and the Sun is simply one of the million million stars in the galaxy known as the Milky Way. Millions of these galaxies populate the universe, and the distances between the galaxies are even more enormous than the distances between stars within a single galaxy. Although the speed of light through space is very great indeed—186,000 miles per second—it takes light many thousands of millions of years to reach an observer on the Earth when the light comes from the most distant parts of the universe.

Not only are the distances between galaxies and stars enormous, we have difficulty in visualizing the magnitude of the immense energy sources

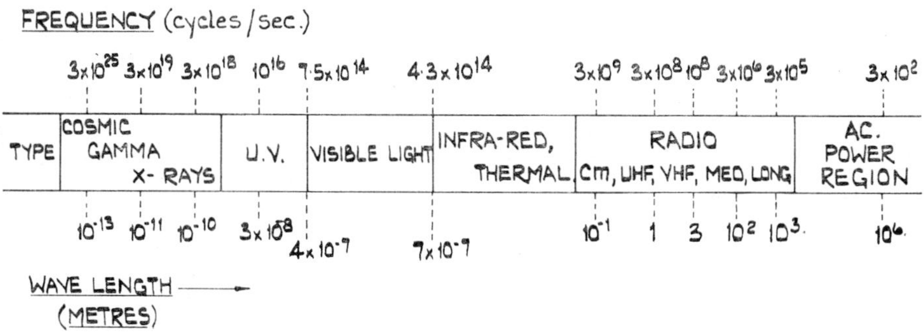

FIG. 2. The spectrum of electromagnetic radiation. The ranges in some of the bands (not shown to scale) overlap. Typical values of frequency and wavelength in the various bands are indicated.

represented in the stars, many of them very similar to our own Sun. Furthermore, in recent years a great variety of different forms of galaxies and stars has been discovered, particularly during the last twenty years. We now know of the existence of quazars, pulsars, and black holes, in all of which energy conversion processes proceed on a massive scale. Many mysteries have still to be resolved in the study of these new objects in space but a new and deeper understanding of the nature of our universe is gradually emerging. On one point there is no uncertainty in the minds of physicists—our universe is abundantly supplied with energy. Since a more detailed discussion is beyond the scope of this book the reader is referred to our section Further Reading on p. 315.

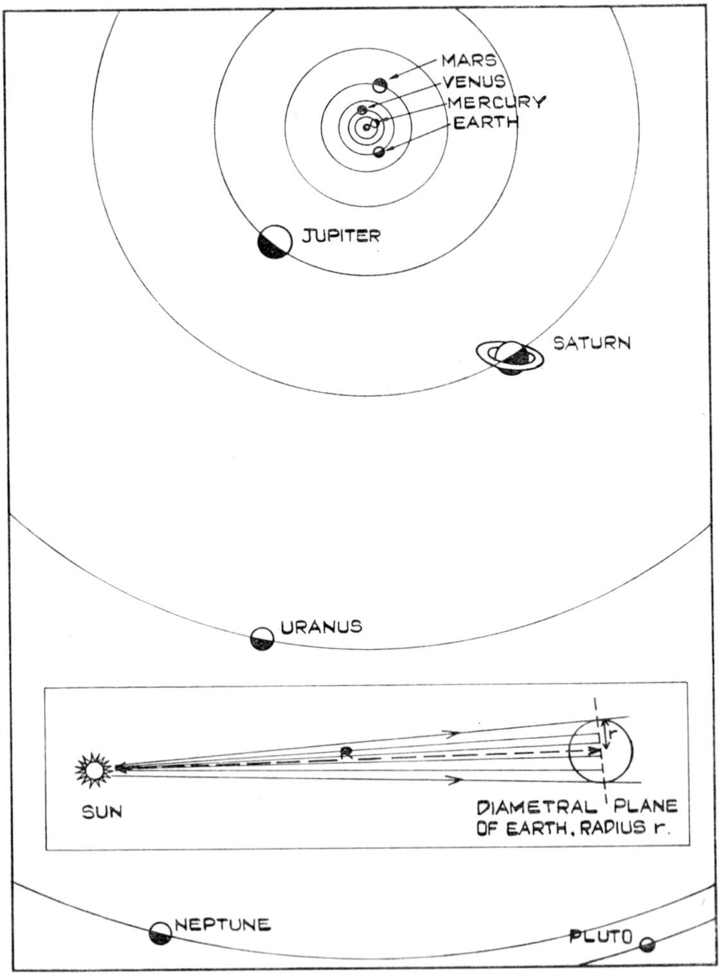

FIG. 3. The Planetary system of the Sun. Insert shows radiation falling on the Earth, radius r, at distance R from the Sun

Chapter 3

Nuclei, Atoms, and Molecules

Nuclear energy is the most abundant form of energy in our universe, but its release by man has given rise to a good deal of controversy. We shall therefore examine this particular form of energy very carefully. In doing so we make no suggestion that nuclear power is more important than other sources of power, nor is pre-judgment of the issues implied in any way.

Classical physics, which posited the atom as the smallest particle of matter, enjoyed immense success in the second half of the nineteenth century, but around the end of the century exciting discoveries, including those of X-radiation and of the free electron, heralded a period of real difficulty. These problems centred mainly on the absorption and emission of radiation by matter (the process by which energy changed between radiant and contained-in-matter forms) and knowledge of the structure of individual atoms was now required. In addition the study of uranium radiation by Becquerel—who observed what was later known as gamma radiation—created insuperable theoretical obstacles. This gamma radiation had its origin in the nucleus of the atom and hindsight allows us to appreciate that our knowledge of nuclear structure was lacking.

Meanwhile studies of heat radiation by Planck and others revealed the need for the introduction of the idea of discontinuity as opposed to continuity which had been assumed to be characteristic of all wave phenomena. Planck found himself obliged to propose the concept of a 'quantum of energy' in order to resolve the discrepancies between theory and experiment in his studies of the emission and absorption of heat radiation. This idea of a quantum, which is applicable to all forms of energy, was a major departure from orthodox ideas but it proved extremely far reaching and in it we have the origin of modern quantum theory. Planck's equation

$$E=h\nu$$

led to the rapid development of quantum physics in the present century. The equation shows that for homogenous electromagnetic radiation of a frequency ν the photons constituting the radiation each have the same energy $h\nu$, where h is Planck's constant of value $6{\cdot}626\times10^{-34}$ J-sec. or $6{\cdot}626\times10^{-27}$ erg-sec. Additionally the frequency ν of radiation is connected to the wavelength λ of the radiation by the simple relationship

$$\lambda=\frac{c}{\nu}$$

where c is the velocity of light in a vacuum (3×10^8 m/sec). To illustrate this we refer to fig. 2. For light at the boundary between ultraviolet and visible light the frequency is $7{\cdot}5\times10^{14}$ c/sec. The wavelength λ for this radiation is given by

$$\lambda=\frac{c}{\nu}=\frac{3\times10^8}{(7{\cdot}5\times10^{14})}=4\times10^{-7}\text{ m.}$$

For our purposes, we can regard a quantum of energy as a packet of energy in the form of radiation and of a defined magnitude which is always the same for a quantum of a given frequency. For penetrating radiations, such as gamma radiation emitted from radioactive substances, the individual quanta have a high energy. At somewhat lower energies we find X-radiation. Ultraviolet and visible light consists of quanta whose energies are usually considerably less than those in X-radiation, and infra-red radiation consists of quanta of still smaller energy than those of visible light. We know also from many refined experiments that all quanta of electromagnetic radiation travel in space with the same velocity, equal to 3×10^8 m/sec. Einstein stressed that the velocity of light was the limiting value of the velocity of any mass as well as of any form of energy.

The Bohr Atom. Following the work of Planck, a major advance in theory came with the work of Niels Bohr whose ideas were based on experiments of Rutherford and others. The striking evidence of large-angle scattering of fast particles, particularly alpha rays, in the experiments, suggested to Rutherford concentration of the mass of each atom in a very small centre or nucleus. Bohr was able to calculate the properties of such an atom. The central massive body is a positively-charged nucleus round which electrons, or small negatively-charged particles, move in well-defined orbits. The charges on the proton and electron are equal in magnitude. The diameter of a typical atom is around $\frac{1}{100,000,000}$ cm (or 10^{-8} cm), and the diameter of the nucleus is roughly $\frac{1}{10,000}$ of the diameter of an atom, about 10^{-12} cm. The nucleus of the simplest atom of all, hydrogen, is given the name 'proton', and in this atom it is regarded as retaining in orbit round itself a single negatively-charged electron.

Bohr was able to show that the orbits in which electrons move depend on 'quantum' considerations and that the electrons cannot circulate round the nucleus except in a number of orbits with particular radii. If an electron is given a definite additional quantum of energy, that is, a small amount of energy of definite magnitude, it can be caused to occupy an orbit farther from the nucleus. If it loses energy, always in a quantum of a particular magnitude, it transfers to an orbit closer to the nucleus. This important idea—that there are stationary states or levels of an electron separated by defined amounts, or quanta, of energy—meant that it was possible to calculate the atomic orbits and the corresponding energy states. Reasonably satisfactory explanation of complex phenomena, such as the emission of quanta of visible light or of X-radiation from atoms, followed. If an electron in any orbit of an atom is supplied with an amount of energy greater than the binding energy of the orbital electron, it may sometimes absorb all of the incident energy and become entirely free of the atom. The atom is then said to be ionized and it carries a single unit of positive charge. The free electron has a kinetic energy equal to the incident energy less its initial binding energy.

The Bohr model for the simplest atom, hydrogen, is shown schematically in fig. 4a. His theory was particularly successful when applied to calculation of the energies of the orbits of the hydrogen atom.

Bohr's findings resulted largely from his work with radioactive substances and helped to explain some features of such substances which are described below (pp. 20–21). An excellent exposition of atomic structure and energy levels is to be found in *Science Foundation Course* (S101) Unit 10 and 11 (Open University, 1979).

Emission and Absorption Spectra. Bohr's theory also proved consistent with work on spectra of electromagnetic radiation emitted from excited atoms. When atoms of the same element—say, hydrogen—emit radiation they usually do so in such a fashion that a line spectrum, a series of different but precisely defined frequencies of light radiation characteristic of the particular atom, is observed. Indeed, in practice, the line spectrum of the atom is a sure guarantee of the presence of that particular kind of atom—it represents a label or fingerprint of the atom. Vice versa, we can have an absorption spectrum, which is indicative of the selection and the absorption, in precisely discrete amounts, of the energy falling on the atom and capable of absorption by it.

Photoemission. If sufficiently energetic photons fall on certain surfaces, particularly metals, they produce what is known as the photoelectric effect—the emission of electrons, called photoelectrons, from the metal surface. Einstein observed that extremely small intensities of radiation could cause this phenomenon and, by applying Planck's quantum theory with Bohr's hypothesis of atomic structure, concluded that radiation could not be spread in a uniform way in space but must be concentrated in small packets which were sometimes able to bring about the liberation of an electron from the surface of the metal. In the photoelectric effect we have an interesting example of processes of energy conversion, in this case light being converted, in effect, into electricity. The reader will find that much of our discussion of energy problems throughout this book will centre around conversion problems. Attempts to utilize the photoelectric effect on a commercial scale are discussed in Chapter 10.

Identity of Mass and Energy. The exact nature of mass and energy, and the relationship, if any, between them, long puzzled scientists. In the early twentieth century the work of Planck, Rutherford, and Bohr showed they were closely connected. Subsequently Einstein propounded his theory of relativity, and this not only profoundly affected scientific thought but established certain relationships of far-reaching significance. The part of Einstein's theory which is truly basic to an understanding of energy takes the form:

$$E = mc^2.$$

Here c is the speed of light, one of the constants of nature, and hence the energy E is, in a very real sense, simply the same as the 'relativistic' mass m. Where the mass m is moving at a velocity v much less than c (the limiting velocity of any object) we can assume that E is measured by the mass m_0 of the object at rest. The mass m itself is a function of its velocity v and given by

$$m = \frac{m_0}{\sqrt{1 - \frac{v^2}{c^2}}}.$$

At sufficiently small values of v, the mass m can be taken as rest mass of value m_0 since in this case v^2/c^2 in the above equation is negligible. In effect, Einstein proves that energy and mass are essentially identical in nature, but extremely different manifestations of the same thing. Among limiting examples of the conversion of mass to other forms of energy is the explosion of plutonium in an atomic bomb. Among the converted forms are blast, heat, gamma radiation, and neutrons. The neutrons themselves have some of the original mass of the plutonium.

Most of us think we have a reasonable understanding of mass. We are inclined to think that we 'feel' mass, but actually we are usually feeling the weight, that is, the force of gravity on the mass. Mass, as such, is simply substance, or stuff—indeed it is a particular condensed form of energy—while weight is a particular property that mass acquires when it is placed in the gravitational field of the Earth. Gravitational force is the attractive force which exists between all masses in the universe. While its theory need not concern us here, its practical manifestations will have a place in our considerations of energy, for the force of gravity exerted by the Earth gives rise to hydroelectric and tidal power and other practical energy systems.

Today we realize that large masses can be made effectively weightless. When a space vehicle is in orbit round the Earth it is in a sense weightless, and it requires no expenditure of energy by an engine to keep it moving at high speed. In this example the force of gravity is counteracted by centrifugal force.

Duality of Matter. As we have mentioned, while radiation could be considered to be wave-like in nature, there were phenomena which forced the conclusion that radiation was at the same time like a stream of photons, or energy packets. In the 1920s it was shown that electrons, until then regarded as particles, could behave like radiation waves. In diffraction experiments for example, practically identical results were obtained with X-rays and electrons. Any particle, for example an electron, was at the same time a wave-packet, a kind of singularity in the energy field. Both matter and radiation appeared to possess, at one and the same time, properties typical of both waves and of particles. A number of textbooks (see Further Reading, p. 315) provide an account of the theory of this apparent duality of matter and radiation.

Neutrons and Protons. The Bohr model presents us with a picture of the atom as a sub-microscopic solar system, the nucleus as sun and the electrons as planets. While it is a gross over-simplification of the nature of the nucleus/electron system constituting an atom of any element, it is extremely helpful in understanding atomic phenomena.

However, in 1932, it was discovered that the nucleus itself was composed of two particles, an uncharged one named a 'neutron' and a positively charged one named a 'proton'. Soon it was realized that they could be regarded as the 'building bricks' of our 92 natural elements. Later it became known that there is a relationship between the two particles—the free neutron is the most simple of the radioactive elements and it can decay into a proton. The neutron is indicated by the symbol n_0^1, where n is the symbol for neutron, the upper suffix showing the mass and the lower the charge, in this case zero. The radioactive decay of a neutron can be represented symbolically by the equation

$$n_0^1 = H_1^1 + e_-$$

H_1^1 represents the proton; the small symbol e_- represents an electron or beta particle, the rest mass of which is only $\frac{1}{1840}$ of the mass of the proton. We say that the free neutron is a beta-emitting particle. Most elements, however, are entirely stable. In other words, neutrons in partnership with protons need not alter or decay.

Isotopes. Assemblies of protons and neutrons make up the nuclei of all natural elements. The number of units of positive charge on any nucleus is the number of protons in that nucleus (as the unit is the amount of positive charge on a single proton). The nucleus is surrounded by orbital electrons whose total negative charge equals the positive charge on the nucleus, with the result that matter is electrically uncharged. Each element has a characteristic number of protons in the nuclei of its atoms, the atomic number, and it determines almost completely the chemical nature of the element. In general the number of neutrons and protons in a nucleus are approximately equal. An element is said to have isotopes when it exists in two or more forms containing the number of protons characteristic of the element but a different number of neutrons. Isotopes usually have closely similar chemical properties. For example, uranium has isotopes of mass 235 and 238 but it is very difficult to separate them by any chemical means. The heavier the atom the greater the tendency or capacity for protons to combine with a somewhat larger number of neutrons. When the number of associated neutrons becomes too large, radioactivity is usually present, showing that a tendency for some rearrangement of the 'building bricks' exists. Thus heavy natural elements, among them the isotopes of uranium and also thorium, show radioactivity.

Usually hydrogen has a single proton as the nucleus, but if this proton is arranged in a special intimate combination with a neutron the nucleus then has nearly twice the weight of a proton and is actually the nucleus of the isotope of hydrogen known as deuterium. The deuterium atom (fig. 4b) shows schematically one proton and one neutron in a nucleus of almost exactly two units in weight. The single positive proton is counter-balanced electrically with the single negative electron. By adding one more neutron we can make tritium (fig. 4b). Chemically, the element is still hydrogen, but in fact each atom is almost three times as heavy as an atom of the normal form of light hydrogen. The three isotopes of hydrogen (hydrogen, deuterium, and tritium) are so different in weight that they appear to differ to some extent in chemical behaviour. In this respect their behaviour is not truly typical of that of the isotopes of heavier atoms.

Atomic Number and the Periodic Table. As most people are aware, scientists use a system of letters to symbolize the atoms of each element; thus H represents an atom of hydrogen, O of oxygen, and Fe (from the Latin

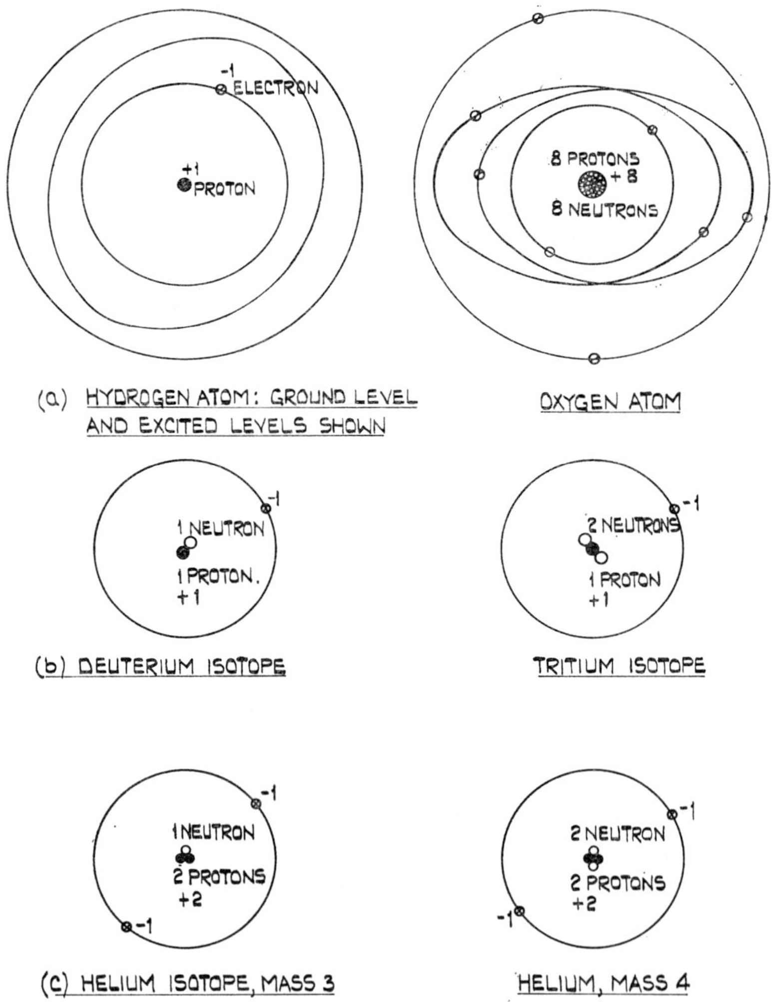

FIG. 4. Schematic representations of atoms of hydrogen, oxygen and helium

ferrum) an atom of iron. Further information about the atom is given by the numbers which follow it. The subscript number is the Atomic Number, that is the number of protons in the nucleus. The superscript number denotes the Atomic Weight, which can be thought of as the sum of the number of protons and neutrons in the nucleus. Thus, for example, the light form of hydrogen is written as H_1^1, deuterium as H_1^2, and tritium as H_1^3.

Ninety-two elements with consecutive atomic numbers compose the universe, varying widely in abundance and distribution. The lightest is hydrogen (A.N. 1), as we have noted; the heaviest is uranium (A.N. 92). Even heavier elements can be made artificially by means of nuclear reactions.

It has been found that the chemical properties of an element are largely determined by the number of electrons in its outer electron shell. Furthermore, if the lighter elements are considered sequentially according to their atomic number (which is, equally, the number of their electrons) each element resembles chemically the element seven places away from it. The overall pattern persists among the heavier atoms, though with some variation. This natural periodicity is always represented diagrammatically in an arrangement known as the Periodic Table.

Following hydrogen in the Periodic Table is helium, which has two protons in the nucleus. The first of the two known forms of helium, the lighter one, H_2^3, is formed when there are two protons in combination with one neutron. There is a tendency for one of the neutrons in tritium to lose an electron; this is what happens when tritium decays radioactively, giving 'daughter' nuclei of light helium (fig. 4c). The process of change of a neutron to a proton which occurs in the beta decay of the simplest radioactive substance, tritium, typifies the beta decay of many far more complicated nuclei.

The second and normal form of helium, He_2^4, has a very stable nucleus with two neutrons combining with two protons, giving a nucleus of four units of mass and a charge of two positive units (fig. 4c). Sometimes this nucleus of helium is emitted in radioactive decay, for example by plutonium, and the decay is said to proceed by emission of alpha rays. It is evidence of the characteristic strength of the helium nucleus. The normal helium atom has of course an electron shell with two negatively-charged electrons, these cancelling the double positive nuclear charge.

Carbon is the sixth atom in the Periodic Table; the most common isotope has a mass 12 and is therefore written C_6^{12}. We can guess it may be possible to add one or possibly two more neutrons giving heavier isotopes of carbon. Actually, in nature we do find C_6^{13}, a heavier stable carbon isotope, and also C_6^{14} which is a radioactive form of carbon (radiocarbon) indicating that two additional neutrons are too many for complete stability of the nucleus.

Nitrogen is the seventh atom in the Periodic Table and is formed of seven protons with the required seven neutrons, so that the nucleus is indicated by the symbol N_7^{14}. Oxygen (fig. 4a) is the eighth and in this case we need eight protons and these, together with eight neutrons, give us a nucleus of mass 16, that is O_8^{16}. Obviously, the atoms of nitrogen and oxygen need seven and eight electrons respectively to be neutral; in both atoms two are in the inner orbit, with five in the case of nitrogen and six in the case of oxygen in the second outer orbit.

Radioactive Decay. Uranium, the heaviest natural element, has 92 protons in its nucleus. Such a large number of protons tends to combine more stably with a greater, rather than an equal, number of neutrons. There are no less than 146 neutrons in the most common form of uranium, the isotope of mass 238 which is represented by the symbols U^{238}_{92}. Even so U^{238}_{92} is not completely stable and the nucleus gradually disintegrates, the transmutations occurring in a long series of decay processes.

In natural radioactive substances these decay processes usually take place through the emission of one of three different types of radiation: alpha particles, beta particles, and quanta of gamma radiation. Alpha particles, as has been noted, consist of the close combination of two neutrons and two protons. Beta particles are fast electrons. The third, gamma radiation, differs from the other two forms, consisting of high-energy penetrating photons of electromagnetic radiation rather than of particles. However, like alpha and beta radiation it is of nuclear origin, and is in that sense distinct from X-radiation which arises in relatively high-energy disturbances of orbital electrons.

The rate of radioactive decay is constant for each radioactive element and is stated in terms of the 'half-life' of the atom. By definition, half-life is the time taken for half of the material to transform into the so-called 'daughter' element. A long half-life corresponds to a slow rate of decay and, vice versa, a short half-life implies a high rate of decay.

Among the radioactive elements first discovered uranium and thorium were notable. Additionally, potassium and rubidium were known to have unstable isotopes which slowly decayed. The modes of decay are usually by emission of an alpha particle or a beta particle, sometimes associated with the emission of gamma radiation. Illustrative examples of such decay processes are

$$Ra^{226}_{88} \rightarrow Em^{222}_{86} + He^{4}_{2}$$
$$Bi^{210}_{83} \rightarrow Po^{210}_{84} + e_{-} + \nu$$
$$K^{40}_{19} \rightarrow Ca^{40}_{20} + e_{-} + \nu.$$

In the first of the above we note that radium decays into the gas, radon (known also as radium emanation), and an alpha particle. In the second, bismuth is shown decaying into polonium by emitting a negative electron (or beta particle) and a neutrino denoted by ν; the neutrino is an extremely light particle with zero charge, highly penetrating and virtually non-ionizing, that is, with almost negligible capacity to interact with matter through which it passes. The last of the three equations shows the way an isotope of natural potassium decays into calcium by the emission of a beta particle and a neutrino. The same element can decay into argon rather than calcium.

Spontaneous Fission. The imbalance of protons and neutrons in the heavy nuclei leads also to a completely different form of decay, not described before. This is decay by spontaneous fission. Whereas in the radioactive decay described previously, the element emits atomic particles and decays into an element close to it in atomic number, in the fission process two nuclei of intermediate mass—say, A and B—are formed. These product nuclei do not require the excessive number of neutrons held by the uranium and, on average, about two or three neutrons are simultaneously liberated.

$$\text{U} \rightarrow \text{A} + \text{B} + \text{(say) 2 or 3 neutrons.}$$

A and B are formed with much excess kinetic and excitation energy and this energy represents an important and useful part of the total energy release in the fission process. Although the much more frequent mode of decay of U^{238}_{92} is by emission of an alpha particle, that is

$$U^{238}_{92} \rightarrow Th^{234}_{90} + He^{4}_{2},$$

we can stimulate the less probable process of fission and so make use of the natural tendency to undergo fission by adding, or attempting to add, a neutron to uranium. This triggering of the fission process is extremely effective and the uranium nucleus which 'captures' a neutron is very likely to undergo fission; thus

$$\text{U} + \text{n} \rightarrow \text{A} + \text{B} + \text{(say) 3 neutrons.}$$

We are, as it were, taking advantage of the in-built weakness of the structure of the uranium nucleus when we design a fission reactor or an atomic bomb.

The absence of electrical charge on the neutron is a matter of great importance. Because it has no charge the neutron can approach without difficulty extremely close to the positively-charged nucleus of any element, whereas the probability of very close approach, and subsequent absorption or capture, of charged bombarding particles is extremely low.

Absorption of a bombarding particle is usually catastrophic so far as the stability of the target nucleus is concerned, so emission of a particle or radiation or both generally follows. Since such disintegration can be used to gain insight into nuclear structure the discovery of the neutron, and its use in these techniques, provided fresh impetus to nuclear physics.

The process of fission was discovered in 1939 and by 1942 it was established that a self-sustaining chain reaction was practicable. This meant that man had a machine which produced very large fluxes of neutrons, and that these could be used to make large amounts of new radioactive materials. In the fission process itself the heavy fission products can take different forms but they are always radioactive. The amount of this radioactive material in nuclear reactors constitutes in itself one of the problems of design. The energy released in radioactive decay can keep the core heated even after shut-down, and cooling is required to remove this heat. In the absence of sustained cooling the core could melt and result in serious difficulties.

Moderation of Neutrons. It was soon realized that even extremely slow neutrons could penetrate into nuclei, and entirely new possibilities in transmutation experiments emerged. Neutrons could, moreover, be easily rendered slow (or 'thermal'), and it was found that not only did slow neutrons penetrate nuclei, but in some cases the penetration probability could be tremendously increased by slowing the neutrons to thermal or near-thermal values. Thus what is called the cross-section for the process of neutron absorption might be increased one thousand fold as compared with the value observed for fast, or unmoderated, neutrons.

We should be clear about what we mean by a thermal neutron. The average energy of a molecule or atom of a gas maintained at room temperature, say around 20 °C, is very small, much less than one electron-volt (the amount of energy used in moving a single electron charge through a voltage difference of one volt, written eV). The thermal neutron is one that has an energy about the same as a hydrogen molecule would have if it were in free air at normal room temperature. Neutrons emitted in fission generally have an energy of approximately one million electron-volts, whereas a thermal neutron has an energy of less than one tenth of one electron-volt.

The reduction in energy of fast neutrons is readily achieved by surrounding the source with material containing a lot of light atoms. The lightest is hydrogen, but we can use deuterium, helium, lithium, beryllium and carbon effectively. The process of slowing down, which is known as moderation of the neutrons, is analogous to the slowing down of a billiard ball struck so as to collide with an assembly of other stationary billiard balls. On average, each collision leads to a halving of the energy so that ten or so collisions can reduce the average energy to only one thousandth of the initial value, and in about twenty collisions the neutron could be reduced in energy by a factor of a million. So a very energetic neutron soon becomes 'thermal' in moving through hydrogenous material. For example, the passage of fast neutrons through a few inches of paraffin wax, rich in hydrogen, brings about a major reduction in energy. In practice, there are usually a number of competitive processes which can occur in the moderating material and when these are quite probable the material is unsuitable. Even hydrogen can absorb a fraction of the neutrons passing through it, transmuting to heavy hydrogen.

In the years immediately following the discovery of the neutron, the high probability with which slow neutrons entered nuclei and created new nuclei meant that many new unstable nuclei could be created. The ease with which neutrons produce radioactive substances is a factor of real significance in the utilization of nuclear fission.

Man has succeeded in these ways in producing a very great variety of unstable nuclei or radioactive elements. Where he has sought to study a still more diverse range of elements he has employed special accelerators. The first one, used successfully in 1932, was a high-voltage generator constructed by Cockcroft and Walton. The cyclotron was introduced by E. O. Lawrence. Later we had machines such as the betatron, synchrotron, and linear accelerator. With such machines a great deal of information about nuclei has been amassed and an extraordinary variety of stable and unstable nuclei has been observed. In the last two decades extremely high-energy machines have been successfully constructed and with their help new insight into the nature of nuclei and nuclear forces is available. We leave the interested reader to pursue this subject for himself. There is a good account in *Science Foundation Course* (S 101), Units 30 and 31 (Open University, 1979).

Fusion of Light Nuclei. Studies of the stability of nuclei show that apart from some fluctuations nuclei increase in stability as the mass increases to about that of iron. Beyond such an intermediate mass the stability decreases gradually with increasing mass. This is equivalent to saying that, in theory, energy can be released in processes of fusion of light elements to yield heavier ones, at least up to the point at which iron is the product. The reverse process (fission) in which heavy elements are broken into lighter ones, has been used to yield energy in fission reactors (see chapters 14 and 15). In principle it is possible to construct fusion reactors, and the 'hydrogen bomb' is an example of the explosive release of thermonuclear energy by such reactions. It has not yet been found practicable to release such fusion energy in a controlled way and in self-sustaining amounts and this is the main aim of continuing fusion studies (see chapter 17).

It is easy to establish theoretically that fusion energy can be released. If we consider the reaction in which four protons are used to make one helium nucleus (that is, four atoms are transformed into one atom) we find that such an elementary 'burning' of hydrogen to yield helium produces vast amounts of heat. In theoretical work, such a conversion of hydrogen to helium is used to explain the means by which many of the stars release energy, though it is believed the process is the outcome of a chain of nuclear reactions in which other light nuclei participate as catalysts. The final result can be expressed simply as

$$4\mathrm{H}_1^1 \rightarrow \mathrm{He}_2^4 + 2\mathrm{e}_- + \text{energy}.$$

The yield per unit of reacting mass in such fusion is about six times the amount per unit mass in fission. There are, however, many difficulties in constructing suitable apparatus and in order to overcome such difficulties we shall see that other reactions which proceed more readily than hydrogen to helium conversion are preferred.

Present Views on Nuclear Structure. We have tended to describe matter as consisting of atoms, and the nuclei of the atoms as consisting of close-packed assemblies of protons and neutrons with electrons circulating around the nuclei. It is important to realize that this is a rather gross simplification. We should not think of the particles like neutrons, protons, and electrons as rigid spheres. They are 'fuzzy' in a sense, and they have a wave-like property as well as a particle one.

The proton and neutron are now thought themselves to be combinations of quarks held in a kind of 'glue'. The study of the forces in nuclei has advanced considerably, and physicists are looking for the connection between nuclear forces and the electromagnetic forces which have their origin in electrical charge in the nucleus. Even more exciting advances have been made in attempts to connect these forces with the force of gravity. In the 'black hole' concept, it is imagined that matter in a collapsed star is so highly concentrated that the strong nuclear forces can be overcome by the immense gravitational force that results from a tremendous concentration of matter. A mass such as that of the Earth is imagined to be concentrated to a mass about the size of a pea. An atom approaching such a black hole is subject to a gravitational force so huge that the electrons are simply torn from it and disappear into the hole, followed by the nuclei. It is believed that such a hole emits nothing, that even quanta of radiation cannot escape. Thus, since no light escapes, the object is invisible, its position being deduced by other means than visual observation. S. W. Hawking (*Scient. Am.* **236**, 1977, p. 36) has put forward the view that in some circumstances even an approaching nucleus could be torn apart. Thus a proton could be disintegrated, with part of it disappearing into the hole and part of it escaping. Much of this recent work has still to gain acceptance but among the ideas that emerge clearly, both in the theoretical work and in the experimental work with extremely high-energy accelerators, is the fact that mass or energy in extreme concentration can transmute into a great variety of forms. In all such changes the total electrical charge must always remain balanced. Nigel Calder provides a stimulating account of recent work in his book, *The Key to the Universe* (British Broadcasting Corporation, 1977). An excellent account of modern ideas in relativistic physics is given by Robert M. Wald in his book *Space, Time, and Gravity* (University of Chicago Press, 1977).

Molecules. We have discussed nuclei and their characteristics mainly to pave the way to the study of nuclear energy. However, much of the energy used today is chemical in nature and derived from burning fossil fuels. Here molecules are involved in exothermic (heat releasing) chemical changes. As these can be very complex it is not proposed to discuss them in any depth, but molecular structure needs to be briefly explained.

A molecule is a structure composed of two or more atoms bonded together by rearrangements among their outer electrons. Hydrogen is the simplest of the atoms and in fig. 4a atoms of hydrogen and oxygen are shown in schematic form. The electrons fall into shells which, in the case of the lighter elements, are completed when occupied by 2, 8, and 8 electrons and for somewhat heavier elements when occupied by 2, 8, 18, and 8 electrons. As we noted in our discussion of the Periodic Table the number of electrons in the outermost of the electron shells of each element strongly influences its behaviour. This happens because atoms tend to rearrange themselves so that their outer electron shell is full. By way of simple illustrations molecules of hydrogen and water are shown in fig. 5. Two atoms of hydrogen share the two electrons that are available to form a pseudo-shell of two in the molecule, and when two atoms of hydrogen combine with one atom of oxygen a complete outer pseudo-shell of eight electrons is made. Energy is released when hydrogen and oxygen combine; on the other hand this bonding energy has to be supplied if the water molecule is to be dissociated. The reaction is:

$$2H_2+O_2 \rightleftharpoons 2H_2O.$$

If we try to break up water by heating it we have to raise the temperature to around 2,500 °C.

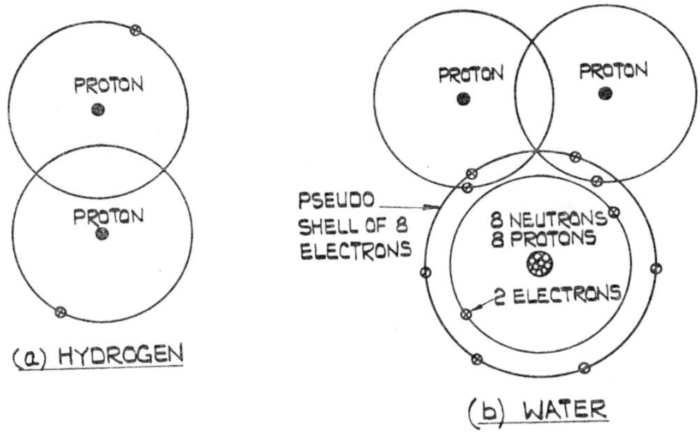

FIG. 5. Schematic representations of hydrogen and water as molecules

Molecules of the form of water owe their stability to covalent bonds, that is, the outer (or valence) electrons of the adjacent atoms are shared. In some instances ionic bonds hold molecules together. These are due to transfer of an electron from one atom to another, and electrostatic attraction between oppositely charged atoms exists.

Reaction between atoms to form molecules, and between molecules themselves, are, as we have seen, really reactions between outer electrons. They are described as chemical reactions and the electron changes are said to involve chemical energy. Nuclear changes involve energies exceeding the chemical by a factor of about 1 million.

The number of molecules that can be made is enormous, and the physical forms of molecular materials may vary considerably. Many compounds can be produced in crystalline form and the shape of such crystals depends in a general way on the shape of the constituent molecules. We shall find in chapter 10 that certain crystals can behave as semi-conductor devices and also as photovoltaic cells. On the other hand some amorphous semi-conductors have been made. In amorphous material a regularity or periodicity in the spacing of atoms or molecules is observed but it is much less well defined than in crystals; glasses are one form of amorphous material.

A large class of compounds of special significance are those containing carbon compounds, other than the metal carbonates and oxides and sulphides of carbon. These are known as organic compounds because they were first found in living matter. They mainly consist of hydrogen and carbon atoms though not infrequently oxygen or nitrogen atoms are also contained in the molecules. They are particularly important in any study of energy because the major fossil fuels—coal, oil, and natural gas—are largely composed of organic molecules, mostly hydrocarbons of the general form C_mH_n. The diversity of such organic compounds is very striking. They may be relatively simple; for example, in the case of methane (the main constituent of natural gas) one carbon atom is combined with four hydrogen atoms. The gas burns readily, combining with oxygen and giving carbon dioxide and water:

$$CH_4+2O_2 \rightarrow CO_2+2H_2O.$$

Among other organic compounds typical of the general class are methanol (methyl alcohol or wood alcohol) denoted by CH_3OH; ethane CH_3CH_3; butane $CH_3CH_2CH_2CH_3$; ethylene CH_2CH_2; and, somewhat more complex, benzanthracene $C_{19}H_{14}$.

Chapter 4

Energy in the Universe

It is believed that all of the matter in the universe was concentrated to extremely high density about 18 billion years ago, and that the universe as observed today is the outcome of the explosion of this dense matter. Both matter and radiation spread out through space after the 'big bang' of all primeval matter and the expansion still continues. Later local concentrations occurred and these constitute the immense galaxies including the Milky Way, the galaxy in which our Solar System is situated. In *The Key to the Universe* (p. 17), Nigel Calder presents some cosmological data in a vivid way.

TABLE 4.1

AMOUNT OF MATTER IN THE UNIVERSE	
Atoms in one gram	say 10^{24}
Grams in one star	say 10^{33}
Stars in one galaxy	say 10^{11}
Galaxies in the universe	say 10^{10}
So atoms in the universe	about 10^{78}

Around 10^{21} stars exist in space, and many pour out as much energy as the Sun, itself a vast energy source. The amount of matter and thus, in the light of Einstein's proof that energy and mass are equivalent, the amount of energy in the universe is enormous. A fraction of it takes the form of electromagnetic radiation. The average temperature of space calculated by measurements of the radiation is 3 Kelvin, three degrees above absolute zero at 0 Kelvin (-273 °C), the lowest temperature theoretically possible. The total of this radiation is immense but its density is very low. The distances between stars are fantastically large and most conveniently measured in light years. One light year is the distance traversed by light in a vacuum in one year. The speed is 3×10^{10} cm per sec or 186,000 miles per second. A year is roughly 3×10^{7} seconds, so one light year is approximately $3 \times 10^{10} \times 3 \times 10^{7}$ cm, that is about 10^{18} cm or approximately 10^{13} km. The nearest star to the Sun is about four light years away, that is about 100,000 times the distance between the Sun and the Earth, which is a mere eight light-minutes (about 93,000,000 miles).

The radiation of other stars has helped us to deduce that the universe is expanding. The speed at which distant stars and galaxies recede from an observer on Earth gives rise to a 'red shift' of lines in the spectrum of the received light; in other words the lines appear nearer the red, or low frequency, end of the spectrum than one would expect. The red shift is greater in value the more distant the source. Edwin Hubble studied such evidence of recession of stars in the universe and it is from calculations based on modern measurements of spectra that we believe that the 'big bang' occurred about 18 billion years ago. The light from the most distant stars has taken times of this order of magnitude to reach the Earth.

It is thought that the Earth consists of material which circulated originally around the Sun, under the influence of its gravitational pull, and that the orbiting matter aggregated. The time of revolution about the Sun is $365\frac{1}{4}$ days. The Earth itself has a small planet, the Moon, which experiences the gravitational attraction of both the Sun and the Earth with the Earth-Moon force the dominant one. The Earth itself rotates and the time of one revolution defines the duration of one day. Compared with the Sun the Earth is small and cold though some of the original gravitational energy, that produced in the process of compression of the matter during the formation of the Earth, keeps the centre much hotter than the crust. The decay of natural radioactive elements concentrated largely in the crust of the Earth contributes substantially to the energy of the Earth itself. The centre consists possibly of a near-spherical mass of solid iron surrounded by liquid iron, though the centre may contain lighter atoms including hydrogen.

Abundance of Elements in the Solar System. Theory suggests that in the aftermath of the 'big bang' light nuclei would be most abundant and that from the light nuclei heavier would be formed later. As we mentioned briefly in the last chapter this fusion provides a great deal of energy until we come to

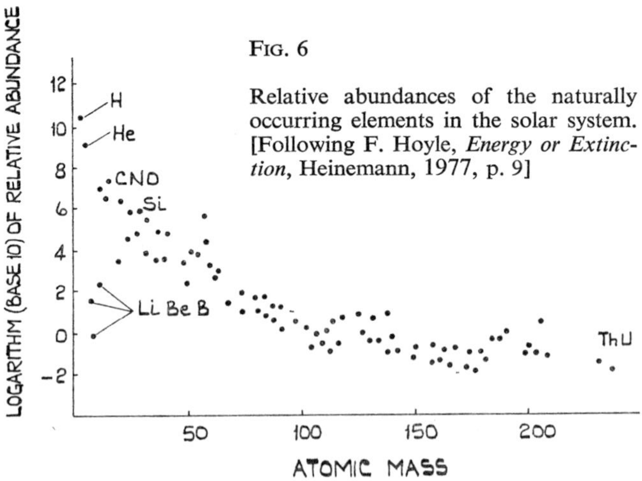

FIG. 6

Relative abundances of the naturally occurring elements in the solar system. [Following F. Hoyle, *Energy or Extinction*, Heinemann, 1977, p. 9]

the formation of nuclei of mass approximately that of iron. Thereafter it is necessary to provide energy in order to build heavier nuclei. If conditions were favourable all nuclear reactions would proceed to the point where only iron and nuclei of about the same mass as iron remained. In reality many factors exert influence on the nuclear reactions that can occur, and as a result the abundance distribution of the nuclei naturally occurring in the solar system is shown in fig. 6. The scale of abundance is logarithmic, that is, in powers of ten. The lightest nuclei—hydrogen and helium—are the most abundant; carbon, nitrogen, and oxygen very approximately less abundant by a factor of $\frac{10^{10}}{10^7}$, i.e. by a factor of 10^3; and silicon less abundant by an additional factor of approximately 10. The very heavy elements—such as uranium and thorium—are less abundant than hydrogen by a factor of roughly $\frac{10^{10}}{10^{-2}}$ or 10^{12}, that is, they are less abundant than hydrogen by a factor of one million million. Following the 'big bang', elements like plutonium would be present but such unstable elements have long since decayed and now only exist when they have been man-made by neutron bombardment. On the Earth, hydrogen and helium are in less abundance than shown in the figure, and carbon, nitrogen, and oxygen are likewise reduced. Lithium, beryllium, and boron are deficient as compared with other light nuclei. On the other hand, in the crust of the Earth, the abundances of thorium and uranium are enhanced compared with those shown for the solar system.

Energy Balance of the Earth. The outstanding characteristic of the universe is abundance of energy. The three principal forms are gravitational, nuclear energy, and radiation, with gravitational energy predominating. The energy flows on the Earth are embedded in the energy flows of the universe, and the surface of the Earth has a dynamic energy balance between energy

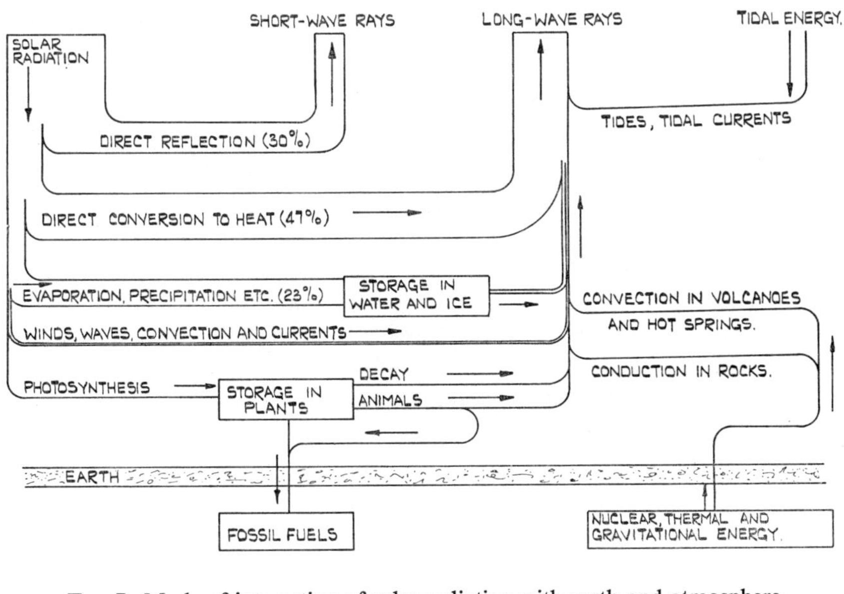

FIG. 7. Mode of interaction of solar radiation with earth and atmosphere. [From M. King Hubbert, *Scient. Am.* Sept. 1971, p. 62]

D

input and energy output. The input consists of solar radiation, gravitational (tidal) energy, and geothermal energy generated by the decay of radioactive material within the Earth and conducted to the surface. The radiation input exceeds the total of the other two by a very large factor, about 5,000, and geothermal exceeds tidal by a factor of roughly 10. For a more comprehensive review of energy in the universe the reader is referred to Freeman J. Dyson (*Scient. Am.*, Sept. 1971, p. 51). As far as useful energy is concerned, the Sun is the only body that significantly influences the Earth. Its major role and its general relationship with terrestrial energy sources are indicated roughly to scale in fig. 7 and summarised in numerical form in table 4.2.

TABLE 4.2

BUDGET OF ENERGY INPUTS AND OUTPUTS FOR THE EARTH
(From M. King Hubbert, *Scient. Am.*, Sept. 1971, pp. 60–87)

Inputs		Outputs	
Nature	Amount in 10^{12} watts	Nature	Amount in 10^{12} watts
Solar radiation	173,000	Directly reflected	52,000
		Converted to heat and radiated	81,000
		Used in evaporation precipation and largely temporarily stored in water and ice	~40,000
		Winds and waves	370
		Photosynthesis (partly retained)	40
Terrestrial (nuclear) thermal, gravitational)	32·3		~32·3
Tidal	3		~3

We observe that the terrestrial energy plays a minor but significant part in the total scheme. We shall explore the geothermal input in more detail in chapter 5 and also examine the process of conversion by vegetation (chapter 11). The constituent elements of the budget are approximate averages and not applicable to a specific part of the Earth's surface. They should, however, convey to the reader some appreciation of the wide range in the scale of the sources of energy available to us and particularly those due to solar radiation, incident and converted (chapters 8–12).

Chapter 5

Non-Solar Sources of Energy on the Earth

There are only two sources of energy on Earth which are not ultimately derived from the Sun. They are the energy dissipated by radioactive elements decaying within the Earth, and tidal energy which originates principally from the gravitational pull of the Moon.

A. Radioactive Elements

Since the radioactive elements are the fuel for fission reactors the quantities available determine the course of our whole nuclear programme. Furthermore, the heat they generate *in situ* is, as we shall see, the principal source of geothermal energy.

In Chapter 4 we explained why radioactive elements are not abundant. The most notable of those that survive (nearly all in the crust of the Earth) are uranium and thorium. Measurements of the decay of such elements are used to deduce a good deal about the time for which some of the rocks of the crust have remained undisturbed (see table 5.1). We are now certain that the Earth has existed in something like its present state for 4·5 billion years.

TABLE 5.1

NATURALLY-OCCURRING PARENT RADIOACTIVE ELEMENTS

Radioactive parent element	Primary modes of decay	Stable end product	Half-life of parent ($T_{\frac{1}{2}}$)
U^{238}_{92}	$U^{238}_{92} \rightarrow Th^{234}_{90} + \alpha$	Pb^{206}_{82}	$4{\cdot}5 \times 10^9$ yr.
U^{235}_{92}	$U^{235}_{92} \rightarrow Th^{231}_{90} + \alpha$	Pb^{207}_{82}	$7{\cdot}1 \times 10^8$ yr.
Th^{232}_{90}	$Th^{232}_{90} \rightarrow Ra^{228}_{88} + \alpha$	Pb^{208}_{82}	$1{\cdot}4 \times 10^{10}$ yr.
Rb^{87}_{37}	$Rb^{87}_{37} \rightarrow Sr^{87}_{38} + \beta$	Sr^{87}_{38}	6×10^{10} yr.
	A^{40}_{18} + X-rays	A^{40}_{18}	
K^{40}_{19}	K^{40}_{19} ↗ ↘		$1{\cdot}4 \times 10^9$ yr.
	$Ca^{40}_{20} + \beta$	Ca^{40}_{20}	

Before radioactivity was discovered by Becquerel in 1896, Lord Kelvin calculated (1871) the age of the Earth to be 100 billion years, about fifty times younger than today's accepted value. He assumed the Earth had been a hot sphere which had gradually lost its heat by radiating it into space. While this part of his work is correct, he failed to appreciate that the Earth, and particularly the crust, was being continuously supplied with energy by the decay of the radioactive substances contained within it. This heat production largely offset the heat loss arising from conduction through the Earth's crust, followed by heat loss by radiation from the surface into space. The cooling has thus been much more gradual than Kelvin expected, and his work provides us with a vivid demonstration of the fact that radioactive substances in large quantity can constitute enormous sources of heat or other forms of energy. Schemes to convert the heat of the crust to useful energy are discussed later in the chapter.

While problems arise from the existence of radioactive substances in the Earth they are one of the many features which indicate its suitability as a home for plant and animal life. The Earth may well be the only planet in the solar system on which any type of life can be sustained. As scientists improve their methods of using interplanetary space vehicles to collect material from other planets, and of making apparatus to carry out measurements on the planets themselves, we may be able to establish if any living organisms exist anywhere in the solar system except on Earth. The existence of the atmosphere and oceans, the contained energy and the large flux of energy from the Sun make the Earth unusually suitable for sustaining life as we know it.

The decay chains of uranium and thorium finishing in isotopes of lead are long and complicated. Uranium occurs in numerous minerals such as pitchblende, uraninite, carnotite, autunite, and davidite. It is also found in phosphate rock, lignite, and monazite sands. It has no fewer than fourteen isotopes and all are radioactive; the three naturally-occurring ones are U^{238}, U^{235}, and U^{234} (only 0·0058 per cent abundant). The most common isotope of radium, one of the daughter elements, is Ra^{226}_{80} of $T_{\frac{1}{2}}=1{,}620$ yr; about 1 g is found in 7 tons of pitchblende. Radium itself produces radon gas, a dangerous radioactive material. A total of eighteen radioactive isotopes of radon are known and the one formed from radium is Rn^{222}_{86} which is α-emitting with a value 3·82 days for the half-life. The isotope Rn^{220}_{86} emanates as a gas from thorium and is called thoron. It has a half-life of 54·5 seconds and is an α-emitter. Radon-219 emanates from actinium (formed in the uranium-235 decay chain) which is itself a very active substance, about 150 times more active than radium. The name actinon is given to radon-219 and it has a half-life of 3·92 seconds, decaying with both alpha and gamma emission. These natural radioactive gases are always present in minute amounts in the atmosphere.

Uranium and Thorium Reserves. The table shows the half-lives of U^{238} and U^{235} are very different. During the 4·5 billions of years that the Earth has existed the amount of U^{235} has decreased until today it is only 1·5 per cent of the original amount for the decay has persisted for a time exceeding six half-lives of U^{235}. The U^{238} has decayed during the same time to half its initial value since its half-life is in fact 4·5 billion years. U^{235} now composes about ·7 per cent of the total amount of natural uranium, and it is clear that its low abundance is the result of the more rapid decay of this particular isotope. Calculation shows that when the Earth formed, U^{235} was about 20 per cent abundant in uranium.

The natural sources of nuclear fuel have been studied rather extensively since about 1940, and the findings are well summarised by S. H. U. Bowie, *Phil. Trans. R. Soc.*, **A276**, 1974, p. 495. He explains why estimates of available tonnages differ widely. He stresses the importance of price in defining 'ore reserves', a term which is applied to material that can be commercially viable now or in the near future. On this basis material available in 1974 at a price of less than \$20 per kg of uranium oxide, U_3O_8, in concentrate form is classified as ore. Resources include material in price ranges above this level, and the amount of estimated resources increases with increase of price.

Uranium is concentrated in abnormal amounts in various parts of the Earth's crust, so figures for the overall crustal average are not helpful. Bowie states that over 70 per cent of uranium reserves are in quartz-pebble conglomerates and sandstones, making estimates of tonnage reasonably reliable. With the lowest commercially viable, or 'cut-off', grade of 50 parts per million of uranium metal, about 10 below the lowest grade readily mined at present, many millions of tonnes of uranium can be made available. Normal granites with an average uranium content of about three times the average crustal abundance, say 5 parts in 10^6, need not be considered in the near future.

Estimates of world reserves of uranium (non-communist countries) in the price range below \$20/kg U_3O_8 are shown in table 5.2. At this price the total is 866×10^3 tonnes U (1 tonne$=$0·984 ton). Bowie considers about 100,000 tU can be added to this total to represent additional reserves in Australia. It is

TABLE 5.2

ESTIMATED WORLD RESERVES OF URANIUM (NON-COMMUNIST COUNTRIES) PRICE RANGE $<$\$20/kg. U_3O_8
(From OECD/NEA and IAEA Report, 1973)

Country	Established reserves in 10^3 tU	Estimated additional resources in 10^3 tU
Australia	71	78·5
Canada	185	190
France	36·6	24·3
Gabon	20	5
Niger	40	20
South Africa	202	8
US	259	538
Others	53	52
Totals	866	916
Additional Australian	100	

clear that with a cumulative uranium requirement of 170×10^3 tU in 1977 and an estimated cumulative requirement of about 946×10^3 tU by 1987 the reserves could prove inadequate within a decade or two unless the price rises markedly. At a price around $60/kgU, uranium may be extracted from sea water.

The distribution of uranium is shown in table 5.2 but other countries with reserves of more than 1,000 tU include Argentina, Central African Republic, Greenland, Italy, Japan, Mexico, Portugal, Spain, Turkey, Yugoslavia, and Zaïre. The combined total for these countries is put at 52,700 tU. Clearly the element is widespread but concentrated in a relatively few countries.

Thorium is not yet considered as useful a nuclear fuel as uranium but it could in due course prove important. The information shown in table 5.3, though less reliable than that for uranium, shows that it could be obtained in comparable amounts.

TABLE 5.3

WORLD RESERVES AND RESOURCES OF THORIUM (NON-COMMUNIST COUNTRIES) PRICE RANGE $< \$20$/kg. ThO_2
(From S. H. U. Bowie, *Phil. Trans. R. Soc.*, **A276**, 1974, p. 503)

Country	Established reserves in 10^3 tTh	Estimated additional resources in 10^3 tTh
Brazil	1·2	31·8
Canada	80	80
Arab Republic of Egypt	14·7	280
India	—	300
South Africa	20	—
USA	52	265
Others	8	—
Totals	176	957

The important ore minerals are monazite, thorite, uranothorite, and brannerite, and in recent years production has been mainly as a by-product of uranium production.

B. GEOTHERMAL ENERGY

A great deal has been written on the subject of geothermal energy, and a most instructive review of research and development is given in the UNESCO publication entitled *Geothermal Energy* (ed. H. C. H. Armstead) which is volume 12 of the series on Earth Sciences (Paris, 1973). Among the contributions is a notable one, 'Basic Theories', by Sir Edward Bullard. Bullard stresses that in recent years studies of plate tectonics have resulted in a major revolution in geology and that new ideas about the history of the Earth, and the nature of motions within it, have been proposed and widely accepted.

The great variety of rocks at the surface of the Earth is evident to all.

Sometimes sediments such as clays, sandstones, or limestones occur, often ancient 'shields' of granite are exposed, and lavas which poured from volcanoes can be observed. It is the sediments which contain all the oil and most of the mineral resources of the Earth. Beneath the sediments of the continents lies a support which is mainly granite. The granite may be formed in different ways as the result of mechanical and chemical processes, including the freezing of molten material and changes in sediments brought about by heat and by liquids and gases from below. The granite contains a large proportion of silica ($Si\ O_2$) in the form of crystalline quartz and much of this has been separated from more basic rocks, containing less quartz, usually deeper in the Earth. Sometimes this basic rock appears near the surface in the shape of basalt dykes cutting the granites. The mountains are composed of masses of sediments which have been squeezed, heated, and distorted and penetrated by molten rocks from below. The flow of water and ice transforms their shapes in the course of time. The oceans are vastly different from the continents. The rocks there are basalts and no granites lie below the sea floor. The under-ocean mountains are due to volcanic action and largely basaltic.

Some 35 kilometres beneath the continents there is a boundary known as the Mohorovičić discontinuity, or simply 'Moho'. This boundary marks the lower limit of the granites and basalts which form the 'crust' of the Earth and below it there is a marked change of composition of the predominant materials. The Moho discontinuity lies only about 5 km below the ocean floor and is on average, therefore, 10 km below the surface of the seas.

Under the crust lies the 'mantle'. The physics of the mantle leaves some questions open. However it appears to behave as a solid in the short term, of the order of a day, and in the longer term, say 10,000 years, it behaves more like a liquid whose viscosity varies with the stress. The mantle, with gradually

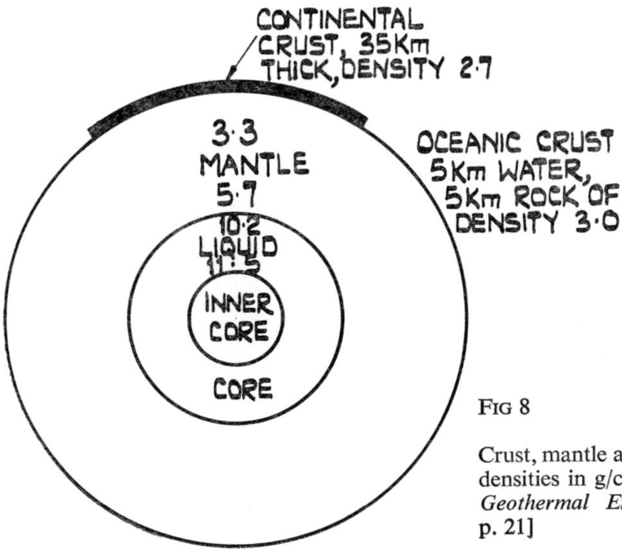

FIG 8

Crust, mantle and core of the Earth with densities in g/cm³. [From E. Bullard in *Geothermal Energy*, UNESCO, 1973, p. 21]

changing structure, extends to a depth of 2,900 km and below that is found a much denser and liquid core which has a thickness of about 2,000 km. It is thought it consists mostly of liquid iron. The central sphere, of radius 1,350 km, may be compressed and solidified iron. Its temperature might be as high as 4,000 K. The radius of the Earth is about 6,370 km. A sketch of the cross-section of the Earth is given in fig. 8.

Heat Flow from Underground. Heat flow from within the Earth is obvious in areas of volcanic activity, of geysers, fumaroles, and boiling mud pools. Measurements carried out in non-volcanic areas show that heat flows upwards from within the Earth, and typical examples of such measurements from Bullard's paper are given here (table 5.4).

TABLE 5.4

TYPICAL MEASUREMENTS OF HEAT FLOW

(From Sir Edward Bullard in *Geothermal Energy*, ed. H. C. H. Armstead, UNESCO, 1973, p. 21)

	Depth m	Greatest temperature °C	Mean gradient °C/km	Mean conductivity 10^{-3} cal/sec cm °C	Heat flow 10^{-6} cal/ cm² sec
Gerhard-minnebron (South Africa)	3,022	46·4	9·5	13·5	1·28
Tocketts (Yorkshire, England)	906	35·7	27·7	4·26	1·18
Adams Tunnel (Colorado, US)	940	24·8	24·1	8·00	1·93
E. basin of North Atlantic	4,670	0·116	25·4	2·28	0·58
Central Valley of Mid-Atlantic Ridge	4,109	0·87	315	2·07	6·52

The heat flow q from within the Earth to the surface, setting aside special circumstances affecting the general pattern, depends on the rate of increase of temperature with depth, or thermal gradient r, and on the thermal conductivity K of the rocks. Heat flow is given by the product, that is,

$$q = Kr.$$

From the table we can deduce that the heat flow can vary fairly markedly. Anomalies in the measured values occur at plate margins, and the upward flow of heat is approximately constant save in exceptional sites. Where poor conductivity does occur we expect a high temperature gradient and vice versa. Away from exceptional locations some hundreds of land measurements and some thousands of sea values have been observed, and a plot of the distribution of the results yields a consistent average value of close to 60 mW/sq.m or $1{\cdot}4 \times 10^{-6}$ cal/sq.cm sec. (fig. 9).

The principal systematic differences in the heat flow should be mentioned. The continental shields give low values around 40 mW/sq.m; mountain ranges tend to give high values of about 85 mW/sq.m; and on the crests of ocean ridges values as high as 340 mW/sq.m are measured.

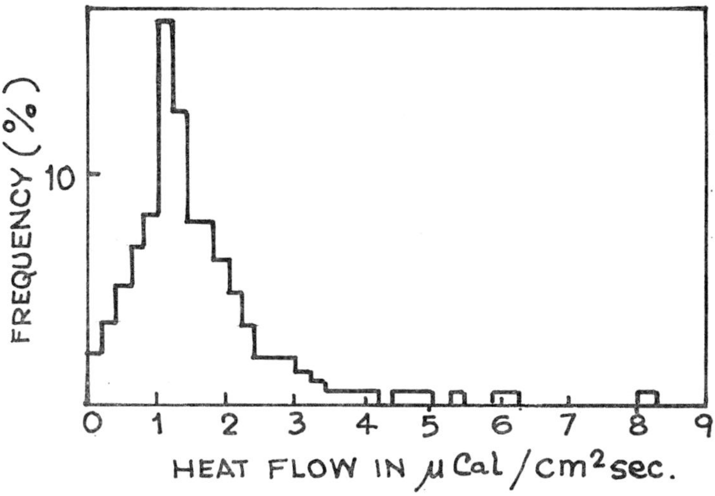

FIG. 9. World heat flow, excluding geothermal areas. [From E. Bullard in *Geothermal Energy*, UNESCO, 1973, p. 21]

We can assess the total output of terrestrial energy with 60 mW/sq.m as the mean value of the flow for the whole Earth. The area of the surface of the Earth $=4\pi r^2$ and, with $r=6{,}370$ km or $6{\cdot}37\times10^6$m, the total $=60$ mW $\times 4\pi \times (6{\cdot}37\times10^6)^2=30\times10^6$ MW. Taking the output of a typical electrical generating station as 1,000 MW, the geothermal output equals that of 30,000 such stations. The total energy available is of the order of 10 times the total world power consumption at present, but we must remember that much of it is probably completely ruled out as a result of practical difficulties and because as a source it is very dilute. Actually it is unlikely to be more than a useful addition to the total required supply. It is, for the most part, virtually inexhaustible, but in those areas where the geothermal energy is usefully released it is, as we shall see, not renewable.

Bullard shows that the total time for which the flow of energy from within the Earth has continued makes it a most impressive source. The Earth is about 4,500 million years old and some rocks are found to be 3,500 million years old. Even with the present-day average flux value of 60 mW/sq.m (which has been declining slowly for some billions of years), in a period short compared with the age of the Earth, say in 1,000 million years, the heat emitted would correspond to the burning of a volume of coal with a surface area equal to that of the Earth and a depth of 20 km. This calculation rules out the possibility of chemical reactions in the crust producing the required amount of energy, and establishes that the energy must be of nuclear origin. The energy is ascribed to the decay of known natural radioactive elements mainly contained within the crust. Table 5.5 shows the principal elements involved but the decay of the many daughter elements of both isotopes of uranium and of thorium is the source of a large fraction of the total.

TABLE 5.5

HEAT PRODUCTION OF LONG-LIVED ISOTOPES AND PRODUCTS
(From Sir Edward Bullard, in *Geothermal Energy*, ed. H. C. H. Armstead, UNESCO, 1973, p. 22)

Isotope	Stable end-product	Half-life in 10^9 years	Proportion of isotope, %	Heat generation cal/g yr.
U^{238}	Pb^{206}	4·50	99·27	0·70 } 0·73
U^{235}	Pb^{207}	0·71	0·72	0·03 }
Th^{232}	Pb^{208}	13·9	100	0·20
K^{40}	Ca^{40}, A^{40}	1·31	0·012	$\cdot 27 \times 10^{-6}$

Using the heat generation figures given in table 5.5., Bullard deduces that the natural radioactive elements in a thickness of 14 km of granite would produce a flow of 10^{-6} cal/cm²sec; and he concludes that the upper crust contains more of the radioactive elements than the lower levels of the crust. There is, in fact, experimental evidence to support such a conclusion, for the heat flow in areas of granitic basement rocks in the US has been shown to be proportional to the radioactivity of the rocks.

It is interesting to speculate on the possibility of utilizing this long-lasting source of Earth energy. Clearly the average value of the temperature gradient matters crucially. Measurements of the temperature as a function of depth for the Gerhard-minnebron bore in the Transvaal are shown in fig. 10. The data are observed temperatures, and the full line is the variation expected for constant heat flow, based on the observations at the greater depths. The gradient is actually less than could be expected in the upper 4,000 feet of the bore, and this reduction in gradient is ascribed to the circulation of water in the fissures, removing part of the heat. Nevertheless the average gradient for the bore is around 10 °C per km. At many parts of the Earth's surface a gradient of two or three times this value can be achieved, and it is not unreasonable to assume 20 °C increase per kilometre of depth. For 3 km penetration a temperature rise of around 60 °C might be obtained. Even if we could use this temperature difference to advantage in a thermal engine, it would still have a comparatively low efficiency, because from thermodynamic theory we know that for such an engine operating between two reservoirs, one at temperature T_{hot} and one at temperature T_{cold}

$$\text{Thermal Efficiency} = \frac{(T_{hot} - T_{cold})}{T_{hot}} .$$

If, then, the surface temperature is 300 K, and the gradient in a hole of depth 3 km gives a T_{hot} of 360 K, then

$$\text{Thermal Efficiency} \leq \frac{(360-300)}{360} \leq 17\%.$$

This is a very favourable case and in practice the efficiency is likely to be less, say 10 per cent. However, efficiency of this order is still worthwhile and we must look at the other major problem—the difficulty of the low thermal conductivity of rocks. This low conductivity implies major problems in removing the heat from the rocks at the bottom of bores. Obviously the heat sink is large, but its size combined with the low conductivity means the heat transfer does not proceed well. Schemes aimed at overcoming the difficulty have been advanced. Among the more unorthodox of these proposals is the use of high explosive, possibly nuclear explosive devices, to create cracks and fissures in the rocks which would then expose a large surface area to a liquid, probably water, that could be pumped down to the large volume of fissured, fragmented rock. After percolating through the rocks the water would return to the surface, via a second bore hole, at an increased temperature. It is at this date (1977) difficult to imagine that the heat of the Earth can be extracted in this fashion for useful applications except where conditions are favourable and temperature gradients above average. F. J. P. Clarke (*Times Higher Education Supplement*, 22nd July 1977) takes a pessimistic view. The

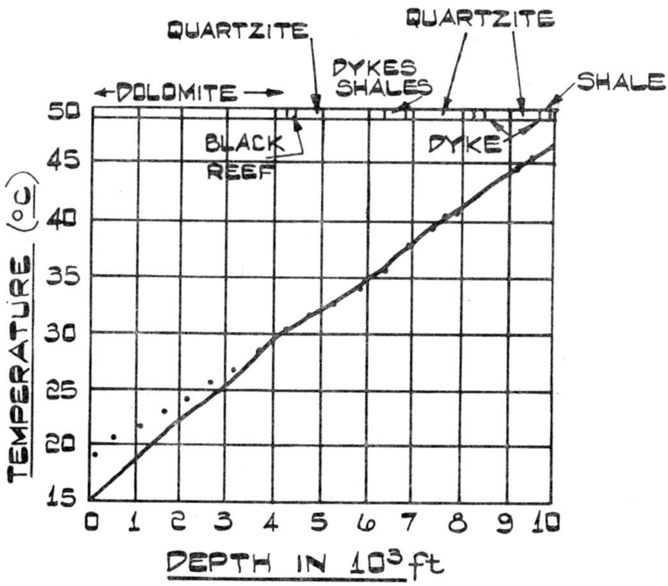

FIG. 10. Variation of temperature with depth at the Gerhard-minnebron bore in the Transvaal. [From E. Bullard in *Geothermal Energy*, UNESCO, 1973, p. 21]

practicability of such methods is, however, the subject of experiments at Los Alamos and, if these proved successful, a great increase in the number of geothermal sites that could be economically exploited would occur.

In more orthodox schemes of heat extraction, fluids would circulate through existing cracks or pores in the rocks. Water injected at sufficiently high pressure might produce much increased fracturing and so enable more heat to be extracted but work on such a method has not yet been attempted.

Special Areas of Energy Release. We have until now considered the Earth as a whole and dealt with its average properties. There are regions of the Earth, however, that do not conform to the average pattern, and the activity of volcanoes is evidence of the release of enormous amounts of energy in a variety of particular locations. Armstead has discussed (*Geothermal Energy*, op. cit., p. 15) the nature and occurrence of the special areas of the Earth where geothermal energy is readily made available. Sometimes the thermal gradient rises well above the modest average value and may indeed be as much as one hundred times higher. Armstead defines the heat in these regions as 'geothermal energy' but, as the reader knows, we apply the term more widely and generally in this chapter. The special regions are usually, but not always, associated with volcanic activity and earthquakes. The most important of the earthquake zones more or less follows the periphery of the Pacific Ocean and a second zone runs along the middle of the Atlantic Ocean—the mid-Atlantic Ridge. This second zone has an easterly branch running through the Mediterranean and the Middle East into Tibet, while yet another branch of this easterly branch of the mid-Atlantic Range follows the direction of the Great African Rift and the Red Sea.

Geothermally active areas likely to prove the most promising and fruitful for energy exploitation tend to be within earthquake zones but are not necessarily close to volcanoes. Two of the most famous geothermal fields, at Larderello in Italy and The Geysers in California, are at large distances from the nearest volcanoes. The earthquake zones are those with crustal weaknesses, and it is this that enables the deep-seated heat to rise near to the surface of the Earth. The exploitation of this heat on a small scale is entirely feasible and exploitation has in fact a long history. The Romans used geothermal heat in baths, and Icelanders, Japanese, Turks, and others have done likewise in modern times. The Maoris in New Zealand have adapted the heat to a number of everyday uses. Perhaps most notable of these applications is that at Larderello where the Italians have extracted boric acid from natural steam jets since the eighteenth century. Later, electric power was generated and at present a group of power stations have a total output of 390 MW. The Italian initiative in the exploitation of the energy is likely to be followed by much more widespread utilization of the natural heat for the production of electrical power, for district heating, air-conditioning, and for a variety of industrial purposes. The application to horticulture in both the USSR and in Iceland is particularly valuable in the relatively harsh climates.

Armstead takes an optimistic view of the future of geothermal energy and among other applications he mentions desalination as noteworthy. He considers dual or multi-purpose plants are much to be preferred. He stresses that with adequate risk capital available the power can be delivered cheaply, and that it provides a steady constant source which is in some ways even more reliable than its principal rival, hydroelectric power. Other experts note that the capital investment can be high for a utility which may prove to have a relatively short useful life. There are additional difficulties such as environmental pollution to complicate applications for permission to develop sites.

Published plans for generating electricity from geothermal sources have been summarized by T. Leardini (see table 5.6 below) and also by J. B. Koenig (in *Geothermal Energy; Resources, Production, Stimulation*, ed. P. Kruger and C. Otte, Stanford University Press, 1973) who gives an excellent outline, on a worldwide basis, of developments in almost all the promising sites presently in use.

The expansion data of Leardini given in this table should be compared

TABLE 5.6

PROPOSED INCREASES IN WORLD GEOTHERMOELECTRIC POWER
(From T. Leardini, *Phil. Trans. R. Soc.*, **A276**, 1974, p. 523)

Country	Site	MW output (Definite)	MW output (More tentative)	Intention
Italy	Travale-Radicondoli	15		Start July 1973
	{Serrazzano extension	15		Start in 1975
	{Upgrading existing units		45	In near future
	Alfina	15		Start in 1976
US	The Geysers			
	units 9 and 10	106		Start in 1973
	units 11 and 12	212		Start in 1975
	Salton Sea and Brady's hot springs	60		Start in 1978
New Zealand	Broadlands Field	120		Start in 1976
Mexico	Cerro Prieto	75		Start in 1980
Japan	Hachimantai	10		Start in 1974
	Onikobe	25		Start in 1975
	Hatcobaru	50		Start in 1975
	Hatcobaru		150	In near future
	Katsukonda	50		Start in 1977
	Katsukonda		150	In near future
USSR	Kamchatka	50		Start in 1980
	Kamchatka		50	In near future
El Salvador	Ahuachapan	30		Start in 1975
	Ahuachapan	60		Start in 1977
Philippines	Leyte	30		Start in 1978
	Leyte		70	In near future
Indonesia	South Sulawesi		250	In near future
World Total proposed increases		923	715	

with the situation regarding geothermoelectric power generation at the end of 1972. At that time the predominant sites were those in Italy and the US. A list of the important stations is given below in table 5.7.

TABLE 5.7

PRINCIPAL SITES OF GEOTHERMOELECTRIC ENERGY IN 1972
(From T. Leardini, *Phil. Trans. R. Soc.*, **A276**, 1974, p. 514)

Country	Sites	Electrical power in MW	Sub-totals in MW
Italy	Larderello	365·1	390·6
	Monte Amiata	25·5	
US	The Geysers	298	298
New Zealand	Wairakee	192	202
	Kawerau	10	
Mexico	Pathé	3·5	78·5
	Cerro Prieto	75	
Japan	Otake	13	33
	Matsukawa	20	
USSR	Pauzhetka	5	5·75
	Paratunka	0·75	
Iceland	Namafjall	3	3
		World Total=1,010·85	

We conclude that within a decade power generation could increase markedly. The sites are, however, largely those which have been well established and their performance can in some ways be guaranteed. It would be more reassuring if a considerable number of new sites were being developed. This fact may be significant in the light of the disparity in the estimates for the future of those with experience in geothermal energy resources. For example, Leardini draws attention to the estimates given by J. C. Denton and D. D. Dunlop (*Geothermal Energy*, ed. Kruger and Otte, op. cit.). Estimates for the US range between pessimistic values for the whole country between 5,000 and 19,000 MW and optimistic values as high as 400,000 MW for the Western US only, in each case by the year 2000, and with a lifetime of the resource of 100 years (see also US Department of Interior, *Assessment of Geothermal Energy Resources*, 1972, Washington D.C.).

The reasons for disparities in the opinions of experts may perhaps be traced, to some extent, to the differences in the nature of the fields that have been employed and may be employed for power generation in the future. We discussed at the beginning of this chapter the possibility of exploiting the vast amount of energy in the interior, most of it at considerable depth and presenting major obstacles to exploitation. If the Los Alamos experiments are successful, and if a reasonably large proportion of the more promising areas of hot dry rock can be found near the energy market, that is, reasonably close to centres of population, the future for geothermal energy would then offer particularly bright prospects. At the moment we ought to assume that it is only in the areas where thermal energy is available close to the surface that we can be sure of successful exploitation. It is pertinent to remark that these special areas of geothermal activity have finite life-times which may be very long, fifty years or more according to the level of utilization, but they are not renewable energy sources in the sense that the term is usually applied. The dry rock scheme would possibly offer almost unlimited duration. It is a little difficult to understand that a geothermal source can be depleted in the same way as a seam of coal but consideration of the formation of the source helps to explain its limitations.

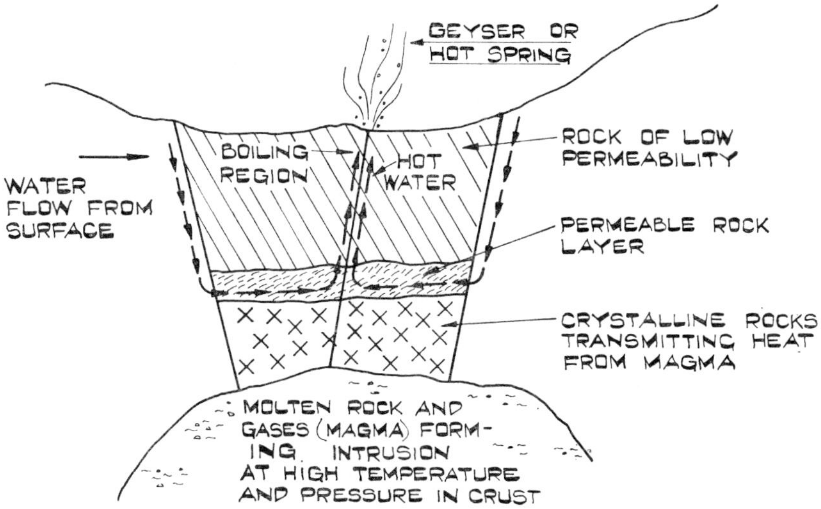

FIG. 11. Geothermal region in cross-section. [Based on L. J. P. Muffler and D. E. White in *Perspectives on Energy*, O.U.P., 1975]

The structure and behaviour of typical geothermal fields have been described by G. Facca (*Geothermal Energy*, ed. H. C. H. Armstead, op. cit.). He finds that there is general agreement that the source of heat, until now exploited in known fields, is a magmatic intrusion into the Earth's crust, the intrusion having a temperature of 600 to 900 °C and at depths of 7 to 15 km. He notes all these 'commercial' fields are in regions of past or present volcanic activity. In active volcanoes, intrusions of magma reach the surface through a large fault system and, in compact hard rock, faulting can provide a channel for the upward flow of magma while clay and other plastic rocks may flow into the fault space and seal it from above. In these cases the magma intrudes to the boundary between the hard and plastic rocks, giving rise to a region of cryptovolcanism. This cryptovolcanism can occur in areas devoid of recent volcanic activity and is often in geological areas of thick plastic formations. Two well-known major fields of dry steam type, namely Larderello and The Geysers, are of this kind. It is clear from this description of this type of field that the size of the intrusion is the main factor determining the life-time of the site as an energy source.

In fig. 11 the cross-section of a geothermal area is shown schematically. Most of the water supply in the majority of fields is from outside, that is, it is rain water initially. It appears the bulk of this rain water is heated conductively as it passes into and through permeable base rock. As hot water is drawn from bores or surface vents for heating or power production, the hydrological balance of the system must be maintained by an adequate inflow of new water. The rate at which the input water is made available and heated determines the level of output of the site.

There are a number of types of fields and the characteristics of several are given in table 5.8.

TABLE 5.8

CHARACTERISTICS OF SELECTED GEOTHERMAL FIELDS

(From P. Kruger and C. Otte, *Geothermal Energy; Resources, Production, Stimulation*, Stanford University Press, 1973, p. 24)

Field	Reservoir temp. in °C	Reservoir fluid	Heat content in cal/gm	Average well depth in metres	Fluid salinity in ppm.	Mass flow in 10^3 kg/hr	Non-condensable gases, %
Larderello	245	Steam	690	1,000	<1,000	23	5
The Geysers	245	Steam	670	2,500	<1,000	70	1
Makukawa	230	Mostly Steam	550	1,100	<1,000	50	<1
Otake	200+	Water	~400	500	~4,000	100	<1
Wairakei	270	Water	280	1,000	12,000	—	<1
Broadlands	280	Water	400+	1,300	—	150	~6
Pauzhetsk	200	Water	195	600	3,000	60	—
Cerro Prieto	300+	Water	265	1,500	~15,000	230	~1
Niland	300+	Brine	240	1,300	260,000	~200	<1
Ahuachapan	230	Water	235	1,000	10,000	320	~1
Hveragerdi	260	Water	220	800	~1,000	250	~1
Reykjanes	280	Brine	275	1,750	~40,000	~400	~1
Namafjall	280	Water	260	900	~4,000	400	6

More generally the sources can be classified according to the US Task Force Report of the National Science Foundation on Project Independence (1974) as follows:

A. Vapour-dominated hydrothermal systems (e.g. The Geysers)

B. Liquid-dominated hydrothermal systems

  1. High-temperature type (temperature >180 °C)
     with (*a*) Low-salinity (<20,000 ppm)
     (*b*) High-salinity (>20,000 and often >100,000 ppm)
  2. Moderate-temperature systems (100–180 °C)
  3. Low-temperature convection systems (<100 °C)

C. Geopressurized sedimentary basins with additional energy in the form of dissolved methane (located in Gulf Coast and Western states of the US for example)
D. Hot dry rock formations (for instance in western US)
E. Near 'normal' thermal gradient environments (20 to 40 °C/km depth)
   1. Sedimentary basins with water-filled pores
   2. Crystalline 'basement' rocks of low porosity and little or no associate water (very widespread).

As the authors remark, the list of principal sources from A to E fall into a scale of desirability and ease of development. The list in a sense sums up the situation so far as immediate future prospects for geothermal energy are concerned. It is an excellent source when it is clearly and demonstrably available for exploitation, and it will make a small percentage contribution to energy sources in a number of countries, but it will be a long time, probably more than thirty years, before it can be transformed to an energy source of widespread availability in view of the need to solve onerous problems in engineering the extraction of thermal energy from deep within the crust. The extraction of energy where it lies close or comparatively close, say 1 km, to the surface is a much simpler business though not without hazards, particularly in the economic sense. There can be risks for the capital investor due to the real possibility of uncertainties in the small-scale geology of the site. J. R. McNitt, C. J. Banwell, and G. E. Sigvaldson, in separate articles, have described some of the methods employed in exploration (*Geothermal Energy*, ed. H. C. H. Armstead, op. cit.) and explained the reason for some uncertainties.

A practical demonstration that the more marginal reservoirs can be utilized economically for generating electric power and for other uses, such as the commercial recovery of minerals, would expand the recoverable geothermal resources many times. In fact the estimate of Donald E. White of the US Geological Survey puts the stored thermal energy in the major geothermal areas of the world at about 4×10^{20} joules (see *Geothermics*, Special Issue 2, 2 vol., 1971). At a conversion factor of 25 per cent the production of electrical energy would amount to 10^{20} joules or 3 million megawatt-years. This energy is depletable but over a period of fifty years would amount to an average annual production of 60,000 megawatts, closely comparable to tidal power (we note it corresponds to sixty generating stations each of output 1,000 MW). The sub-surface area of the Earth involved in this scale of exploitation is well under 1 per cent of the total.

It is clear that a number of serious questions remain unanswered at the present time (1977), and Peter J. Smith (*Nature* **257**, 1975, p. 10) has emphasised their importance. The life-time of dry steam reservoirs is unknown; the interactions between heat flow, fluid flow, mass transport, and tectonic stress changes are obscure. The prevention of clogging and corrosion of the pipes and machinery by minerals such as silica is not yet entirely satisfactorily resolved. Pollution and other side effects are discussed in the next section.

All these difficulties bring back into the picture the extraction of heat from dry rocks, which in principle can be done almost anywhere in the world as long as sufficiently high temperatures are reached at depths down to which drilling is economic. The Los Alamos work along this route is advancing steadily. It is indeed a route advocated in 1927 by Hodgson though the Los Alamos approach involves fracturing of the rocks at considerable depth. It has been stated (A. W. Laughlin, *GeoTimes*, 20th March 1975) that general improvements in drilling techniques in recent years have made it possible to make a test drilling to about 3,000 metres with hydraulic fracturing at almost 2,000 metres, and it is intended that this will be repeated at the bottom. In several other areas of the world search continues for sites of dry rock heat at much less than the expected depth for the temperatures required—in a sense for anomalous sites. A site of this kind has been located in the Gulf of California and it may prove to be extensive. Here heat flow of 30 μcal per cm^2 per sec. has been measured, a heat flow ten or more times higher than might have been normally expected. It appears to be due to a basaltic intrusion within 100 metres of the surface. It is an extremely interesting example and, if uncomplicated by the presence of hydrothermal fluids, justifies an optimistic outlook on the future prospects of exploitation at other sites where the thermal energy can be drawn from regions of high heat flow existing close to the surface.

In spite of what has been said about the possibility of complicating features arising at many of the readily observable sites, the fuel cycle for the hydrothermal extraction of the heat for electricity production is technically straightforward. The generation must take place at the site itself to avoid energy losses and the geofluids are first separated into steam and liquid phases. The ratio of these varies from up to 100 per cent steam in vapour-dominated systems to up to 100 per cent liquid in water-dominated. In the former, the steam supply is provided by a gathering network of wells. The steam is passed through a condensing steam turbine-generator system and the exhaust enters cooling towers. Around 80 per cent of the steam mass is evaporated in the cooling towers and the condensate with the minerals within it injected into the less productive steam wells. In liquid-dominated systems a fraction of the geofluid is flashed to steam for electric power production and the remainder of the fluid used to provide a hot-water supply. This supply can be used in both domestic and industrial applications or simply rejected. The schematic form assumed by such a plant is shown in fig. 12.

Environmental Considerations. In recent times there has been strong opposition to the detrimental environmental and ecological consequences of many industrial enterprises. There is no doubt the effects of the Industrial Revolution during the last two hundred years have been serious and many environmental dangers and hazards have been faced and not always successfully overcome. Yet the economic arguments for improving the standard of living remain strong, especially for the developing countries. We shall examine environmental considerations in more detail in chapter 26. Here we would note that serious consequences could ensue from widespread adoption of other sources including some that would appear to be attractive renewable resources such as wind, wave, and tidal energy. In the case of geothermal energy the adverse effects of development are often very obvious.

While the activities related to power production at geothermal sites are localized and do not involve mining, fuel processing, transportation and handling facilities, other undesirable side-effects such as the release of hydrogen sulphide give rise to difficulties for a range of some miles around. There can be very considerable gaseous and particulate emissions, land modifications, subsidence seismic hazards, surface-water and ground-water pollution, biological effects, noise problems, and adverse social effects. The Project Independence study of 1974 drew attention to the lack of safe disposal techniques for toxic and saline liquid wastes containing sometimes arsenic, boron, and chloride salts. Among the gases that can be released are carbon dioxide, methane, and ammonia as well as hydrogen sulphide.

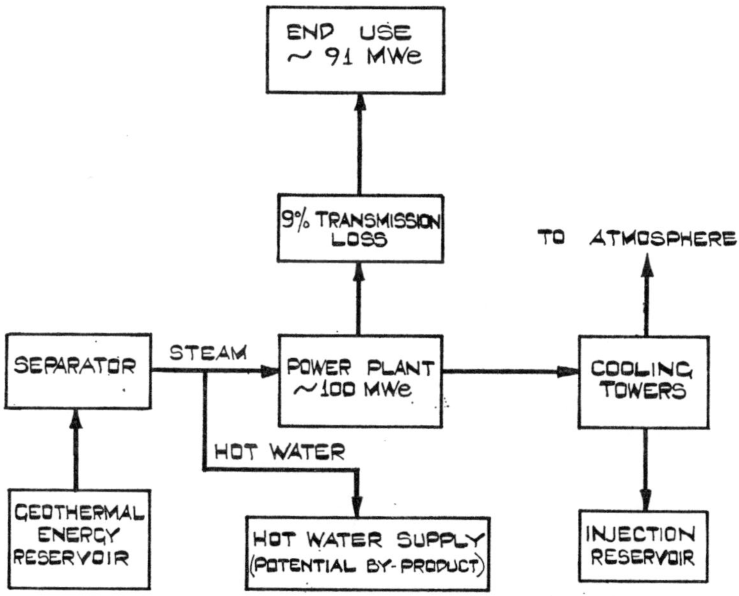

FIG. 12. Fuel cycle for a geothermal plant of 100 MW electrical output

Morton C. Smith of Los Alamos has called attention to the possibility of extensive subsidence arising from the removal of large masses of sub-surface fluid. This subsidence clearly can cause serious economic and legal complications, probably sufficiently grave to delay the rapid exploitation of many potentially useful sites. Subsidence has been observed in the Wairakei field in New Zealand and it can obviously prove to be a serious disadvantage affecting a range of structures including railroad beds and airport runways. Drilling is sometimes a cause of severe noise disturbance and it causes contamination of sub-surface water. In spite of the lack of ready-made answers to such problems the Project Independence analysis appears to be optimistic in tone and an estimate of a generating capacity in the Imperial Valley of California of as much as 100,000 MW of electrical power, over a period of fifty years, is mentioned. The environmental and ecological problems can probably be largely resolved with sufficient research and development effort, and capital, devoted to the tasks.

Potential for the Future. Before concluding this subject it is desirable to refer to the views of A. L. Hammond (*Science* **177**, Sept. 1972, p. 978) which represent a balanced view of the geothermal prospects as seen prior to the great upsurge of interest in all energy matters triggered by the OPEC policy of 1973 on oil production and pricing. The low pressures and temperatures at which plants like those at The Geysers operate, result in the conversion efficiency of thermal to electrical power being about 22 per cent, which is relatively low, but this does not hinder the Pacific Gas and Electric Company from achieving at The Geysers plant their cheapest source of electricity. Hammond remarks that The Geysers might produce between 1,000 and 4,000 MW of electrical power but uncertainties in the time of depletion hinder more rapid expansion of the output, planned to increase at a rate of 110 MW per annum for some years. He goes on to quote the potential generating capacity in the Imperial Valley of California, based on the work of Robert Rex and others, as being up to 20,000 MW. From examination of the work of other authorities he arrives at the conclusion that, on a conservative basis, as much as 100,000 MW of generating capacity could be developed with vigorous efforts by the year 2000 and that the prospects appeared to be worth the price. While this resource could not replace fossil fuels or nuclear power it amounts to a considerable additional resource.

A more recent expression of expert opinion conceived in an apparently less favoured country, the United Kingdom, is given in *Geothermal Energy: The Case for Research in the United Kingdom*, Energy Paper No. 9 of the Energy Technology Support Unit at Harwell. The author, Dr. John Garnish, argues convincingly for a positive, if cautious, approach to the prospects of geothermal resources in the UK. R. Oxburgh discusses Garnish's paper in his own contribution to the subject (*Nature* **262**, 1976, p. 526) and stresses once more that in most situations geothermal energy must be regarded as depletable, defining it as heat mining. He remarks that the rate of convective removal of heat from the rocks of a hot source is orders of magnitude faster than the heat can be replaced by conduction from below, and that in most cases the exploitation rate which is designed into the plant could result in exhaustion in some twenty years. On the other hand, in a minority of sites, near active volcanoes for instance, the transport of heat to the surface by magma could match or even exceed the extraction and prevent depletion.

The views of experts would thus appear to amount to cautious optimism for the future, with more exciting possibilities opening up if intensive effort is devoted to research, development, and industrial exploitation in the next twenty or so years. If such effort proved successful a contribution of about 2 per cent to the present total world energy consumption could be successfully provided within twenty or thirty years.

C. Tidal Energy

Tidal energy has not found extensive application but it may be used on a larger scale in the future. It is a good example of a renewable source which may yet be turned to good account.

The oceans are subject to the gravitational force of the Sun, the Moon, and the Earth itself. Being liquid the water can move readily under the influence of these forces. The attraction of the Moon, moving in its orbit round the Earth, is slightly more than twice that of the Sun because, although its mass is very much less than that of the Sun, it is far closer to the Earth (about 240,000 miles as compared with 93,000,000). The rotation of the Earth itself introduces a 24-hour period but in fact two tides occur per 24 hours, roughly one per 12 hours. The reason for the 12-hour period is made evident by fig. 13. The attraction of the water towards the Moon, on the side of the Earth nearer to the Moon, causes the water to be raised upwards in its direction. However, as the angular momentum of the Earth and its oceans must remain constant, an equal and opposite outward movement of water on the other side of the Earth must take place. Thus high tide exists on opposite sides of the Earth as indicated and, in consequence, low tides are caused elsewhere. The Earth takes 12 hours to rotate from one high tide position to the next, so likewise low tides are at about 12-hour intervals. The actual extent of a tide depends on the position of the attracting body with respect to the equator and on the relative positions of the interacting bodies. The Sun further

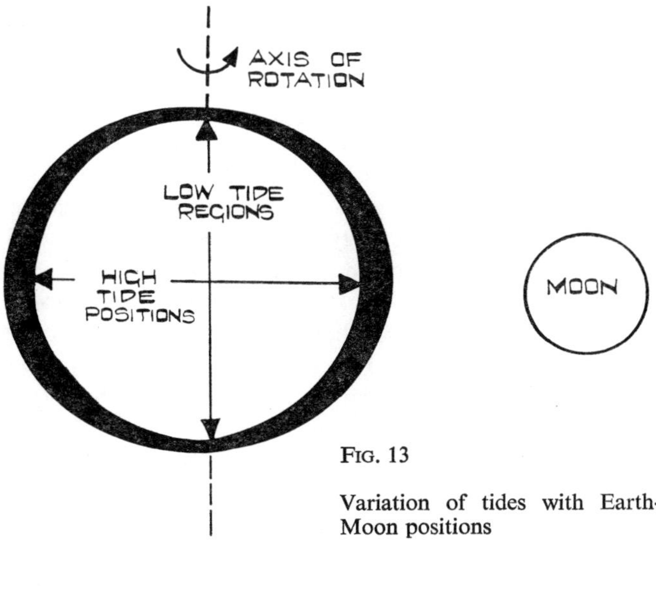

Fig. 13

Variation of tides with Earth-Moon positions

complicates matters, as it may add its attractive force to that of the Moon or operate in the opposite direction. It is perfectly feasible to calculate future positions of the Sun and the Moon and the forces they exert, but difficult to assess with accuracy the effect on the oceans at different parts of the coasts. Water depth and shape of coast are among influencing factors.

The energy represented in tidal action is very large and, if it could be harnessed efficiently, would theoretically yield enough to meet a worthwhile part of man's energy requirements. Unfortunately, there are not many accessible sites at which large amounts of this energy can be conveniently harnessed; presently (1977) tidal power schemes provide only a very minor element of the total energy consumption. In the inhospitable regions of the ocean, such as the Barents Sea, a large part—perhaps 80 per cent or more of the world total tidal energy—is uselessly dissipated in frictional losses concentrated where the water meets the shores. Where water has to enter and leave some parts of coasts in large volume but through narrow channels the

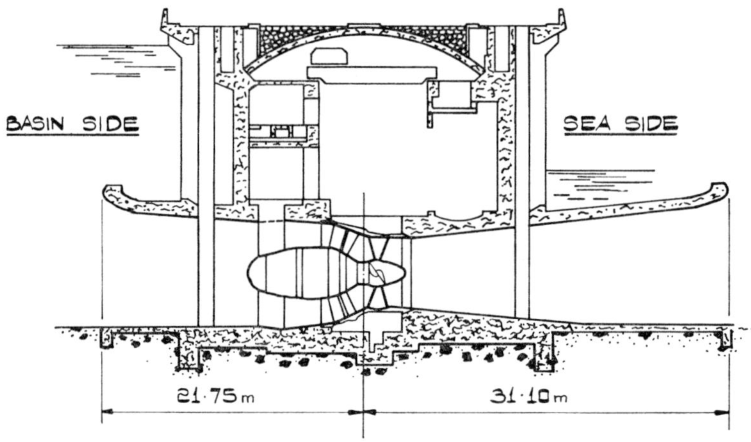

FIG. 14. Rance tidal power station for 240 MW. [From K. R. Vernon, *Phil. Trans. R. Soc.* **A276,** 1974, p. 490]

rate of flow can be very high; for example, 7 knots at some locations on the west coast of Scotland. The English Channel and certain adjacent waters represent good sources of tidal energy with high tides, of about 15 m, occurring in the Severn Channel for instance. Another interesting site is the Solway Firth. While there has been serious study devoted to the harnessing of tidal power in Britain, as with all such schemes the capital cost is a deterrent and the ecological effects are difficult to assess. The main financial difficulty is that the large amounts of water raised by tidal action have to be trapped in dams that are not too extended or difficult to construct. The number of such large reservoirs which can be suitably located is small. The water from the reservoir is released in controlled fashion and, in falling to a lower level, drives a water turbine which is coupled to an electric generator. One such tidal station is in operation at La Rance where tides of over 13 metres occur. The dam is $\frac{1}{4}$ km long and even in the favourable conditions exploited in this instance the construction cost per kilowatt proved high. The mean capacity of the La Rance plant averaged over the year is 65 MW. A cross-section is shown in fig. 14. Rated at 240 MW the turbine is reversible and can generate electricity with flow in either direction.

Another problem is that of matching output to demand. Tides occur at times which are not related to the peak of demand so storage of some kind is vital to success. One dam being inadequate in this respect, thought has been given to multiple reservoir configurations. Alternatively, electrical power from other stations can drive the turbines of the tidal plant as pumps and so raise the water into the dam so that it can be used later when electrical energy is required. This is an example of pumped storage of power, a process which has become increasingly popular in recent years in hydroelectric power installations, since it is an important way of matching more economically electrical generating capacity and demand.

Thus there are no serious scientific and technological problems in harnessing tidal power, though sophisticated techniques might have to be employed to time the output, but suitable sites are rare. It is thought that a two-basin system could prove successful in a Severn Estuary plant and that as much as 4,000 MW could be generated for the daily high demand period of about 12 hours. This would add about 7 per cent to the UK electrical energy total of 1977 or about 3 per cent of the total UK energy consumption. Such a contribution is well worthwhile and in 1978 the UK government allocated further funds for feasibility studies. Few locations, however, are as favourable as the Severn. In the Bay of Fundy, which is considered to be the best tidal power source for Canadian use, the average height of the tides is 11 metres and although this is promising it is able to provide only a small fraction of the total energy requirement of Canada. Other possible sites for practical development are located in the White Sea, the Sea of Okhotsk, the Timor Sea, and off the cost of Argentina. One estimate of the total output of potentially useful sites (see K. R. Vernon, *Phil. Trans. R. Soc.* **A276,** 1974, p. 489) proves rather disappointing, amounting to around 350 TW hours per annum which is not much in excess of the electrical energy consumption in the UK. A still more recent and reduced estimate by the World Energy Conference of 1977 indicates that tidal power at all the possible sites could supply the world with between 13 and 20 GW, which is the equivalent of between thirteen and twenty large electrical generating stations and only a small fraction of total world needs. Obviously therefore tidal energy can at most supply only a very small percentage of the total world energy requirements. This fact does not in itself mean such an energy supply should be neglected. If the high capital costs can be met the supply is assured and the life of tidal stations could be very long.

With the limitations of gravitational energy as illustrated in tidal possibilities in mind, we can turn to solar radiation which far exceeds all other sources available to us.

Chapter 6

The Sun—The Primary Energy Source

The Sun is the source of all the energy represented in sunlight itself, in winds, in waves, in dams and rivers, and stored after conversion in all our natural or chemical fuels such as wood, coal, natural gas, and oil.

Formation and Form of the Sun. The Sun is thought to be a second-generation star, that is, it is formed from the ashes of earlier stars which had burned out. Sunlight itself provides evidence of this as spectroscopic examination of the radiation reveals the presence of substantial amounts of heavy elements, such as iron, within the structure of the star. It is believed that condensations of interstellar matter took place and, as the gravitational force exerted by the condensed matter increased, it attracted more and more of the matter in the surrounding volumes of cloud. Later the matter reached an amount such that the interior pressure and temperature were sufficiently high for nuclear burning to start. In 1938 Bethe and von Weizsäcker worked independently on theoretical explanations of the nuclear transmutation processes which were capable of sustaining such a furnace. They concluded there were two main nuclear cycles, in each of which the most important step was the proton-proton reaction. This fusion reaction, which was discussed in chapter 3, results in the formation of helium, with a consequent loss of mass and an energy release related to the mass reduction. In the second of the cycles, the carbon cycle, the same final result was achieved in a more complex way. In this second cycle heavier nuclei—carbon, nitrogen, and oxygen—were involved in nuclear catalytic fashion to bring about the conversion of protons to helium nuclei. It is believed that the Sun as a furnace is representative of the very large number of less massive stars in the universe. The energy released in such stars generates an immense outward pressure, and such a self-disrupting force is counterbalanced by the strong gravitational attraction of the bulk of the material on the layers of gas composing the outer parts of the stars. (The containing gravitational force is negligible in man-made experiments so in these the thermonuclear energy is released explosively, as for example in the 'hydrogen' bomb. Conversely containment of reacting material in a vessel in which thermonuclear energy is released, as in laboratory fusion experiments, presents major problems, and is the main reason for the preference for fusion reactions which proceed more readily than the proton-proton one.)

The scale of thermonuclear energy release is impressive; for the Sun the rate of release is $3{\cdot}86 \times 10^{33}$ ergs per second, so, using Einstein's relationship

$E=mc^2$, we find the mass loss per second is no less than 4,700,000 tons per second, that is, $1{\cdot}4 \times 10^{14}$ tons per annum. A mass of more than 100 million million tons is consumed in proton to helium transmutation. The scale of mass to radiation change is truly astronomical. As these reactions consume about 10^{-11} of its mass each year the Sun has enough fuel to continue at its present rate of burning for roughly ten billion years.

During most of this long life-time the Sun will continue to be a glowing ball of gases, approximately 865,000 miles in diameter and consisting at present of roughly 70 per cent hydrogen and 28 per cent helium. It is entirely gaseous and, while it rotates with an axis almost aligned with that of the Earth, the period of rotation, as observed from the Earth, is 27 days at the solar equator and over 30 days near the poles.

The rotating ball of gases forms a mass of $2{\cdot}19 \times 10^{27}$ tons, which is 330,000 Earth masses. The specific density is 1·41, considerably higher than that of water, but only one quarter that of the Earth. However, the density does vary—from an extremely high value at the centre of the Sun (about 100 times that of water) to an extremely low value at the apparent surface of the Sun (about ·0003 that of air at sea level on Earth). The Sun appears as a brilliant disk and the boundary of its spherical surface consists of highly luminous gases known as the photosphere. The temperature of the Sun as we proceed inwards from the photosphere, which is at about 6,000 K, increases enormously and reaches between 15,000,000 and 20,000,000 K at the centre. Above the photosphere the temperature first declines from 6,000 K to about 4,000 K and then increases rapidly to above 1,000,000 K in the outer atmosphere.

The photosphere emits the greater part of the radiant energy and the 'effective' temperature of the Sun corresponds to its temperature. The well-established laws of radiation based on experiment allow the determination of the temperature of the photosphere in much the same way as we would determine that of an ordinary furnace. We visualize the Sun acting as a perfect radiator, or black-body. The distribution of the radiant energy with wavelength varies for an emitting source, of the form contemplated, in the fashion illustrated in fig. 15. The visible part of the radiant energy changes in relative intensity (more of blue as against red for example) as the temperature rises from 3,000 K to 6,000 K, and the distribution curve can be used as a reasonably accurate means of determining the temperature of the source. The peak of the curve moves towards shorter wavelengths as the temperature is increased. Although the radiation from the Sun observed at the Earth's surface has gaps in the spectrum, the curve which best fits the observations made near the centre of the disk is that for 6,000 K with the peak emission intensity in the blue-green at about 4,700 angstrom units (one angstrom = 10^{-8} cm). The 'effective' temperature may in fact be a little above 6,000 K as the photosphere may be a somewhat imperfect radiator.

The Sun and our Earth. Much of our knowledge of the Sun derives from examination of the light and other forms of radiation which it emits. As we have said, the visible light of the Sun resembles that of a black-body heated to about 6,000 °K, but detailed spectroscopic study of the radiation shows numerous spectral lines and bands to be present; these supply invaluable information on the nature of the source. Work at optical observatories has been reinforced with radio antennae measurements and by employing orbiting solar observatories, high-altitude rockets and balloons, which permit measurements uncomplicated by the effects of the Earth's own atmosphere, and in particular facilitate the study of X-rays and ultraviolet radiation. Recently infra-red telescopes have been successfully deployed.

For the purposes of this book the importance of the Sun rests mainly on its capacity to provide us with energy. The magnitude of the flow of solar energy to the Earth can be appreciated by considering the value of the *solar constant*. This is defined as the rate of flow of energy reaching the top of the Earth's atmosphere, 2 calories per sq.cm per minute or 333 calories per sq.m per second. Now 1 kilowatt (kW)=239 calories per second, so the solar constant can be expressed as $\frac{333}{239}$ kW per sq.m, or 1·4 kW per sq.m. There is considerable absorption in the atmosphere and the maximum flow of energy in clear sunny weather at ground level over the major part of the US, for instance, is around 1 kW per sq.m while the average is 0·18 kW. Opinions differ about the significance of such a flow. Some contend it is insufficiently concentrated to supply all our needs in practical form. Others are impressed

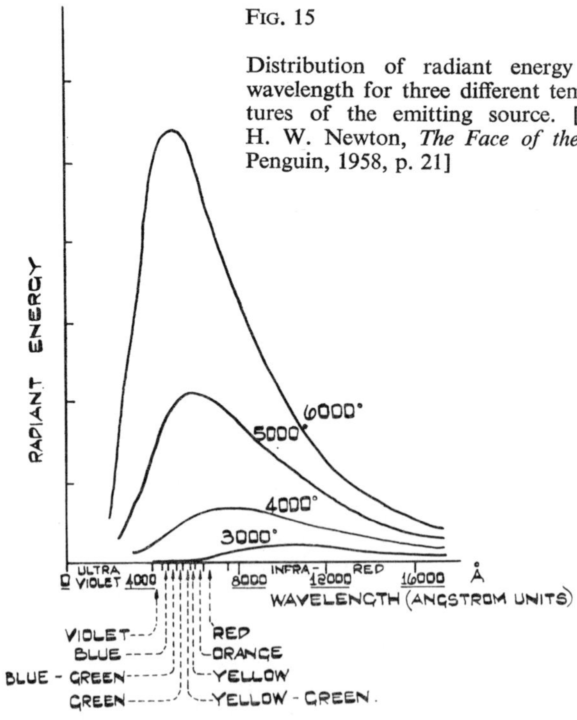

FIG. 15

Distribution of radiant energy with wavelength for three different temperatures of the emitting source. [From H. W. Newton, *The Face of the Sun*, Penguin, 1958, p. 21] by the total magnitude, which is $2{\cdot}4 \times 10^{14}$ hp for the Earth as a whole. For those who appreciate electrical units more readily, $2{\cdot}4 \times 10^{14}$ hp is $2{\cdot}4 \times 10^{14} \times 0{\cdot}746$ kW, or $1{\cdot}79 \times 10^{14}$ kW (1 hp$=0{\cdot}746$ kW). This total energy supply can be expressed in gigawatts as $1{\cdot}79 \times 10^{8}$ GW. A large modern electrical generating station usually has an output of around 1 gigawatt (GW); the total solar energy flux on Earth is therefore approximately equivalent to that of $1{\cdot}79 \times 10^{8}$ electrical generating stations, that is, almost 200 million large power stations. On a thousandth part of the Earth's surface the energy received equals that of 200,000 generating stations, more than enough to keep us provided with all the energy we need.

The Earth, 93 million miles from the Sun, has a diameter of only some 8,000 miles so it intercepts a tiny fraction of the total solar energy flux, about $0{\cdot}5 \times 10^{-9}$ of it (one half of one billionth of the total). This fraction is evaluated by applying the inverse square law which tells us that the amount of radiation intercepted varies as the inverse of the square of the distance between the source and a surface perpendicular to the rays. If we look at the inset of fig. 2 we note the total radiation flux of the Sun passes through the spherical surface of area $4\pi R^2$ at distance R of 93,000,000 miles from the Sun. The plane representing the Earth has an area of πr^2 with $r=4{,}000$ miles. The fraction collected by the Earth's surface is therefore $\pi r^2/4\pi R^2$ or $0{\cdot}25 \times r^2/R^2 = 0{\cdot}46 \times 10^{-9}$. Conversely, the total received at the Earth allows us to estimate the total energy flux from the Sun—it is two billion times greater than the flux incident at Earth and is thus $2 \times 10^{9} \times 2{\cdot}4 \times 10^{14}$ hp which is $4{\cdot}8 \times 10^{23}$ hp. The Sun is an impressive furnace and, rather surprisingly perhaps, it is extremely steady so far as total output is concerned. Measurements made during some fifty years show its output varying by only a few tenths of 1 per cent.

Solar energy reaches us in many forms. We can appreciate that visible light and infra-red radiation reach us in vast amounts influencing the atmosphere, the land, and the oceans and creating trade winds, ocean currents, thunderstorms, hurricanes, and so on. The bearing of these major effects of the Sun on conditions on the Earth is related in various ways to our energy needs and to our ways of meeting those needs, including of course the conversion of solar energy, the use of wind for energy generation, the use of river flow—particularly hydroelectric power—and the extraction of energy from waves.

Additionally the Earth and its atmosphere receive ultraviolet (UV) rays and X-rays. These ionizing forms of radiation produce the Earth's ionosphere. The ionosphere fluctuates in step with variations in the amount of incident UV and X-radiation. Indeed short-term outbreaks of solar activity can cause serious disturbances in high-frequency communications and, a more general effect, the heat generated by increased intensity of UV and X-rays can give rise to an expansion outwards into space of the atmosphere of the Earth. This effect has been substantiated by satellite research, as has the phenomenon of solar wind. The Earth's magnetic field is affected by the wind; some of the solar wind particles can be caught up by the extended magnetic field producing Van Allen belts of radiation encircling the Earth. Other trapped particles of the solar wind can give rise to aurorae when accelerated into the Earth's atmosphere. Finally, some less frequent flares of the Sun are of such a nature that they cause the ejection of particles of extremely high energy, actually of penetrating cosmic ray type, and, together with proton storms, the energetic particles can be a health hazard to astronauts—in fact to everybody at ground level though to a much reduced extent.

The Earth's Atmosphere. The atmosphere consists of four main parts, the troposphere extending from the Earth's surface to about 20 km height, the stratosphere between 20 km and some 50 km, the mesosphere extending from 50 km to about 90 km, and the thermosphere at still greater heights. The temperature variations of the atmosphere with height form the basis on which this subdivision rests, and the way temperature varies with height is shown in

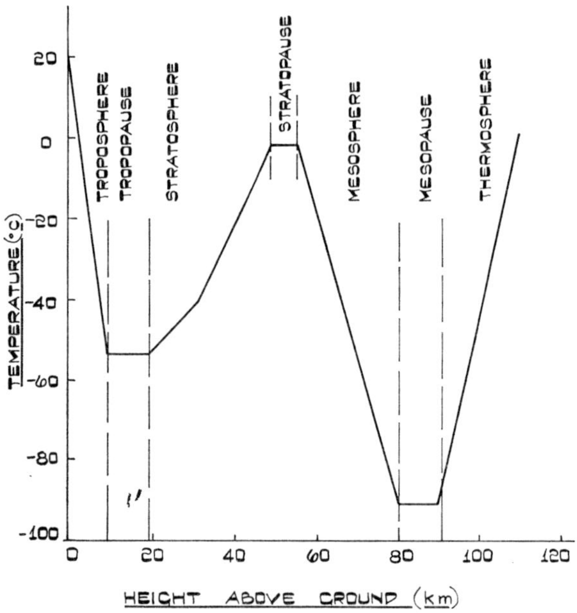

FIG. 16. Structure of the atmosphere, and variation of temperature with height above ground

fig. 16. The changes are striking and the causes complex. They can be related, however, to the way the atmosphere absorbs and reflects short-wave and long-wave radiation. From most points of view, and certainly in considering energy, the troposphere is the region of most importance.

The fact that a thermal energy balance exists allows us to apply the Stefan-Boltzmann Law to deduce the temperature of the top of the atmosphere and, in turn, the temperature of the Earth's surface. The Law states that

$$S = \sigma T^4$$

where S is the energy radiated per sq.m, T the absolute temperature, and σ, the Stefan-Boltzmann constant, is $5{\cdot}67 \times 10^{-8}$ watts/m^{2}K^4. It is assumed the Earth plus atmosphere behaves as an idealized black-body.

Now the total rate at which radiation falls on the upper atmosphere is 1,400 watts/sq.m. This radiation falls on a circle and the outgoing radiation emerges from the surface of the sphere representing the upper atmosphere.

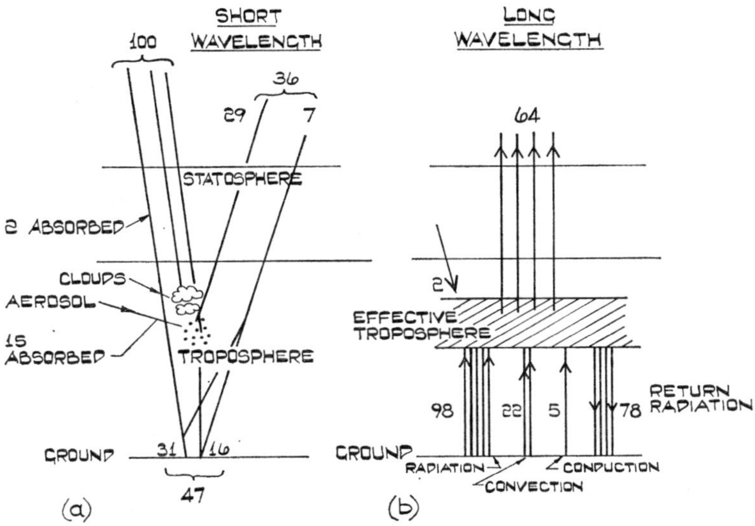

FIG. 17. Behaviour of short and long wavelength radiation in the atmosphere.

(a) For every 100 energy units of short wavelength radiation entering the top of the atmosphere, 2 are absorbed in the ozone and 36 are reflected as short wave radiation, most of it from clouds and particles in the troposphere. A total of 47 is absorbed at the Earth's surface, with 15 absorbed in the atmosphere at lower altitudes.

(b) The long wavelength radiation due to Earth heat contributes 98 + 22 + 5 (total 125). The effective troposphere returns 78, giving net energy loss of 125 − 78 = 47 units of energy. This is returned as shown in (a), in the form of short wave radiation. Likewise the 64 units emitted to space, added to the 36 units of short wavelength reflected as in (a), equals 100 units incident.

The outgoing radiation must have a surface density one quarter that of the incoming (ratio of area of circle to area of sphere

$$= \frac{\pi r^2}{4\pi r^2} = \tfrac{1}{4}$$

where r is the radius). Thus allowing for the albedo (the proportion reflected), and the directional factor, we find

$$\begin{aligned} &\text{Total radiation emitted/(sec)(m}^2\text{)} \\ &= 0{\cdot}64 \times \tfrac{1}{4} \times 1{,}400 \text{ w/m}^2 \\ &= 224 \text{ w/m}^2 \quad \text{and hence} \end{aligned}$$

by Stefan-Boltzmann Law
$224\ \text{w/m}^2 = 5{\cdot}67 \times 10^{-8}\ T^4$
or $T = 251$ K or -22 °C.

The temperature of the edge of the atmosphere calculated here is much below the average temperature of the surface of the Earth, and this difference illustrates the thermal blanketing role of the atmosphere, a natural effect of prime importance.

The behaviour of the short- and long-wave radiation in the atmosphere is shown in fig. 17(a) and (b). The troposphere transmits, absorbs, and re-radiates the energy of short and long wavelength so that a balance of energy is achieved. The more important components in the energy accounting are shown in table 6.1. All the values in the table are averages for the components which in practice may vary markedly from point to point on the surface of the Earth depending on vapour content, dust content, latitude, and nature of the surface of the Earth or sea.

TABLE 6.1

RADIATION BUDGETS OF THE EARTH AND COMPONENTS OF THE ATMOSPHERE
(derived from data of R. G. Barry and R. J. Chorley in *Atmosphere, Weather and Climate*, Holt Rinehart and Winston, 1970)

Zone	Input	Output	
Outer edge of atmosphere	100	reflected	36
		re-radiated	64
Stratosphere	2		2
Troposphere			
short-wave	15		
long-wave	91	space	57
in form of stratospheric	2	surface	
convection	22	(as back-	
conduction	5	radiation)	78
	135		135
Earth surface			
short-wave	47	atmosphere	98
long-wave	78	convection	22
		conduction	5
	125		125

We know that in the same units of energy flux 98 are reradiated from the whole surface of the Earth where 64 are absorbed at the edge of the atmosphere. According to the Stefan-Boltzmann Law T varies with S, the flux, so that it is proportional to $S^{\frac{1}{4}}$. If then for a flux of 64 at the upper edge of the atmosphere we find $T=251$ K, then for a flux of 98 at the surface of the Earth the corresponding temperature is

$$T=251\ (\tfrac{98}{64})^{\frac{1}{4}}\text{K}=279\ \text{K (or 6 °C)}.$$

This mean temperature of the surface of the Earth is not accurate and the value generally regarded as consistent with available measurements is 15 °C and not 6 °C. Many explanations of the discrepancy can be advanced, including the assumption that the Earth acts as a black-body radiator which is possibly unjustified. The mixture of incoming short-wave radiation and outgoing long-wave radiation is a further complication.

It is desirable to examine the matter of wavelength in a little more detail. The energy E of a photon or quantum of radiation is given by

$$E=h\nu$$

where h is Planck's constant and ν is the frequency of the radiation (p. 13). The frequency ν is related to wavelength λ by the formula

$$\lambda=\frac{h}{\nu}.$$

The higher frequencies thus correspond with the shorter wavelengths and as higher frequency means higher energy we deduce that a number of quanta of short-wave radiation are more energetic than the same number of quanta of long-wave radiation. (The inverse relationship $\lambda=h/\nu$ between wavelength and frequency is illustrated in fig. 2.)

The temperature T of any emitter is what determines the most probable wavelength λ_{max} of the radiation emitted. The relationship between the two is stated in Wien's displacement law, thus

$$\lambda_{max}\,T=\text{constant}=2{\cdot}9\times10^{-3}\ \text{metre Kelvin}$$

The curves of fig. 18 show the spectra or energy distributions of radiation for the Sun and for the Earth but not on a comparable energy scale, as the enormous difference between the Sun and the Earth is ignored to show both on the same order of scale. The Sun has a λ_{max} of about 5,000 Å and the Earth has a λ_{max} nearly seven times as long, around 35,000 Å. This radiation is in

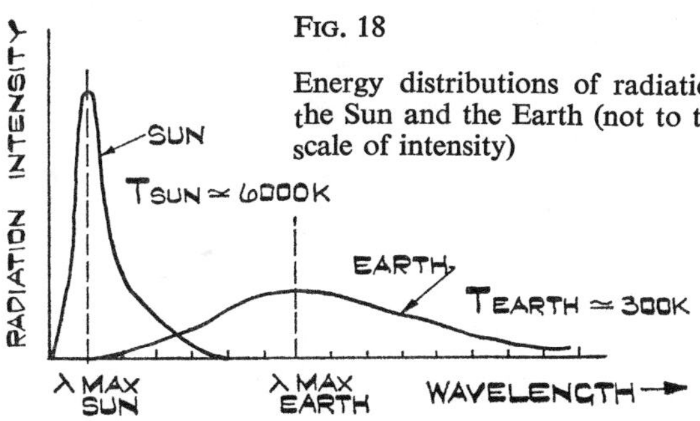

FIG. 18

Energy distributions of radiation from the Sun and the Earth (not to the same scale of intensity) the far infra-red and while the amount is very large it is difficult to apply such radiation usefully. The quantum energy corresponding to λ_{max} is only 0·35 eV.

Chemical Composition of the Atmosphere. Chemically, the troposphere consists principally of nitrogen and oxygen. The chemical composition of the *dry* atmosphere at the surface of the Earth is as shown in the table below:

TABLE 6.2

COMPOSITION OF THE DRY ATMOSPHERE AT THE EARTH'S SURFACE

Gas	Volume Concentration (per cent)
Nitrogen	78·084
Oxygen	20·9476
Argon	0·934
Carbon Dioxide	0·0314
Neon	0·001818
Helium	0·000524
Methane	0·0002
Krypton	0·000114
Hydrogen	0·00005
Nitrous Oxide	0·00005
Xenon	0·0000087
Sulphur Dioxide	traces up to 0·0001
Ozone	between 0 and 0·000007 in summer and between 0 and 0·000002 in winter
Traces of nitrogen dioxide, ammonia, carbon monoxide, and iodine	

There is contained in the atmosphere an extremely important and variable constituent, namely water vapour. The Sun has a marked influence on the variation which has a range extending from 3 per cent in the Tropics to 2×10^{-5} per cent in the Antarctic. The main mass of the atmospheric water vapour is located in the troposphere and the concentration falls rapidly with increasing elevation. If the water vapour in a column of air in a typical part of the atmosphere at temperate latitudes was all condensed at the Earth's surface, it would form a layer of some 1·6 to 1·7 cm in thickness. Information on the water-vapour content of the stratosphere seems not yet entirely reliable, though it would appear from recent studies to be very dry.

The Sun plays a dominating part in influencing the process of evaporation of water over the seas and over land, and evaporation, condensation, and transportation processes become interwoven. Condensation results in cloud formations from which precipitation of water takes place in the shape of rain, hail, or snow. Most of the weather phenomena occur in the troposphere and, on average, some 50 per cent of the Earth's surface is covered by layers of cloud. Such a cover can be seen in the typical photographs of the Earth's surface taken in space flights.

There has been a marked increase in interest in ozone influence on the atmospheric processes; this influence arises chiefly from its power to absorb ultraviolet solar radiation, the main factor increasing the temperature in the stratosphere. The ozone content varies with the time of year (a phenomenon known as the solar radiation factor which has a minimum in autumn and a maximum in the spring), and latitude, increasing from the equator to the poles. The man-made changes in the ozone content are still the subject of research.

The concentration of carbon dioxide can be affected by industrial pollution, though plant growth and ocean absorption are the main factors determining its value. It is definitely established that the carbon dioxide content has increased in the present century; this increase has its origin in the large rise in the burning of fossil fuels. Its effect is uncertain but it may give rise to a hothouse effect and so raise the average temperature of the Earth's surface. The hothouse effect refers to the increase in the amount of heat radiation that is retained within the atmosphere, when it would otherwise escape into space in the absence of gases such as carbon dioxide, vapours like water vapour, and dust. Such a man-made change in climate is possibly unique and must be monitored carefully. Other reference will be made to this problem (p. 266).

While small particles suspended in the air may have an industrial origin they are, for the most part, derived from the dust-storms, volcanic eruptions, and other natural phenomena such as whirlwinds. The small particles which range up to some tens of microns in size (1 micron $=10^{-6}$ m) act as nuclei for condensation of water vapour and so an atmospheric aerosol is created. The aerosol has an important influence on solar radiation and must be considered in any energy balance in the atmosphere, tending as it does to lower the surface temperature.

Radiation Balance of the Earth's Surface. As we saw above, the atmosphere absorbs relatively little of the shorter wavelength radiation from the Sun, but it does absorb the longer wavelength radiation (infra-red rays) which is emitted from the surface of the Earth and which in part is derived from the incoming solar radiation that does reach the Earth's surface. This capacity of the atmosphere to absorb infra-red is important in reducing heat emission from the Earth to outer space. Fig. 17 shows us how the absorption of solar radiation in the atmosphere, chiefly by the minor constituents, is accompanied by scattering brought about both by aerosol particles and clouds. The combination of direct and scattered radiation makes up the total radiation, and the total radiation on the Earth's surface is partly absorbed and partly reflected—the average value of the albedo being 0·36. The heat exchange between the Earth's surface and the atmosphere is known as the effective radiation. The difference between the short-wave radiation absorbed at the surface and the effective radiation is defined as the radiation balance.

The principal source of the heat provided to the atmosphere is the Earth's surface and not the incident solar radiation. In fact the amount of solar radiation directly absorbed in the atmosphere is less than the heat lost from it to outer space (by longer wavelength radiation). The input of energy from the Earth's surface to the atmosphere takes two principal forms—turbulent heat exchange and the heat of water-vapour condensation in the atmosphere.

The heating processes just outlined are not uniform over the surface of the Earth and as a result there is much air movement. Among the factors causing non-uniform heating is the marked variation in solar energy input with latitude. Thus we find a general atmospheric circulation takes place, this circulation being such as to tend to reduce differences in the heating of the air by horizontal heat transfer. The water vapour from the oceans is carried over dry land by similar movements and the rotation of the Earth also plays a part. The energy input of the Sun is thus converted to some degree to a complicated system of air flows, some almost stable and others extremely variable. The more stable include trade winds (towards the equator in both hemispheres), monsoons (between ocean and continent), and in the temperate latitudes westerly winds (from west to east).

The high-altitude wind phenomena do not appear to be important so far as energy studies are concerned but they are complex and interesting. Conceivably, knowledge of conditions outside what is normally regarded as our atmosphere will, at some time in the longer-term future of energy planning, prove useful in an applied sense.

For the present purposes we are mainly concerned with the conversion, on a substantial scale, of solar energy into wind and waves. We have mentioned the large-scale air flows but there are, in addition, in the lower atmosphere, numerous air circulations which may be variously classified, as, for example, breezes and mountain-valley winds. In some parts of the world there are relatively prolonged periods of winds of comparatively high speed. Thus, off the west coast of Scotland and north-west England we have a high percentage of days with winds of around 15 miles per hour. These winds can be regarded as thoroughly viable energy resources. The practical aspects of employing winds and waves as energy sources will be examined. They are obviously associated to a considerable extent.

Chapter 7

Direct Exploitation of Solar Radiation

Before considering the more complicated issues that arise in using solar energy at ground level, it is interesting to digress a little to mention a scheme for utilizing the radiation in outer space. Schemes both above and on the ground frequently depend on the deployment of photovoltaic cells. These are such important devices for energy conversion that they merit a separate discussion and the reader is therefore referred to chapter 10.

Exploitation of Solar Radiation in Space. It is possible in theory to avail ourselves directly of this 'wasted' energy. Some thought has been given to the use of stationary satellites of high reflectivity which could possibly be deployed as miniature suns, mainly reflecting sunlight on the Earth rather than allowing it to escape into space. An imaginative system has been described in rather more detail by Peter Glaser. The concept may be rather far-fetched but it could offer rewards in the longer term. Glaser visualizes the construction of two large panels, each consisting of arrays of thousands of photovoltaic cells. These panels would be placed in a stationary orbit at about 22,300 miles above the Earth. Conversion of the incident sunlight by the cells at 18 per cent efficiency is assumed, and the electrical power of about 1,500 MW so generated is fed from the two huge arrays (which would themselves absorb 8,500 MW of solar energy) into a microwave generator provided with cooling attachments. The resultant microwave energy of some 1,000 MW, generated with around 66 per cent efficiency, is beamed to a suitable aerial on Earth and there converted to such forms of electrical power as may be required for transmission and distribution to consumers.

Much remains to be done before such a scheme can be put into practice, but it does offer the exciting prospect of trapping solar radiation which otherwise would pass uselessly into empty space, and it gives, in theory, a supply independent of the time of day, or weather, or seasons or indeed geographical location. It could conceivably be the successor to the nuclear reactor even though the project seems a formidable one. A summary of discussions held in Washington in September 1977 has been given by J. Kistemaker (*Physics bull.* **29**, 1978, p. 58).

Exploitation of Solar Radiation on the Ground. At ground level solar radiation varies with latitude, climate, and seasons, but in fact the total radiation received annually over a large part of the more populated part of the globe does not vary by a very large factor. In Northern Europe, in latitudes between 50° and 60°, the total is roughly half that incident at the equator. In Britain the average annual value is 0·1 kW per sq.m and in the US it is about 0·18 kW per sq.m. The variation of insolation (exposure to the sun's rays) for several countries is shown in fig. 19. In practical terms if 1 per cent of the amount falling on Britain were converted to electricity with 10 per cent efficiency all the present electrical requirements would be met. However, 1 per cent is still a large area of land (945 sq. miles), and this example drives home the point that solar radiation is not highly concentrated. Another difficulty is that seasonal variations are marked and in winter and summer the insolation can differ in northern latitudes by a factor of about 6, while at the equator it varies very little between winter and summer. This highlights the importance of the problem of storage of the energy obtained by solar conversion. Unfortunately it is in winter, when solar radiation is lowest, that energy demands tend to be at a maximum. Most industrial requirements do not show very substantial differences between winter and summer, but in the temperate countries of the more northern latitudes domestic consumption

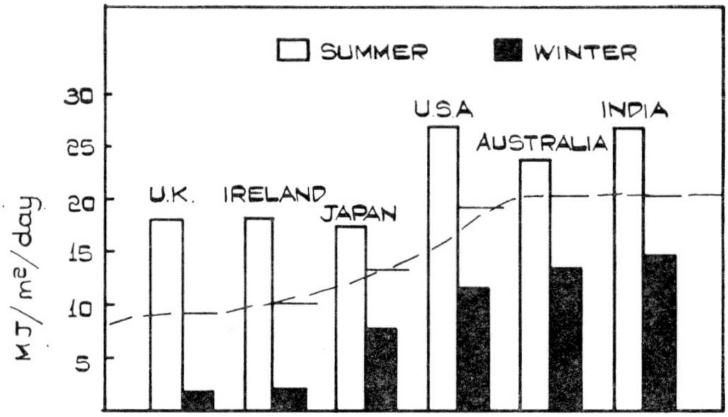

FIG. 19. Summer and winter insolation in several countries. The seasonal variation is more marked than the average annual values, which differ by a factor of about 2 (the curve through the bars indicates averages). [From J. K. Dawson, *Atom* **231**, 1976, p. 13]

reaches a peak in winter when space heating is most required. It may be that at best solar radiation will be a very useful part of a total energy provision. The question of diluteness arises also in considering other renewable converted forms of solar energy, that is, wind and wave power, and must be taken into account.

Petr Beckmann has argued in his book (*The Health Hazards of Not Going Nuclear*, Golem Press, 1976) that solar power is much too dilute to be truly helpful in meeting the energy demands of an industrial society. He takes 1 kW per sq.m (a value appropriate to very favourable conditions) as the input and, adopting a figure of 10 per cent efficiency for conversion, with 50 per cent spacing for the large arrays of photovoltaic cells, he points out that a 1,000 MW solar plant occupies an area of 50 sq. miles. He appears to have miscalculated: an area of 10 sq. miles seems to be required and not 50 sq. miles. Beckmann argues that 50 sq. miles per 1,000 MW is in stark contrast to the few acres required in providing the same energy from a fossil-burning or nuclear plant. His argument is weakened by his error but carries force on other grounds including that of danger of accident to workmen in cleaning the structures covering the extensive installation. He continues to examine the situation for small-scale conversion, for example by homeowners, and deduces the situation there would be very much worse since many of the solar units would be on roofs of homes and cleaning would be even more hazardous.

By implication Beckmann relates the diluteness to the high capital cost of solar plants. The size of the structures required to collect the energy does mean substantial outlay at the start: he quotes a figure of $1,500 per kW (1976) of photovoltaic power. This high cost could conceivably be reduced by the required factor to compete with fossil or nuclear sources but the diluteness problem, though apparently exaggerated, remains as a serious handicap.

Practical Schemes for the Direct Conversion of Solar Energy. Solar energy has found until now very little direct application even for domestic purposes but there is clear evidence that this situation is likely to change rapidly in the near future. We must not overlook the fact, however, that the glasshouse has a relatively long and honourable history in horticulture, and during the past two decades transparent plastics which help to 'trap' solar radiation and prevent a high rate of loss of energy into the atmosphere, especially in windy conditions, have become popular. Without doubt the increase in crop productivity achieved in these ways has proved very valuable.

Very recently architects, building designers, and engineers have constructed multi-purpose blocks, schools, offices, and homes so that all forms of energy, including solar, were used with maximum economy. A good number of solar houses has been built to study the whole problem and also to evaluate the rather complicated question of space heating and air-conditioning. More scientific institutions are becoming deeply involved in analysing the range of problems that arise. The work looks promising and if the designs can be standardised to some extent and made reliable, there could well be a rapid rise in the demand. Recent tests show that many solar panels deteriorate rapidly in use. Some claims by manufacturers have been exaggerated but a new realism both in the scientific and industrial scene is emerging. Work such as that in the EURATOM establishment at Ispra in northern Italy will prove of much value in putting the new solar studies on a sound basis.

Most of the flat solar panels on the market are rather simple affairs with water circulated inside the relatively shallow container. The flat panels are intended for mounting against an inclined wall or roof so that the incident solar radiation is almost maximized. The water is usually taken to a large storage tank and used for hot-water supplies or, in more ambitious schemes, for space heating and even air-conditioning. Sketches of a manufactured system are given in fig. 20 (a) and (b), The sections through the Collector System and a Unit are shown and manufacturing data provided.

Among the most important features in the design of the solar panel is the transparent cover through which the radiation penetrates, thus reaching the water. This has to have good transmission characteristics, low reflectivity, and poor transmission for radiation passing from the water outwards. Some designs of such covers are much better than others. With simple, well-designed panels thermal efficiencies of 40 per cent are possible. The durability of the system is a matter of considerable importance as maintenance costs prove onerous if the design is faulty.

Recently more attention has been given to energy concentration, even in the case of panels for use on the domestic scale, and it is essential for all high-temperature solar installations. Martin Processing use a metallised film made by Imperial Chemical Industries Limited (ICI) and known as Melinex 442. A series of concave reflective surfaces, covered with Melinex, concentrate the Sun's rays on to thin copper tubes in which water can be raised to temperatures up to 70 °C. A special ultraviolet screening and glazing material, lightweight and of high clarity, is used in preference to glass for covering and protecting the units. The system presents few installation problems. The

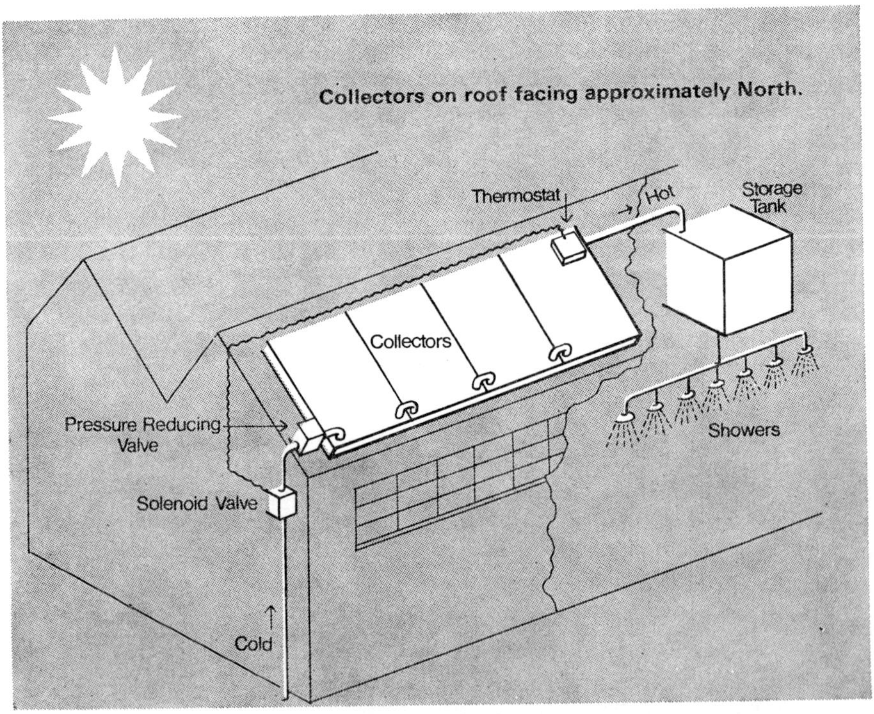

FIG. 20 (a). Solar panels for a hot water supply. A very big well-insulated storage tank, holding two or more days' supply, is not shown (from Hall Thermotank Group publication).

concave surfaces are regarded as useful in 'tracking' the Sun, giving longer effective collection times. These large lightweight panels seem to show that an extremely valuable energy complement can be secured in appropriate climates. The cost is modest and in 1977 was between 10 and 12 US dollars per sq.ft of panel. At this price the investment for domestic supply purposes is considered to be recoverable in about ten years by fuel savings, without allowing for escalation of fossil fuel prices.

To reach higher temperatures in the water of a solar system it is of course even more necessary to concentrate the incident radiation. Radiation falling on an array of mirrors can be focussed on water in a tower to bring it to a temperature of several hundred degrees centigrade. In this way high pressure steam can be provided to drive a turbo-generator and deliver electricity. The high capacity stations require solar towers. Rapid progress has been made recently. A solar thermodynamic power plant at La Pas (Mexico) started operation in 1976 (*Nature*, **260**, 1976, p. 477) and stations of 1 MW output are under construction at Odeillo. Such a station, or heliostat, for 5 MW is being built in the US and plans for 10 MW plants in the US and in France are in preparation. Mitsubishi Heavy Industries plan to build one in Japan to provide 1,000 MW. The optimism prevalent in this field of work is evident when it is noted some experts anticipate that in the US some 6 per cent of the total energy consumption will be by solar conversion in the year 2000 and about 25 per cent by the year 2025.

F. J. P. Clarke (*Times Higher Education Supplement*, 22nd July, 1977) takes the view that while in many more sunny countries the current development work in solar concentration techniques may bring some success in the future, they cannot be used to advantage in the UK and other countries with prevalent cloud cover for much of the year. In such places the flat-plate type

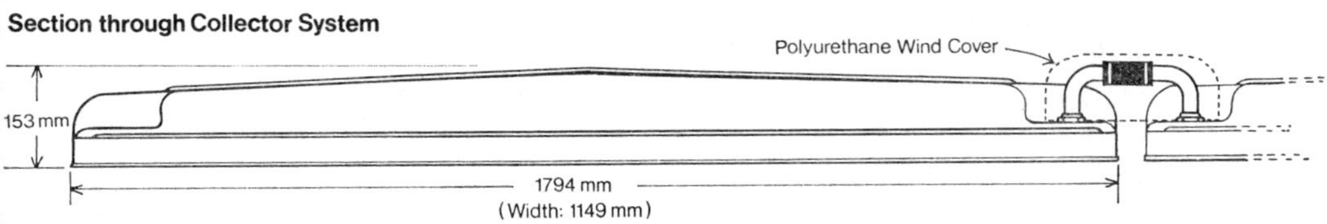

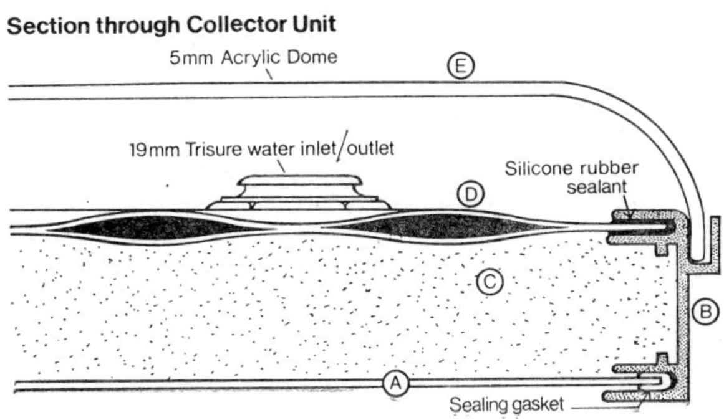

FIG. 20 (b). Details of the solar panel and of a single unit of the system of figure 20 (a)
A is galvanised steel, 0·6 mm thick
B is the frame of extruded aluminium
C is polyurethane foam, 25 mm thick
D shows stainless steel sheets, 0·5 mm thick, resistance welded to form water channels and heater spaces
E is acrylic dome transmitting 90% of solar radiation (Hall Thermotank system)

of radiation collector has to be used and the sunlight falling on typical appliances in this class raises water to about 75 °C. Higher temperatures tend to result in increased heat loss but nevertheless water around this temperature is eminently suitable for many domestic purposes. Passed into water tanks, and from there to various heat exchangers, it can be made available for domestic hot-water supply and for space heating or even air-conditioning. In the warmer climates the increased summer flux of solar energy gives the opportunity to provide air-conditioning, while in winter it can be used for space heating and supplying hot water.

It is clear that there will be useful energy economies arising from both simple plate-type collectors and concentrating-type collectors of sunlight. The rapidly increasing amount of study devoted to improvement of collectors, to study of reliability and maintenance, and to capital cost and installation problems would seem to assure rapid progress in this field. Independent assessment of these features is proceeding, and work such as that at EURATOM in Ispra will be most valuable. The problem of the scale of investment is one requiring attention. Clarke estimates that solar energy, used directly in well-engineered and well-installed forms of flat-plate systems, might contribute by around the year 2000 the equivalent in energy of several millions of tons of coal per annum in the UK. This would be around 1 per cent of the annual total coal equivalent consumption in the UK of 325 million tons in 1976. If such a worthwhile contribution to fuel saving can be made in what is a rather unattractive climate, it would seem that continued attack on the engineering and scientific problems would be amply rewarded.

In the UK domestic space heating consumes 20 per cent of the total primary energy demand (domestic hot water requires some 7 per cent) and the total space heating requires around 30 per cent of the total primary energy supply. Success in space heating by solar radiation could certainly, in many countries such as the UK, be guaranteed to provide a relatively large fraction of the total energy requirement. Here again it should be noted that reasonably compact heat storage is extremely important. It is no exaggeration to say that all the sources that are considered to promise well as alternative renewable sources would be very much more attractive if compact storage of heat, or energy in any usable form, could be provided. The need for convenient storage is self-evident when we remember that many humans live in flats in densely populated cities. It is not only difficult to store energy at an adequate level in such circumstances—it is also difficult to see how the flat collectors can be placed to collect radiation in sufficient amount. The country dweller with sufficient roof and garden space is much better placed to avail himself of solar energy.

Chapter 8

Solar Energy Conversion: Wind, Wave, and Ocean Thermal Power

Wind Power. The energy in winds is essentially derived from the energy radiated from the Sun. The atmosphere provides the mechanism whereby conversion occurs; the seas, likewise, provide the means of transforming solar energy to wave energy. Obviously, the total derived energy in winds will be less than the original solar supply, but in wind energy there can be local concentrations, even though in general wind energy is also dilute. Much the same can be said of wave energy; indeed both of these derived sources may be large and concentrated enough to be useful. At the World Energy Conference of 1977 the agreed estimate of the total energy in winds, in the whole atmosphere, was about $2,600 \times 10^{12}$ gigawatt-hours per year. Roughly one quarter of this, that is 650×10^{12} gigawatt-hours, was the estimate of the energy in wind over land areas. The Conference estimates of the usable wind power were, however, much less attractive, ranging widely and falling between 1 and 20 million megawatts. Calculation shows that the total wind energy in the atmosphere is a million times more than the usable energy, if the estimates are roughly correct. The lavish provision of wind power in nature means that, in spite of this reduction factor, we could still have available about 10 million megawatts, or 10,000 comparatively large power stations each of 1,000 MW output. This is in excess of the present-day total of generated electrical power. Clearly the energy is abundant and we have a problem not of availability but of practicality before us.

Man has made use of wind energy for a long time and Holland is famous for the decorative, if cumbersome, windmills that adorned the landscape. In more modern industrialized countries, including of course Holland, they have played only an insignificant role. However, since the 1950s there have been numerous attempts to use windmills on an industrial scale. Such efforts have in general raised grave doubts about wind as a viable large-scale resource, and it has usually been argued that wind systems do not merit substantial research and development programmes. Winds are fickle over most parts of the Earth's surface, so that the intermittency of a supply of electrical energy produced by a rotor or propeller is a substantial difficulty. The cost of energy storage, for such is implied in a viable wind-based energy scheme, must be taken into account in assessing costs, and on present knowledge is likely to represent a considerable addition to the capital investment required. It is possibly the intermittency of wind, and hence supply, that has resulted in little attention being given to this source until very recently. For a long time it was thought that the answer was to use the surplus current to electrolyse water, thus producing oxygen, and hydrogen which is a valuable fuel. This method has not yet been found attractive in large-scale practice. There are nevertheless regions of the world where wind is prolonged and reasonably steady and Martin Ryle has stressed the important fact that in Britain the average wind power is greatest in winter when the energy need is at a maximum. This factor reduces considerably the gravity of energy storage difficulties.

At the beginning of the twentieth century Denmark made great use of wind power and some 33,000 generators each produced on average some 35 kW of electrical energy. A total of 1,155 MW was thus obtained from 33,000 installations and this output equates roughly with that of a single large fossil-fuelled or nuclear-fuelled modern power station. The grand total is not impressive, but distribution from such a large number of small local power units is a relatively trivial exercise and this is a factor in favour of smaller, more local, supply points. At the opposite end of the spectrum a very large wind rotor built in Vermont, US, produced 1·25 MW. The twin-bladed propeller was 175 ft in diameter. In the late 1950s a large 100 kW wind generator was built in the Orkneys, but mechanical breakdown occurred soon after the start of operation and work was abandoned on this installation, just as in Vermont, when metal fatigue resulted in breakdown.

It seems certain that serious work on wind power came to a halt because of the low cost of fossil fuels. However, even in the Orkneys, now closely involved in the North Sea oil industry, the fact that wind power is free while oil is becoming more and more expensive may mean that we see renewed efforts to exploit wind power.

The advance in design of both rotors and propellers has led to clearer understanding of the inherent problems. Thus Professor K. Bergey has reached the conclusion that a windmill should not be built to operate at a level directly related to the windspeed; no effort should be made to accommodate very strong winds and the speed of the rotor or propeller is better limited to the value reached in a wind of around 20 miles per hour. Such limitation of the speed of rotation can be ensured with relatively simple governors, or by automatic feathering of the blades, and the mechanical stresses are reduced to levels which are readily tolerated. In certain quarters it is contended that sophisticated technology could discredit wind-driven generators and that their future is more assured when they are built by inexpensive and comparatively simple techniques.

The modern windmill may be best constructed in the way described by George Darrieus who patented his idea in 1927. The windmill essentially consists of two curved blades attached to a vertical axle which rotates as the blades receive the wind from any direction. The blades are aerofoils and they respond well in winds of either low or high velocity. Tests have been done with such windmills from around 8 ft to 16 ft in height, clearly of domestic type. Other windmills have been made with twin-bladed propellers of a size to deliver around 100 kW. The size needed for still larger outputs becomes somewhat daunting—for 3 MW output at about 30 miles per hour wind speed the blades would be about 200 ft across. The Savonius rotor is shown in fig. 21 and an example of a modern windmill design is illustrated in fig. 22.

The energy collection efficiency of a well-designed rotor or propeller may be about 60 per cent. The mechanical losses and other losses in energy conversion would probably result in an overall efficiency of 40 per cent. The capacity installed has to exceed that required for peak demand by a factor of about three, to allow for idle times when stored energy would be deployed. On its own as an energy source such an installation could well result in high capital investment, and a number of estimates by experts seem to confirm such an undesirable conclusion. A Swedish estimate due to the State Power Board puts the cost of constructing a 2 MW plant at £1¾M which is considered to be four times the cost of competitive systems. Comparatively large structures, which are still very low in output compared with fossil-fuelled generating stations and would hence be needed in very large numbers, would also occupy valuable land. In the long term the stations could of course be off-shore or on less valuable land, but they then need a complex transmission system. On the whole the prospect for wind systems seems bleak. The situation would change if a break-through with a small rugged installation of inexpensive design were to be made. Used on a single family scale and as an auxiliary to the more conventional supplies it is possible wind energy could prove valuable.

Recent work, in particular that of Martin Ryle (*Nature* **267**, 1977, p. 111) seems to offer a definite challenge to the more orthodox and established ideas of energy generation by wind. He stresses that, at least in Britain, the fluctuating loads associated with space heating of buildings and other needs have given rise to peak loads about three times the average. He underlines the fact that the short-time high demands that have been met until now with the help of oil and natural gas will not be met at all easily when fossil fuels are rare

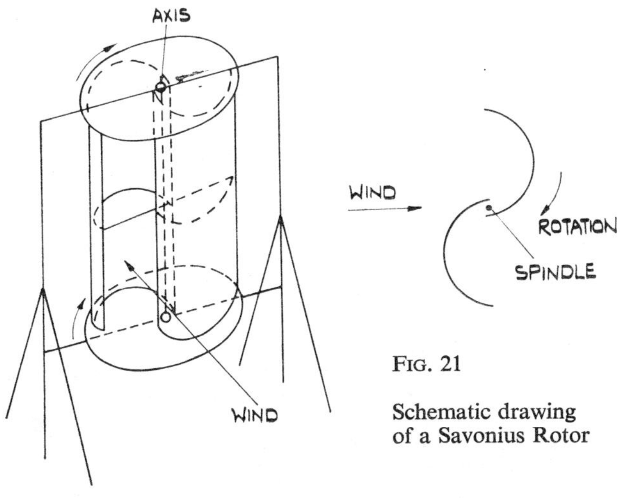

FIG. 21

Schematic drawing of a Savonius Rotor

and electricity is required to meet a much increased fraction of the total demand. He contends that the necessity of meeting the short-time fluctuations, and the winter difference, dramatically affects the analysis in favour of 'renewable' sources of energy such as wind and wave.

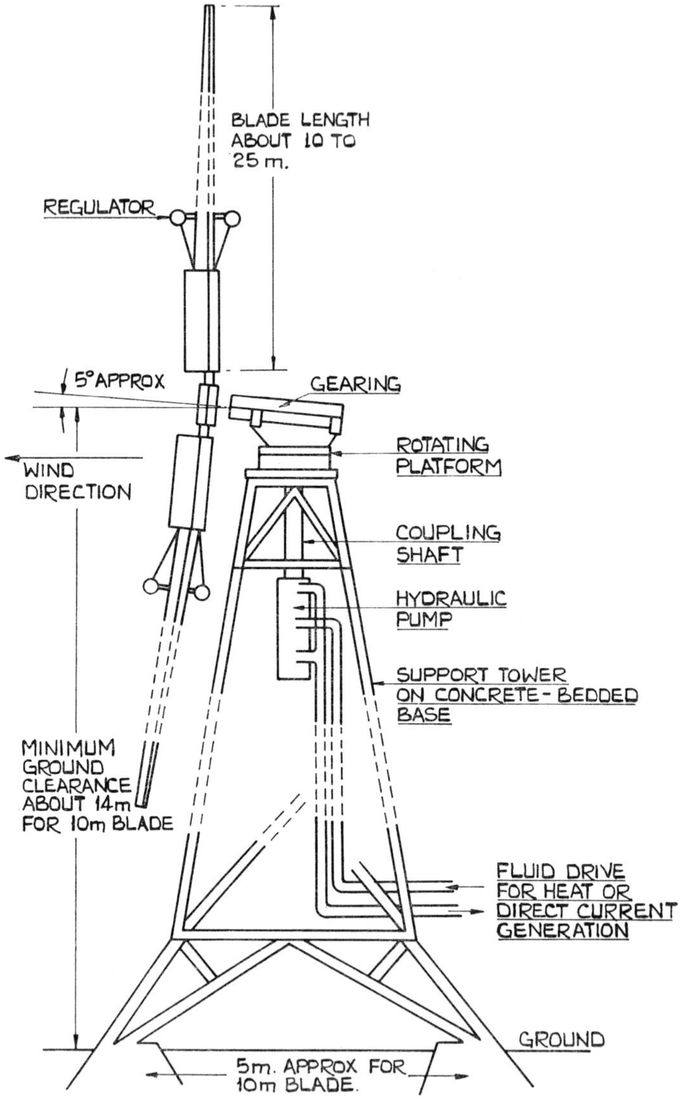

FIG. 22. Main components of a wind generator. For a rotor diameter of say 50 m. about 2×10^6 kWh per annum (approx. 1 MW peak power) can be produced on a good site with a wind speed averaging 30 km/h. The output varies directly with the cube of the wind speed. Self-feathering of blades can limit the speed to say 3 revs/sec. The number of the blades used in different machines usually lies between 1 and 5. In some designs the blade (or blades) rotates in a horizontal plane to improve reliability and structural strength.

The demand fluctuations in gigawatts as calculated by Ryle are given in table 8.1 below. His data for the various modes of consumption, in winter and summer and during the day and night, show the marked disparities that arise and the pronounced difference in different consumer needs. The modes of consumption are analysed carefully to calculate the equivalent electrical energy consumption as, for example, in transport where petrol is estimated to have a mean conversion efficiency of 14 per cent and diesel 21 per cent. In order to arrive at the equivalent electrical loads he assumes that rechargeable batteries or hydrogen fuel cells are among the storage devices in use and that they have an overall efficiency of 60 per cent. His analysis is used to establish that $2{\cdot}10 \times 10^9$ gigajoules (GJ) is the total annual energy requirement, while the present entire electrical output is $0{\cdot}77 \times 10^9$ GJ. The difference of about three in these two totals shows the seriousness of the problem of replacing the fossil fuels, in particular oil and gas. He also considers a 50 per cent increase in coal consumption, and the table shows this can have a marked effect. The large fluctuations in demand remain, for instance 213 GW in cold winter day

TABLE 8.1

DEMAND FLUCTUATIONS (GW) AND ANNUAL ENERGY REQUIREMENTS
(Data from M. Ryle, *Nature* **267**, 1977, p. 111)

	Annual energy requirements (10^9 GJ)	Winter				Summer	
		−3° C		+11° C			
		Day	Night	Day	Night	Day	Night
Domestic:							
Space heating	0·241	49	—	12	—	—	—
Hot water	0·087	1·7	3·3	1·7	3·3	1·7	3·3
Public buildings and so on:							
Space heating	0·233	48	—	12	—	—	—
Hot water	0·038	—	2	—	2	—	2
Industry:							
Plant	0·72	32	14	32	14	32	14
Space heating	0·20	41	—	10	—	—	—
Transport	0·34	5·5	16·5	5·5	16·5	5·5	16·5
Present Gas-turbine generators	0·24	36	—	7·5	—	8·7	—
Total:	2·10	213	35·8	80·7	35·8	47·9	35·8
~150-hr storage			91			42	
Annual mean					66·5		
Assuming 50% increase in coal production	1·91	187	31	79	31	47	31
~150-hr storage			82			39	
Annual mean					60·5		

conditions as against 47·9 GW in mild summer day conditions. Also in the table are entered the consumption rates when a 150-hour storage system is in use. The figures for consumption then change to 91 GW and 42 GW, or 82 GW and 39 GW with increased coal use. Thus storage changes the overall fluctuation from a ratio of 5·95 to 2·17 or 2·10. Whether the whole of the argument is accepted or not the crucial importance of storage of energy for times of 150 hrs (approximately one week) is very clearly demonstrated.

The adoption of a storage system results in a much reduced fluctuation in the required output from electrical generating stations, giving a smoothed-out demand that is tolerably close to the monthly variation of wind power shown in fig. 23.

This result is turned to advantage by Ryle, who reasons that since wind energy is eminently suitable to meet the annual variation this must be reflected in its favour in calculations of cost of generating installations. Table 8.2 below gives his estimates of the capital and running costs of several systems designed to meet the requirement of providing 1 GJ with the time-variations encountered in actual practice and including the use of 150-hr storage.

TABLE 8.2

THE CAPITAL AND RUNNING COSTS (IN 1975 $) OF PRODUCING 1 GJ ANNUALLY, FOR ALL NON-NUCLEAR SYSTEMS COSTS INCLUDE STORAGE.

(Data from M. Ryle, *Nature* **267**, 1977, p. 111)

	Capital Cost	Annual Running Cost	Total for 20 years
Nuclear			
(*a*) no storage	174	3·6	246
(*b*) 150-hr storage	90	2·1	132
Solar panels	85	0·5	95
Wind	49	0·8	65
Wave	(125)	(3)	(185)

The storage cost is assumed to be $2.0 per kWh. The figures are very revealing and show in such arrangements wind power is relatively cheap to install and to use. For solar panels the figures are much less favourable, mainly because the energy available from solar radiation is not well suited on a seasonal basis to the varying demand. In summer the radiation is at its maximum and the average energy need is at a minimum.

In his study of wind energy Ryle adopts for his immediate purpose a 1 MW wind-turbine with vertical axis; there have been models of this type and output used in France and elsewhere. For 1 MW a diameter of 50/60 metres is implied, so the machines are large. For a total output of about 43 GW in winter no fewer than 43,000 machines would be in operation. This gives a vivid picture of the scale of the scheme as envisaged, and Ryle does refer to strong environmental objections. At the same time it is reassuring to know that if there were demanding reasons to rely heavily on wind power, it is feasible to meet a substantial part of our vital energy requirements by this method. Ryle's analysis is for Britain but clearly applies in many countries with somewhat similar climatic conditions. His data for land requirements are criticised by C. F. Clement (*Nature* **268**, 1977, p. 396) who argues that outputs on the scale proposed by Ryle imply windmills covering about 20 per cent of the land area of Britain, a point accepted by Ryle in the same volume (p. 482) but not regarded as damaging to his main conclusions. The investment cost of a practical 150-hr storage system is thought by some experts to be among the more serious difficulties of the scheme. Others have criticised the proposals on environmental grounds.

Wave Power. Wave energy, like wind energy, is a derived form of solar power and again it is a potentially large source. It is not easy to generalize about it as regards diluteness, but perhaps it should be stressed that a derived form can be more concentrated than the original form to which it owes its origin. In some places wave power can be considered relatively concentrated. Off the west coast of Scotland it has been estimated that waves carry an energy flux, averaged over the year, of between 50 and 100 kW per metre. The collector must be of the order of dimensions of the waves, that is, several metres across and deep. This helps to explain the apparently large value of the energy flow—the volume of water must be substantial for a power yield of 100 kW. For a station capable of delivering around 1,000 MW we would need, on the more conservative basis, to extract the energy over a distance of 20 km, roughly 14 miles. Using machines extending for 400 kilometres we would theoretically be easily able to avail ourselves of energy equivalent to 25

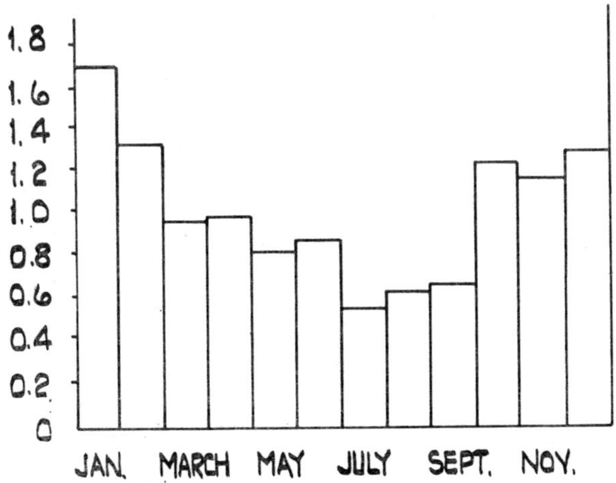

FIG. 23. Variation of the energy available from wind power with month of the year, expressed as a fraction of the mean energy. [From M. Ryle, *Nature* **267**, 1977, p. 114]

million tons of coal, or about 6 to 7 per cent of the UK total energy consumption (in 1977). Such an amount of energy is well worth winning if the price is not too high. The energy obtained from between 800 and 900 miles of collector would be approximately equivalent to the total output of all existing generating plant (60,000 MW) in Britain (1977).

While wave energy is clearly another renewable source available in really worthwhile quantities, exploitation will raise a considerable range of new problems. The system has not been tried in anything approaching prototype scale and so far most of the interest and research has been confined to Britain. The ideas of Salter of the University of Edinburgh seemed to act as a stimulus and four principal schemes have been devised. These are shown schematically in fig. 24.

They are described as Salter ducks, Cockerell rafts, the Russell rectifier,

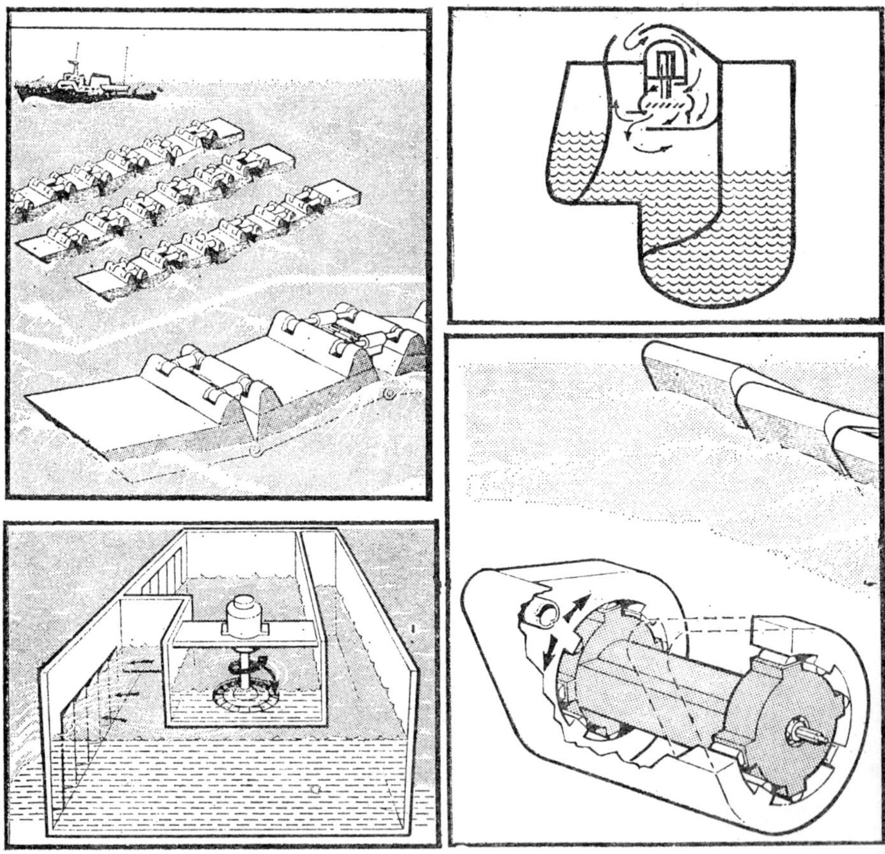

FIG. 24. Four systems for extracting wave energy. From left to right clockwise are a Cockerell wave contouring raft, an oscillating water column, Salter's ducks, and a Russell rectifier. [From F. J. P. Clarke, *Times Higher Education Supplement*, 22nd July, 1977]

and the Oscillating Water Column (OWC). The Hydraulics Research Station (HRS) is busy with the Russell or HRS rectifier and the National Engineering Laboratory is examining the potentialities of the OWC.

Both the duck and raft devices make use of the motion of one component relative to another (an axle or another raft), while the rectifier makes use of a constant head of water, averaging a few metres in height. The OWC compresses air to provide the energy store, which then can be used to drive a turbine.

The low speed of the devices in each case implies large machines for the conversion of the random energy of the waves, and the capacity of these large machines to damp out the fluctuations in the energy supply is an essential element in the design. No serious snags have been uncovered in these early stages of the work so doubtless progress will continue. It is clear from research studies with the Salter duck that it does absorb the wave energy well. In 1978 prototypes of the Salter and Cockerell devices, both at one tenth size, were being tested on Loch Ness and the Solent, respectively.

Some scientists regard wave power as a daunting project. The lack of experience of industrial scale plant in the new environment of rough coastal waters could prove a major impediment to rapid advance. Experience recently acquired in Britain with off-shore oil installations does not appear to be particularly relevant in this completely novel undertaking. Construction may prove hazardous and installation costs high. The assemblies should be placed at a distance of tens to hundreds of miles from the shore in order to collect effectively the energy generated over distances of several hundreds of miles. The transfer of energy to land might well require a solution of the energy storage problem; for example conversion to hydrogen. On the other hand sophisticated means of carrying electrical power, as such, to the shore may be necessary. The effects on shipping in general, and fishing in particular, could be a subject of prolonged debate and argument. If international interest is lacking it could prove a very formidable task for Britain to assume in isolation.

TABLE 8.3

COMPARISON OF POWER OF VARIOUS ENERGY SOURCES
(From F. J. P. Clarke, *Times Higher Education Supplement*, 22nd July 1977)

Type of Source	Watts of power per sq.m
At heat-exchange surface of fossil and nuclear units	50,000
Approximate upper limit of radiant solar energy	750
Annual daytime average	200
Annual average of wind energy in United Kingdom coastal region	400
Average wave energy over several metres of favourable waters	1,000

Further Considerations of Wind and Wave Power. We have already outlined in general fashion the views of proponents and critics of both wind and wave schemes for power generation. A good review of the main problems still to be solved has been given by F. J. P. Clarke. He stresses the dilute or diffuse nature of wind and wave power as compared with fossil or nuclear fuels, and gives figures to support his criticism on this criterion (table 8.3). Although nature does manage to concentrate the solar energy in winds and waves in certain regions, the contrast in the sources is still most marked.

Clarke's review of wind energy brings out three extremely relevant factors affecting the probability of success in providing energy in harnessing the winds. The output of an aerogenerator depends on

(1) the cube of the wind speed
(2) the total time for which the wind blows
(3) the square of the rotor diameter.

Thus only windy sites and large machines should be used. Clarke states that the cost per kilowatt of power produced falls by a factor of 10 in raising output from 1 kW to 10 MW. Until material science is much more advanced it would appear, therefore, that about one megawatt, which corresponds to blade diameter of 50 metres, is the practical limit. Taking 10,000 sites as the estimated number of suitably windy locations this suggests that about 10,000 MW could be generated. This is not an inconsiderable contribution to the UK energy scene. It corresponds to a saving of 6 million tons of coal per annum and is roughly 2 per cent of the total UK energy need. This total could be increased by using off-shore sites, but transmission of electrical power and storage questions become crucially important factors in that case. In actual fact the storage problem underlies the whole question of wind energy exploitation.

Ocean Thermal Power. Solar radiation falls on large areas of the world's oceans and maintains the surface temperatures higher than those at depth. The radiation converted to thermal energy in this way constitutes, in principle, an enormous converted source. Between the tropics of Cancer and Capricorn the near-surface temperature has an average fairly steady value of 28 °C during the year, while at a depth of 2,000 ft or more the temperature remains tolerably constant at less than 3 °C due to inflowing polar currents. Thus warm water in immense amount is held 25 °C above the water underneath.

Theoretically heat engines can extract about 8 per cent of such converted solar energy in the seas to deliver it as electricity from generators. An estimate of C. Zener (*Physics Today*, Jan. 1973) shows one square mile of ocean waters could provide, by virtue of the temperature difference, about one megawatt of electrical power. Clearly the energy is vast in quantity but very dilute even compared with wave energy. About 100 million square miles satisfy the conditions for extraction so the 'warm bath' could in theory yield 100 million MW or 10^5 GW, an amount exceeding total world energy consumption.

As long ago as 1881 Jacques D'Arsonval considered the form that the heat engines might take. For a temperature difference of 25 °C the efficiency is given by

$$\text{Efficiency} = \frac{\text{T surface} - \text{T depth}}{\text{T depth}}$$

At a T depth value of 276 K (about 3 °C) and a depth to surface rise of 25 °C we have

$$\text{Efficiency} = \frac{25}{276} = 9{\cdot}06 \text{ per cent.}$$

In actual practice the efficiency would probably be in the range of 3 to 6 per cent.

Design studies for large engines make the project look promising. In fig. 25 a huge submersible tank is employed and the tank is divided into three

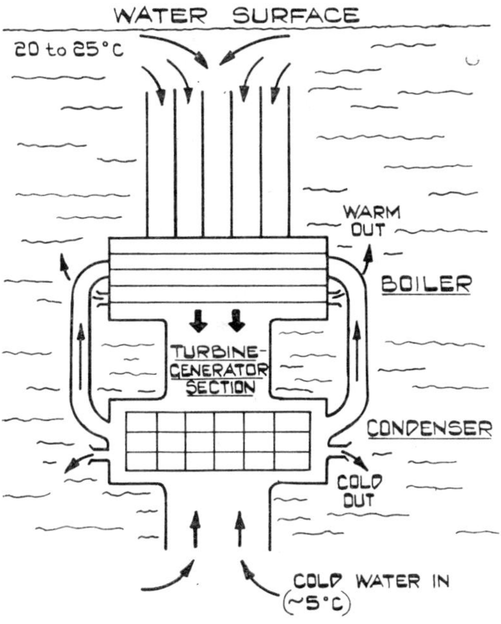

FIG. 25. Sea thermal gradient engine

main parts. The top section is the heater, the turbine is underneath and the condenser is the bottom portion. The warm water of the heater, that is the near-surface water, heats the working fluid and the vapour of the fluid drives the turbine. The vapour is condensed after passing through the turbine and returned to the heater section. The engine has to be extremely large but off-shore oil ventures, including for example the construction and use of drilling and production platforms in the North Sea, have shown that large-scale machines of the required type might be successfully operated. The specification for a suitable working fluid is perhaps the most difficult criterion to meet.

Ideally it has to have a high vapour pressure at a modest temperature. Some of the fluids used in refrigeration such as freon ammonia and propane are among the possibilities.

As in the case of wave power the transport of the energy to the customers presents obstacles. In principle the turbine could drive a generator and the electrical power from this could in turn be used to deliver hydrogen from electrolyzers. Cost analysis for such a system is not unfavourable and the scheme could well merit investment in the construction of a prototype.

Chapter 9

Hydro Energy and Pumped Storage

Hydro Energy. The use of hydroelectric power has been the most notable exception to the general tendency, so evident during the past century, towards a rapidly increasing dependence on fossil fuels particularly oil and natural gas. Tidal power, discussed in chapter 5, can, from the practical point of view, be regarded as a special type of hydro power. Hydro power itself is a converted form of solar energy while tidal power has an entirely different origin in gravitational energy.

The heat of the Sun causes evaporation of water from the seas, lakes, and rivers (and to a lesser extent from land), and this water returning to the ground as rain or snow can be retained in natural or man-made dams at some height above sea-level. The water stored in this fashion has potential energy which can be released, and converted in form, when the dam supplies water to a turbo-generator at a lower level, generally below the level of the bottom of the dam. Maximizing the available head of a river system can lead to the adoption of a wide variety of installations. Stations with large reservoir capacity may be located in the upper reaches of the catchment and less well-regulated stations may be installed further downstream. If the fall of a river is small, and the flow large, a series of run-of-river stations with the generating plant integral with a barrage may be appropriate.

Hydro power usually involves relatively high construction costs but with long life expectancy for the dam the power is among the cheapest forms and certainly among the cleanest. It is a renewable resource, a major advantage in times of rapid increase in the price of fossil fuels.

The contribution to the world production of electricity made by hydroelectric generation is substantial. The United Nations data for 1970 showed that hydroelectric plants generated 24 per cent of the world's total electricity supply, that is $4{\cdot}24 \times 10^{18}$ J or 1,178 TW (one terrawatt$=10^{12}$ watt). In the context of UK electricity supplies this is 4·7 times the total electricity generated in 1970 in the UK. The amount is certainly impressive being of the order of the total electricity generated in the US. In terms of the saving of coal it would represent no less than 600 million tonnes assuming the coal was used to provide electricity. Very large generating plants have been constructed, notably the 4,500 MW Bratsk plant in the USSR. Two more Russian stations have capacities around 2,500 MW and in the US the Grand Coulee dam provides 2,000 MW with the possibility of an additional plant for 3,600 MW at the same site.

An excellent survey of hydro energy production and the potential for development of water-power sources has been made by K. R. Vernon. He shows figures for production in 1961 and 1970 in the developed countries, centrally planned economies, and developing countries (table 9.1). He notes that while in 1970 the developed countries consumed almost three-quarters of the world hydro energy, the rate of development outside them has been twice as fast during the period. He takes the view that this trend, obvious from the table, will become more marked as exhaustion of hydro potential in the developed countries approaches. Much remains to be done in some of the developing countries but the increase of 137 per cent in ten years shown in the table is evidence of rapid exploitation.

TABLE 9.1

WORLD HYDRO PRODUCTION
(From K. R. Vernon, *Phil. Trans. R. Soc.* **A276**, 1974, p. 485)

	1961		1970		% Increase
	10^{18}J	TW hr	10^{18}J	TW hr	
Developed countries	2·07	575	3·05	847	47
Centrally planned economies	0·32	89	0·67	185	107
Developing countries	0·22	61	0·52	146	137
Totals	2·61	725	4·24	1,178	62

As we have already noted from Vernon, in many countries the question of suitable sites for dams seriously affects the rate of expansion of hydro power. It has been stated that by 1985 almost 60 per cent of the available hydro power in the US will be harnessed and that subsequently the available sites will be suitable only for relatively small plants, each providing 50 to 150 MW. These plants could of course still be valuable sources of power but somewhat uneconomic compared with stations in the 1,000 MW range. Perhaps a more serious argument against exhausting all possible sites is the unnatural pattern of behaviour imposed by man on most streams, rivers, and lakes. There is a limit to the degree of artificiality that can be introduced in the environment. In the case of certain more sparsely populated mountainous countries hydro power can supply almost all needs for electricity. Norway is in this position. By contrast in the US only about 4 per cent of total energy needs are provided by hydro power. While hydro power should be able to provide a most valuable additional amount of energy it is not feasible to expand it, at least in the developed countries, to take the place of the fossil fuels.

However, if we look in more detail at hydro resources we note a figure of $82{\cdot}8 \times 10^{18}$ J or 23,000 TW hr was given at the 1968 World Energy Conference (20 times 1970 production). Vernon considers this estimate to be optimistic by a factor of between 2 and 3 and it is his figures that we give in table 9.2.

TABLE 9.2

WORLD HYDRO RESOURCES
(From K. R. Vernon, *Phil. Trans. R. Soc.* **A276**, 1974, p. 486)

	Potential		% developed
	10^{18}J	TW hr	as of 1970
Developed countries	4·68	1,300	65
Centrally planned economies	6·47	1,800	10
Developing countries	18·0	5,000	3
Totals	29·15	8,100	14·5

So far as the developed countries are concerned the bulk of the untapped resources is in Canada, which can, in due course, increase its own supplies and contribute to needs in the US. In Europe it is thought that around 80 per cent of the hydro potential will be exploited by 1980. In the UK most of the hydro resources have been developed. Much the same is true in the US, in Japan, and in Australia.

In the centrally planned economies the main developments will be in Central Siberia on the Angara and Yenisei rivers. In China there are vast amounts of potential hydro power but little is known of the plans for the future.

In the developing countries it appears that Africa and South America have very substantial potential resources. So far as the future of the developing countries is concerned they have a real opportunity to provide a rapidly increasing part of their needs with hydro power, an opportunity which must surely be seized.

Pumped Storage. We have often had occasion to refer to the serious difficulties that arise in attempting to store electrical energy on a large scale. These difficulties contribute in some degree to the increase in dependence on convenient fossil fuels, particularly oil. However, in pumped storage one means of providing an answer to the problem is found. It is really a method of enabling some electricity generating stations (those in the neighbourhood of suitable high level lakes) to be operated more efficiently and usually it is employed for the conversion of thermally generated power to hydro power. Such systems do not contribute to the total energy supply but are fuel conservation devices operating on a large scale. Additionally they provide an opportunity to use large base-load stations of about 1,000 MW output to meet varying demand from customers. Thus large nuclear stations can be used in spite of the fact that it is not desirable to alter that output. The ability to make any kind of large base-load generating station more flexible by introducing pumped storage is economically attractive. In fig. 26 the rate of increase of world pumped storage capacity between 1940 and 1972 is seen to be on an impressive scale.

Half of the total capacity is located in Europe and the remainder mainly in the US and Japan. In the UK a 360 MW plant is located at Ffestiniog in North Wales and a 400 MW plant in Cruachan in Scotland. A station for 300 MW is in construction at Foyers in Scotland and a 1,440 MW station has been planned.

In any storage scheme the primary requirement is the existence of upper and lower reservoir sites in close proximity but with a level difference of not less than 150 metres. The site should be reasonably close to centres of load and off-peak generation of electricity. Usually it is arranged that run-off water, from the catchment area of the site, is collected and used to improve the overall performance of the pumped storage station. In this sense pumped-storage systems can be partly performing as hydro power plants in their own right. In practice, several of them are conventional hydroelectrical plants which did not secure an adequate amount of natural water supply. Good sites are not readily obtained so that recently thought has turned towards the provision of lower level reservoirs by underground excavation. In such cases the upper level reservoir can be at ground level. The level difference of at least 150 m of the two reservoirs is not daunting as an engineering programme. The environmental attractions of such schemes are to be offset against possible increases in construction costs. It is possible that old mine workings could provide the lower reservoir and an attractive sailing lake at ground level be made from the upper one.

It is interesting to note in a little more detail the ways in which pumped storage can be manipulated to help large thermal generating plant. Such plant is usually operated much more economically, both from the customer's point

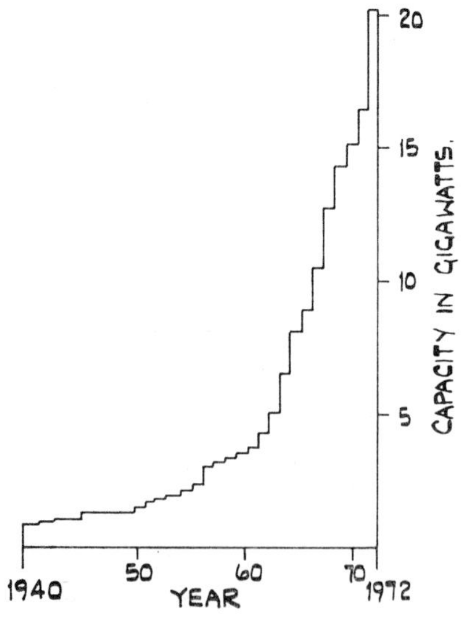

FIG. 26

World pumped storage capacity. [From K. R. Vernon, *Phil. Trans. R. Soc.* **A276**, 1974, p. 488] of view and that of the fuel conservationist, if the load is the optimum one. Then all of the machinery is used according to design. However, there are very substantial variations in the level of demand for electricity, and the differences of supply level are marked during periods of 24 hours and from season to season. Weather conditions can in themselves bring about sudden changes in the load on the supplier. Electrical power is extremely flexible and when customers can switch on or off at will the supplier can encounter serious problems in meeting the fluctuating load. The summer and winter demands on the Central Electricity Generating Board system in 1976/77 are shown in fig. 27. In the course of a single day the demand can be seen to change by a factor of about 2·4, and this large change in customer requirement is the strongest argument in support of pumped storage.

In the past it was often necessary to construct much more electrical generating capacity than was on average required, and this situation obtains today except where pumped storage facilities exist. Even when the plant capacity is available in order to meet the maximum expected demand it is sometimes difficult to respond rapidly to a sudden large increase. Coal-burning and nuclear plants are well fitted to supply a steady output and in level operation they work at their highest efficiency. The sudden increase in load is often handled by switching on additional low-merit plant, frequently diesel oil burning type. Such plant is not only expensive to provide but the power supplied into the transmission network is not generated with the high efficiency of the larger base-load stations. The need for it is overcome if the large high-merit stations can continue to generate throughout the 24 hours,

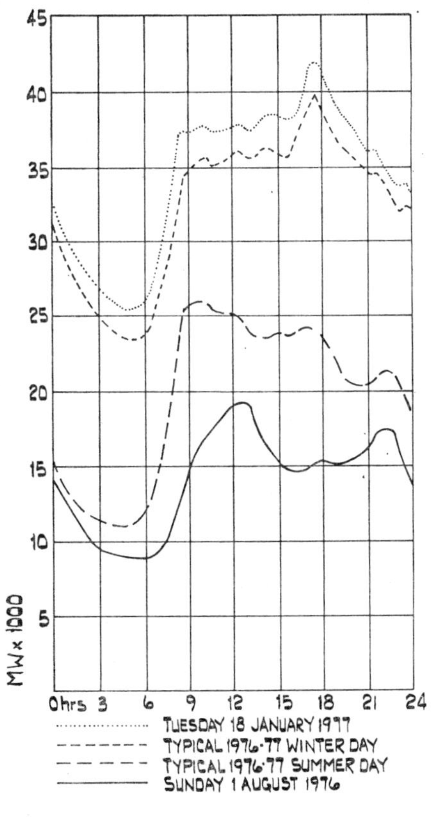

FIG. 27

Summer and winter demands on the CEGB system in 1976–77, including days of maximum and minimum demand. The highest level exceeded 42,000 MW and the lowest less than 9,000 MW, giving a variation factor approaching 5. [From *CEGB Annual Report and Accounts 1976–77*] pumping more and more water to the higher reservoir as the customer calls for less and less electricity at the end of the working day. Hydro power is quickly and readily generated by releasing the water stored in the higher reservoir. The hydro power that is put into store overnight can be used the following day to add to the total amount of electricity generated in order to satisfy the peak daytime requirements of the customers. In overall terms considerable savings in unit costs can thus be achieved, and since the large base-load plants are being used in the most suitable way they perform more reliably and with longer effective life expectancy. The pumped storage component is particularly beneficial when the amount of nuclear power generated in the total system of the supplier exceeds the night load, because nuclear plant is not well suited to rapid change in level of output. The excess of nuclear power is applied during the night in pumping and, in effect, nuclear power can be used to meet peak load and not be limited to meeting steady base load.

The design and construction of the pumped storage part of a comprehensive electricity generating system, which incorporates many base-load stations making use of the same pumped storage facilities, can be complicated. The peak load period is usually comparatively short, say 6 hours, and justifies pumping only at night and at weekends. The larger the nuclear energy contribution to a total system, the more valuable is pumped storage.

Chapter 10

Photovoltaic Energy Conversion

Types of Solids. In recent years the direct conversion of solar energy to electrical energy has received a much increased amount of attention. A sense of urgency is apparent in the research and development effort, but this would be pointless if there had not been great advances in the science of the solid state, particularly in understanding the mode of action of semiconductors and photovoltaic devices. For several decades striking discoveries in this sphere have been made, and many of the discoveries have been of a fundamental nature affecting phenomena beyond the immediate purpose of the investigators. The transistor is the outstanding and most familiar example of the innovations.

Electrical Conductivity and Semiconductors. Solid materials fall into a number of groups, of which the principal are metals, polymers, and ceramics. A fourth group are composites, made up of materials from two or more of these groups. When the atoms in solids are arranged in regular patterns and consequently the materials show a well-defined lattice form of arrangement of the constituent atoms, the solids are described as crystalline.

The electrical conductivity of a material depends on the availability of free electrons within the material. When a voltage is applied across a material containing free electrons the induced flow of the electrons constitutes a current. In metals a large number of electrons can move freely within the solid and hence they are good conductors of electrical current. Ceramics are usually a mixture of metallic and non-metallic atoms and have lattices of the same types as metals. The atoms in ceramics are strongly bound together with ionic or covalent bonds, or a mixture of the two, and they contain few free electrons so electrical conductivity is poor.

The structure of polymers is different from that of metals or ceramics. The molecules are often very large, made up of 10,000 or more units, and the units themselves often consist of several carbon and hydrogen atoms with sometimes a small proportion of other elements. The large molecules of a block of polymer are tangled together in a rather irregular way and the material is said to be amorphous. The amorphous polymers contain essentially no free electrons and are non-conductors. Some polymers do show a degree of regularity of structure and are both exceptional and important members of the class, for they can be used as semiconductors as we shall see. Rubbers and glasses fall reasonably well into the polymer groups so far as most properties are concerned.

The conductivities of solids vary over an extremely wide range and a factor of 10^{23} between values may exist. The values for metals do not differ very markedly from each other, and the value for a given metal is not much influenced by the presence of impurities. On the other hand the conductivity of poor conductors in the ceramic, polymer, and composite groups can be altered substantially by the addition of impurity to the material. In practice the value of the electrical resistivity of a material is often used to classify substances as conductors, semiconductors, and insulators. The resistivity varies over an extremely wide range, for example from $1{\cdot}67 \times 10^{-6}$ ohm-cm for a good conductor like copper, to 2×10^{17} ohm-cm for a good insulator like sulphur.

In our short review of the properties of atoms and molecules (chapter 3) we showed that the electrons had only certain allowed energy values. This is true of isolated atoms but when large numbers of atoms form a solid, large electromagnetic forces between the electrons of the separate atoms are created. The allowed values of quantum energy-change for the individual atoms are then altered and some are raised and some reduced. The electrons in the case of a single isolated atom may have discrete energy levels, as indicated in fig. 28(a), but the energy level diagram for the electrons of a solid formed by a large number of the same atoms can take the form of energy bands and gaps, as shown in fig. 28(b). The two top bands are the important ones in conduction phenomena. The top one is the conduction band and the one immediately underneath is the valence band. When the conduction band is empty the material does not conduct as the valence electrons remain bound in their own band. However, heat can enable some electrons of the valence band to acquire sufficient energy to be excited into the conduction band and the solid becomes a conductor. When an electron is raised from the valence

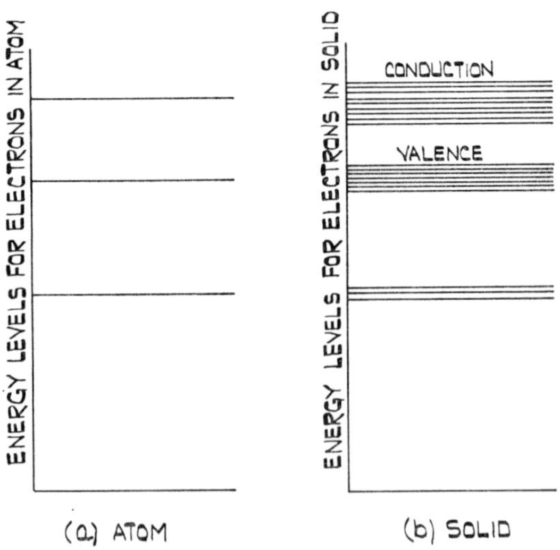

FIG. 28. (a) Energy levels of a single atom. (b) Energy bands and gaps of a solid formed of a large number of the same atoms.

band to the conduction band a hole is left in the conduction band; the hole participates in conduction to the same extent as the conduction electron. We picture the hole moving in the opposite direction to that of the electron.

Materials with conductivities in the range 1 to 10^{-11} siemens are usually classified as semiconductors. If we consider a well-known semiconductor, such as silicon or germanium, we realize the atoms of either of these materials have four valence electrons. In a lattice arrangement of the atoms of these substances the atoms arrange themselves so that each atom has its own four valence electrons and, in effect, one 'borrowed' from each of the four neighbouring atoms. Hence in a piece of silicon crystal all the atoms, save the small number near the surface, have an effectively complete shell of eight electrons. The large changes brought about by the addition of impurities are readily explained in terms of conduction and valence band behaviour. If the impurity has five outermost or valence electrons, for each atom taking a place that would normally be occupied by a silicon atom in a lattice containing only silicon, there is a 'spare' electron. Arsenic is able to serve in this way as an impurity and it can be used to 'dope' silicon crystals. The spare electron is free to serve as a conduction electron in a silicon crystal doped in this fashion. The situation here is different from that arising in a pure solid.

When we are dealing with a pure solid the production of a conduction electron is accompanied by the creation of a hole. In the case of the doped material only the conduction electron is created. For this reason such materials as silicon with arsenic as the impurity are negative charge carriers; they are classified as n-type semiconductors. On the other hand if the impurity atom is able to supply only three valence electrons, and boron is a good example of such an impurity, the lack of a fourth valence electron leads to the creation of a hole in the lattice. This is the same as the creation of an excess positive charge carrier and the materials are classified as p-type semiconductors. Thus for each impurity atom a positive or negative charge carrier is created.

If two pieces of semiconducting material of different types, one p-type and one n-type, are joined together, electrons and holes from each side will move towards the junction between the materials. Holes will obviously move more readily in one direction and electrons more readily in the opposite direction. In effect the resistance of the p–n junction is larger for current in one direction than in the other and the junction acts as a current rectifier. Such junctions are used in great numbers in radios, computers, and other electronic devices.

We are now in a position to consider photon excitation of a semiconductor. As the reader will remember a photon is a quantum of electromagnetic radiation. A photon of sufficient energy can excite an electron of an atom from its normal orbit to an orbit of higher energy. It can even cause ionization of the atom if the electron is made completely free of the attraction of the nucleus. In all semiconductors, but more efficiently in doped ones, a similar process of excitation of an electron can occur when a photon is absorbed. In this case the electron is usually able to remain in the conduction band and move freely for some time. Eventually it can arrive at an atom with a vacant lower energy level and remain fixed, but until then it is able to act as an electron in the conduction band and contribute to the conductivity of the material. The material thus conducts current more readily by virtue of absorption of the photons and is said to show photoconductivity. When the semiconducting materials are illuminated and used to produce voltage across an external circuit and supply current to an external load, they are acting as photovoltaic devices. In this form they represent an important type of energy convertor. The mode of action of a photovoltaic cell is illustrated schematically in fig. 29.

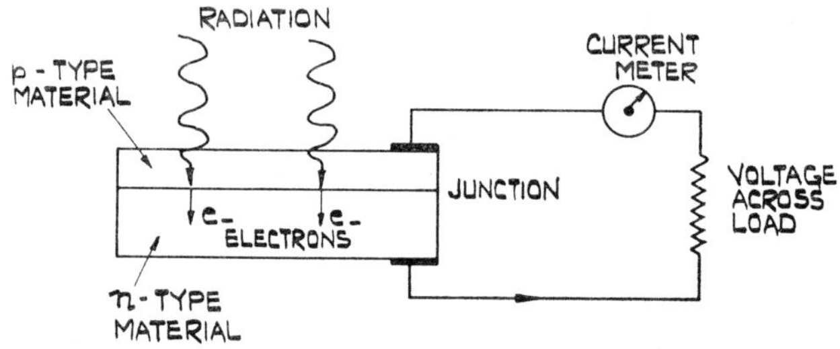

FIG. 29. Response of a photovoltaic cell to solar radiation

Among the first applications of such devices were power supplies in space satellites. In most applications to date the preferred photovoltaic cell has been silicon doped with a small amount of phosphorus, which, like arsenic, has five valence electrons. In a silicon crystal the available electron has an energy close to that at the bottom of the silicon conduction band when it is doped. In this state the absorption of a small amount of energy brings about ionization and a 'free' carrier in the conduction band is created. The possibility of the development of a number of crystalline semiconductors with a variety of electrical properties follows from the incorporation within them of different types and concentrations of impurity atoms.

The most serious disadvantage of crystalline solar energy conversion devices has been the need to grow suitable crystals. Silicon semiconductors incorporating a controlled amount of impurity have proved reliable in performance, but the expense of preparing good quality crystals has proved inhibiting for most applications. The efficiency of conversion is, of course, another critical factor, for it determines the area which must be exposed to sunlight to generate a required amount of energy. In 1977 the cost of generating electricity with silicon solar cells was approximately \$1·50 per kW-hr, around 100 times the cost using fossil fuels.

To overcome this difficulty attention has turned to amorphous solids with a degree of periodicity in their structure. This limited amount of regularity appears to affect their properties in a manner similar to that ascribed to the regular lattice arrangement of atoms in, say, a silicon crystal. Their electrical properties, when they are properly doped, resemble that of doped crystals. David Adler (*Scient. Am.*, **236,** no. 5, 1977, p. 36) has given an excellent account of recent work on both crystalline and amorphous photovoltaic devices. Adler discusses the advances in research on chalcogenide glasses, and also on tetrahedral amorphous semiconductors such as those investigated by Spear and Le Comber who doped silicon which itself was deposited from silane gas, SiH_4. The doped amorphous silicon solar cells have been shown by Carlson and Wronski of RCA to operate at up to 6 per cent efficiency. These results may well prove the starting-point of a major break-through in solar cell technology.

Practical Photovoltaic Devices. In practice, silicon is still the material most frequently used for the production of photovoltaic cells. These have been manufactured in quantity for specialised purposes where cost is not an important factor. Purified and 'doped' with minute quantities of particular elements, such as phosphorus, boron, or arsenic, they are prepared as small crystals and connected in parallel and series as required. The array of cells then provides a practical source of electrical power.

The electrical characteristics of such a cell are typical of most devices. The

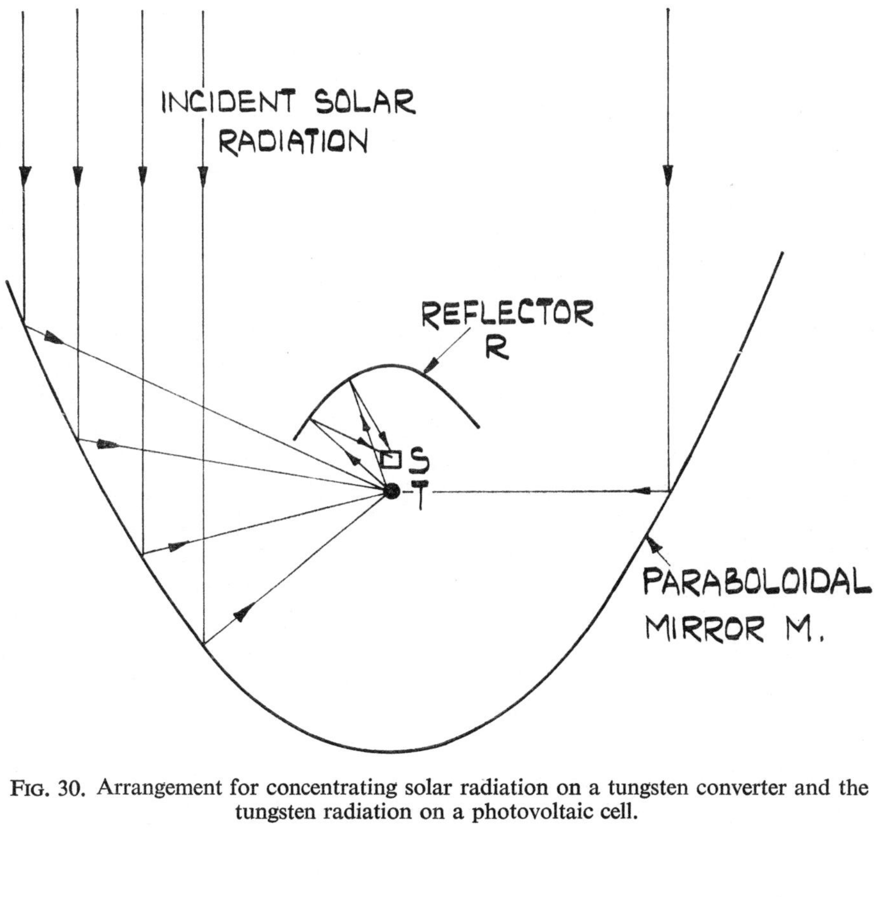

FIG. 30. Arrangement for concentrating solar radiation on a tungsten converter and the tungsten radiation on a photovoltaic cell.

incident radiation appears to be converted with about 5 per cent efficiency in good operating conditions but higher values around 13 per cent have been claimed. The manufacture of crystalline silicon cells is complicated by the tendency of pure molten silicon to be chemically very active and also to absorb impurities in manufacture; such impurities have adverse effects on performance even in extremely small amounts. Furthermore, large single crystals of pure silicon suitably doped are difficult to grow and the process of cutting into thin slices and polishing is a matter of expert skills. The more recent technique of single crystal silicon ribbon is valuable in avoiding cutting losses.

A combination of techniques may overcome the problem of mass producing efficient cells by an arrangement such as that shown in fig. 30. Sunlight is concentrated on a tungsten filament (T), which becomes incandescent. Around 2,800 K it acts as a source of secondary radiation, which in its turn is concentrated on a smaller mirror R and focussed on a cell at S. The type of cell chosen is one with an optimum response to the spectral distribution of the tungsten radiation. In this way efficiency can be raised to 25–30 per cent, the number of photo-voltaic cells kept to a minimum, and maintenance reduced.

Thin film semiconductor developments, which include gallium arsenide cells, copper sulphide/cadmium sulphide cells, and cadmium telluride cells, look promising but possible material supply limitations and life-time considerations may be serious. The description given by International Research & Development Company Limited shows how arrays of cadmium sulphide thin film solar cells can be constructed. The cell is essentially a large area planar copper sulphide/cadmium sulphide hetero-junction with the light incident on the Cu_2S side. The CdS is vacuum-deposited on a 20 micron layer on a metallised high temperature plastic substrate. The upper surface is then converted to a Cu_2S p-type semiconductor with a high absorption coefficient for ultraviolet, visible and near infra-red light. Absorbed light releases electrons which are accelerated across the junction region into the CdS. An integral extension of the metallised plastic substrate forms the negative lead to the cell and the positive connection is provided by a fine-mesh, gold-plated grid cemented to the Cu_2S layer. The whole cell is protected by a clear plastic cover, held in position with epoxy adhesive, and is 0·09 mm thick save at the edges which are provided with additional plastic protection. The standard cell is 2 cm × 6 cm in size and multi-cell arrays can be constructed. They are flexible, light, and resistant to humidity, severe thermal shock, cycling and particle radiation. The power to weight ratio is high as can be appreciated from the voltage-current characteristic curve shown in fig. 31. The peak power output of 72 mW is obtained at 0·4 volts and 0·18 amp. in space sunlight of 1·4 mW/sq.mm with the corresponding figures for terrestrial sunlight of 1 mW/sq.mm being 60 mW at 0·4 volts and 0·15 amp. The efficiencies of conversion in the two cases are 4·3 per cent and 5·0 per cent with power to mass ratios of 129 mW/g and 107 mW/g respectively.

A recent (1977) announcement from IBM, engaged on contract work for the Langley Research Centre of NASA, gives cause for optimism. A solar cell of gallium arsenide together with a thin layer of gallium aluminium arsenide, operates at 18 per cent efficiency, higher than the 13 to 14 per cent optimum values for silicon-based cells.

The recent discovery of a method of making efficient amorphous photovoltaic cells using silicon doped with phosphorus and boron may have paved the way to the production of much cheaper cells. Silane gas is used with phosphine and borane to deposit alternate layers of silicon, each about one micron thick, doped with the two impurities. The junctions between the two layers respond to the incident radiation at about 2·4 per cent efficiency of energy conversion but it appears there are grounds for expecting that around 15 per cent may be achieved. Professor Spear (*The Scotsman*, 12th August 1976, p. 11) is quoted as saying that an output of 70 MW might be obtained from one square kilometre of cells and around 25 MW in overcast conditions, at an estimated capital cost of £1·00 per sq.m.

To date the most spectacular solar energy equipment in use has been the one used in the Skylab work. This was built with four large arrays of doped silicon solar cells; they produced the required power but only after teething difficulties which were found to be of a mundane nature.

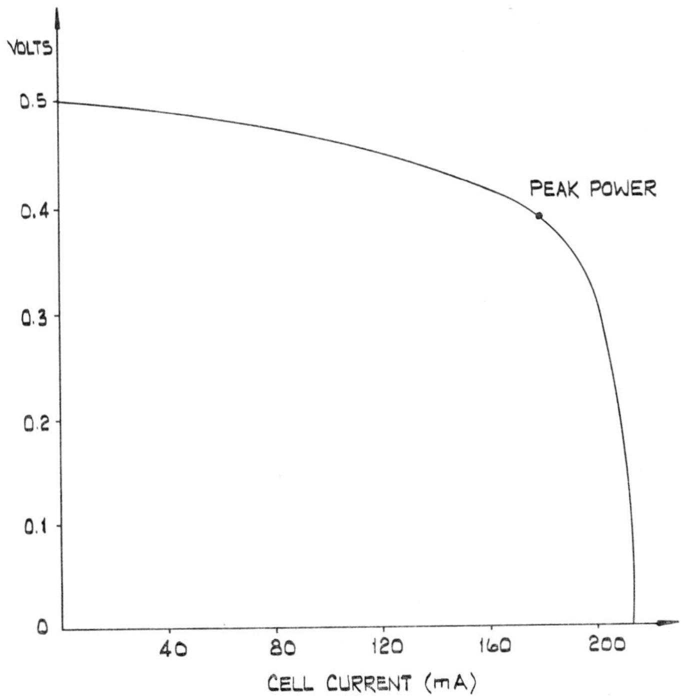

FIG. 31. Voltage-current relationship for Cd S cell in space sunlight. [From International Research and Development Co. Ltd.]

Large-scale Photoconversion Systems. A paper by K. Krebs and E. Gianoli of the Euratom Centre entitled 'Photovoltaic Conversion of Concentrated Solar Radiation' is an excellent study of the recent (1977) advances in concentration of radiation and photoconversion (reference EUR 5723e). They admit that the use of convertors depends on achieving economic production, but think that by the year 2000 some 1 to 5 per cent of global production of electrical energy will be obtained by this means. They believe the elimination of fuel transport costs is important and tends to cancel out the comparatively high cost of convertors. In this connection it is stressed that, in 1977, 2,000 million people were without electricity and, while the cost of conventional electricity is virtually certain to be high, thinly populated areas of the globe require a new form of electrical power. The particular case of Pakistan is noted. Here 20,000 water wells could be constructed and pumped to irrigate land which has become unproductive by virtue of saline deposits. There is clearly urgent need to exploit solar energy in many of the developing countries.

The paper also provides an excellent account of the economic aspects of the photovoltaic cell systems, a discussion of the means of concentrating radiation, and analysis of the electrical characteristics of solar cells. The

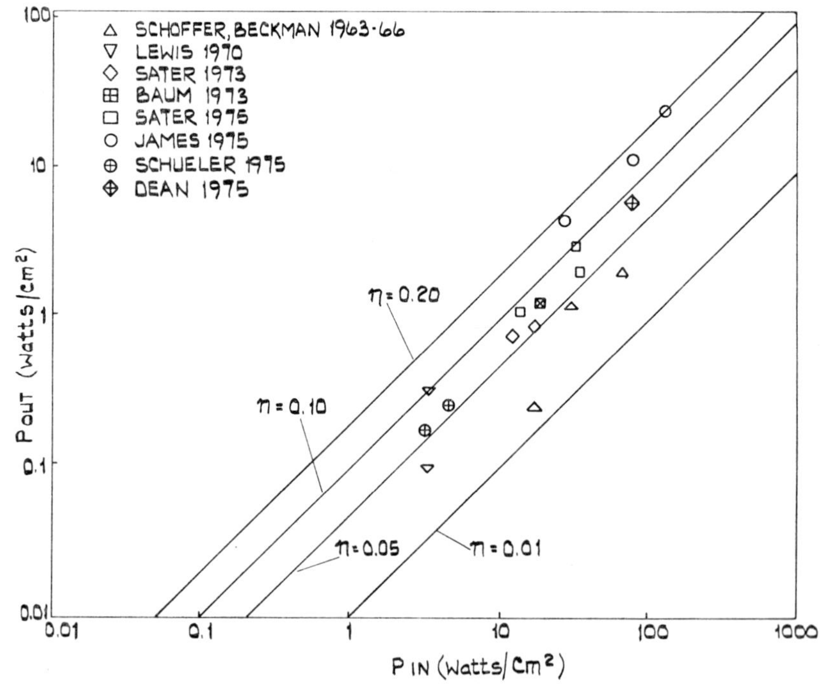

FIG. 32. Output power as a function of input power for various types of solar cells, illustrating operation at high intensity with good efficiency. Note the efficiency can reach 20%. [From K. Krebs and E. Gianoli, EUR 5723e]

behaviour of cells under high intensity radiation is discussed, and effects of temperature and internal resistance investigated. The analysis of the operation of several concentrator arrangements is given including double mirror, parabola, double parabola, Fresnel lens (single and multiple arrays), and different combinations of these methods.

Obviously it is critically important that the solar cells used in concentrator systems should be efficient at high intensity of radiation. The cost reduction in providing electrical power with fewer cells, presently expensive as convertors to electrical energy, can only be secured if the cells efficiently absorb the high intensity of concentrated sunlight and convert it usefully. The results shown in fig. 32 demonstrate clearly that cells can be made that will operate efficiently at least to intensities around 100 watts per sq.cm. Peak sunlight conditions in sunny countries provide 1 kW per sq.m, and 100 watts per sq.cm is $100 \times 10{,}000$ watts per sq.m or 1,000 kW per sq.m. Concentrations of one thousand fold above peak sunlight conditions can therefore apparently be handled in good cells.

The authors consider that the concentration of solar radiation is economically necessary, and they conclude that a cost of 1·5 US dollars/peak watt might be reached. At this level the demand could rise rapidly and ensure that cell production costs fell steeply. As a stimulating example of the success of concentration an intensity of 1,735 suns applied to gallium arsenide cells has yielded 19 per cent overall conversion of solar radiation to electrical output.

The climate of certain countries, such as Australia, the US, and France, all with abundant sunshine, will prove most suitable for early exploitation of solar power. In these countries fairly simple thermodynamic conversion systems may be able to supply low grade heat up to around 120 °C. At temperatures even lower than this the low grade heat can meet most domestic requirements and also be used for desalination, drying, air-conditioning, and refrigeration. The possible weather changes will always introduce uncertainty in the supply until storing is practised, and for low grade heat this usually implies a large volume of space for thermally-insulated water tanks. In the case of solar cell systems it is possible that research will lead to methods of direct production of hydrogen, an excellent multiple-purpose fuel. Even without such a success electrolysis of water to give hydrogen is a possible answer.

The present position in research and development work seems full of promise and, with sufficient resources devoted to the work in the immediate future, a truly valuable contribution to the energy supply can be forthcoming in the next two decades. In the whole gamut of renewable sources direct solar conversion seems to be singularly appealing.

Chapter 11

Energy Conversion in Growth and Agriculture

Energy Conversion in Plants. We have already discussed some of the principal modes of conversion of solar energy and we shall next examine the mode of conversion in plants, which led to the storage of energy in the fossil fuels. Plant life has depended on exploitation of solar energy for hundreds of millions of years. Growth processes achieve the ordered structure of the plant from apparently disordered matter and in doing so, according to the law of entropy, require energy to be supplied from outside. We must consider the mode of conversion of this energy, and the ways in which a proportion of it can become available for our use.

The surface of the Earth is bathed, day by day, in solar radiation. Much of the energy required for growth is supplied as light and heat, converted by green plants through a chain of chemical and biochemical reactions known as photosynthesis, and partly stored. This storage has proceeded for some hundreds of millions of years and it is this fact which makes the energy store, residing mainly in the fossil fuels, uniquely important. Hasty use of the store could adversely affect future generations. Its replacement on anything like the scale on which it has been available until now would appear to be completely impossible but the rapid advances made in modern bio-science suggest there may be ways to minimize the severity of the problems. Biologists are trying today to understand more fully how chemical elements and molecules come to be fashioned into living organisms by virtue of incident solar energy. In the light of such knowledge, they hope to learn to grow the most suitable organisms, plants, and animals, in an efficient manner and so achieve energy storage in the same way as in natural biological processes but much more rapidly.

The manufacture of foodstuffs within green plants has been a critically important part of their evolution. Chlorophyll, the pigment which gives them their green colour, enables them by photosynthesis to combine carbon dioxide from the atmosphere with water from their cells to yield glucose. Photosynthesis itself leads to liberation of oxygen and it is thought almost all of the oxygen in our atmosphere has been formed by the process of photosynthesis. As the availability of oxygen in the atmosphere increased during hundreds of millions of years the process of respiration (use of oxygen as in breathing) replaced fermentation as the commonest means of obtaining energy from the sugar. Photosynthesis is actually the most massive of all known chemical processes and nearly all forms of life depend on it for nourishment.

The process takes place in the following manner. When light falls on a chlorophyll molecule it excites the molecule and the outermost electron of the electronic system moves into a higher energy orbit. Subsequently, as this excited electron returns to its normal orbit in the molecule, the absorbed energy is transferred and used to split a water molecule into hydrogen and oxygen. By a complicated series of chemical reactions the hydrogen is in turn used to form glucose with the aid of the available carbon dioxide. The glucose is thus a store for the energy of the solar radiation. In many plants the glucose energy is taken up by a chemical called ADP (adenosine diphosphate) to form ATP (adenosine triphosphate). Sometimes the ATP is formed directly rather than indirectly by the previous formation and consumption of glucose.

We have probably said enough to establish that all plants which can bring about photosynthesis are using and storing part of the enormous flux of radiant energy reaching the Earth from the Sun. In a straightforward way man has been well aware of this for a long time—he has observed that most plants burn readily and yield considerable quantities of heat in doing so. Wood featured prominently as a fuel in industrial ventures of the past. The related and now dominant fossil fuels, coal, petroleum, and natural gas, are derived from plant and animal life and it is the energy of solar radiation trapped by such animal and vegetable life in the past that is now released by man.

Use of Photosynthesis for Commercial Energy Production. Biologists regard ATP as packaged energy available to drive living processes—the universal energy source for plant life. The ATP releases energy in any suitable situation when energy is needed, becoming ADP as it does so. If man could find a means of imitating the ATP formation process then he could store solar energy at will. By bringing about its release from ATP this energy could be used for industrial and domestic purposes. If we could even release hydrogen (the intermediate step as we have seen in the forming of glucose) then the hydrogen could itself be our fuel. No successful prototype machines of this type have yet been constructed but it is conceivable that one might emerge as a result of the rapidly increasing research effort in this fascinating field. However, it will not be easy to imitate the step-wise processes that are the secret of success in natural photosynthesis.

David Gates establishes the fact that photosynthesis must be a step-wise process in an instructive way (*Scient. Am.*, Sept. 1971, p. 89). He considers the energy content of a molecular bond and multiplies this by the number of molecules per mole, a constant known as Avogadro's number which is 6×10^{23}, to get the bond energy of the substance per mole (where the molecular weight is the number of times one molecule of a substance is heavier than one twelfth an atom of carbon, and where the mole is the weight in grams equal to the molecular weight of the substance). If the same method is used to calculate the mole equivalent of blue light (wavelength $4{,}500 \times 10^{-8}$ cm), it turns out to be 64 kilocalories per mole. For ultraviolet light (wavelength $= 2{,}250 \times 10^{-8}$ cm) it gives 128 kilocalories and for infra-red (wavelength $= 9{,}000 \times 10^{-8}$ cm) 32 kilocalories per mole. The energy needed to break molecular bonds can be calculated likewise; for a single bond between two carbon atoms it is 82·6 kilocalories per mole, for a double bond 145·8, and for a triple bond 199·6. It is clear from these figures that only ultraviolet light has sufficiently high energy to break the bonds, yet light involved in photosynthesis is blue and red so a step-wise mechanism using a number of less energy-consuming processes is indicated.

Reasons have been advanced for optimism in attempts to improve the efficiency of natural photosynthesis. In considering photobiological energy conversion D. O. Hall (Inaugural Meeting of the UK Section of the International Solar Energy Society, London, January 1974) considered the main photosynthetic process of synthesis of carbohydrates from carbon dioxide, CO_2:

$$CO_2 + \text{water} \xrightarrow[\text{chlorophyll}]{\text{sunlight}} (CH_2O)_n + \text{oxygen}.$$

Looking on this as the conversion of energy-poor compounds (CO_2 and H_2O) to energy-rich compounds (carbohydrate and oxygen) he notes the process was developed by the blue-green algae about three billion years ago and is still almost the sole source of fixed carbon and of oxygen. Animals, plants, and bacteria use the reverse of the reaction to produce energy from stored carbon compounds. He remarks, however, that only about 0·2 per cent of the solar radiation is used by the whole of the plant life of the Earth and less than 1 per cent of the fixed carbon so produced is consumed as food, leaving tremendous scope for improvement.

On the other hand he shows that theoretically the photosynthetic fixation of carbon dioxide can be about 27 per cent efficient. In practice only between 6 and 8 per cent of the light can be used to make fixed carbon compounds (or roughly one-third of the possible 27 per cent) and then only when algae plants or tropical grasses, like sugar cane, are carefully cultivated. In normal crop conditions around 1 per cent of the light is the upper limit of conversion in temperate climates, so improvements are certainly possible.

The figures given by Hall for the possible photosynthetic conversion of solar power are interesting. If, for example, hydrogen gas was produced using solar energy at 10 per cent overall efficiency, the equivalent of the projected US gasoline consumption for the year 2000 would be obtained from a collection area of 20,000 square miles with negligible effect on the heat balance of the Earth. By way of contrast, a coal-burning power plant delivering 1,000 MW would require an area of approximately 100 sq.km to be stripped of coal to a depth of one metre to provide the equivalent supply for a period of twenty-five years only. The figures are most impressive and provide the case for considerably increasing effort in maximizing photosynthetic energy studies. Possible methods of improving the efficiency of the process are outlined by J. Zelitch in *Science* **188,** 1975, p. 626.

Hall describes the process of capture of photons of light by chloroplasts, the membranous bodies containing chlorophyll, electron carriers, and enzymes which give green plants their colour and which also occur in algae. Most bacteria contain hydrogenase, an enzyme which catalyses the reaction

$$2e_- + 2H^+ \rightleftharpoons H_2.$$

By adding hydrogenase to isolated chloroplasts and illuminating the mixture, protons (H^+) formed by the photolysis of water can be combined with the electrons released by the pigment system to liberate hydrogen, H_2. This means that solar energy provided to a chloroplast and hydrogenase system can continuously generate hydrogen H_2 and oxygen O_2 from water

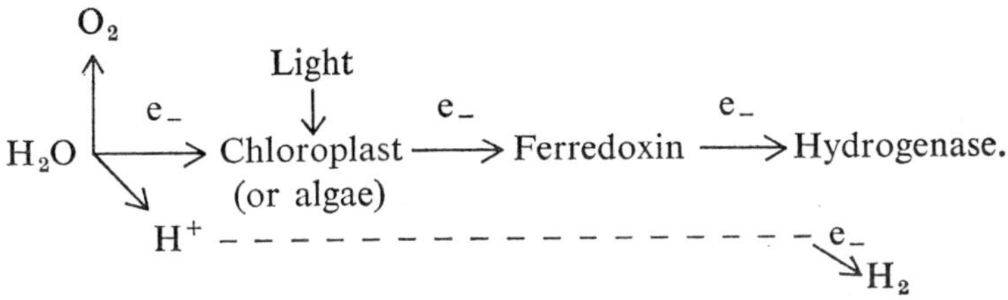

Ferredoxin is an electron carrier protein contained in chloroplasts. The final outcome of the processes of plant-bacteria type is the production of a most versatile fuel, hydrogen. We shall have reason to look at the possibilities of the hydrogen economy later.

Cesare Marchetti has considered the possibilities of an energy plantation (see, for example, his paper entitled 'Energy Plantation Revisited: Milking the Forest', March 1977, published by IIASA, Laxenburg). He advocates, for the less developed countries, a 'soft' technology which implies little sophisticated hardware but requires a combined knowledge of energy systems, plant biology and physiology, and genetic engineering. He believes a feasible system could produce 2·5 kW of high quality energy per person, which is roughly half the present-day level in the European countries. The first problem Marchetti discusses is the efficiency with which plants use the low percentage of carbon dioxide ($\sim$320 ppm) in the atmosphere. Recent investigations of photosynthesis suggest that some plants are more wasteful of energy than others, the difference being due to the two types of leaf cell involved. One type of plant known as 'C_3', which includes wheat, spinach, tobacco, and hay, uses a less efficient chemical cycle to collect carbon dioxide than the so-called 'C_4' plants, maize, sugar cane, and sorghum. As a result C_3 plants grow less quickly than C_4 plants; J. L. Monteith concluded that the efficiency values are 1·4 per cent and 2·0 per cent respectively (*Expl. Agric.* **14**, 1978, pp. 1–5). Marchetti believes that in both C_3 and C_4 plants some re-engineering may be required. Insects can induce the formation in plants of galls; he considers it possible to bio-engineer galls so that sugars from the sap of trees are converted to hydrogen which can be piped to a store. Hydrogen is preferred to methane (CH_4) as it requires no supply of carbon, and therefore depletes the biosphere to a lesser extent.

In effect Marchetti proposes a transformation of solar energy to a transportable and storable fuel simply by using a special kind of forest equipped with modest hardware. He acknowledges the need for further work but argues that most of the steps are understood and that a practical demonstration would stimulate research in the more difficult area of increasing photosynthetic efficiency.

The papers by both Hall and Marchetti provide good lists of references to the rapidly expanding volume of literature on this intriguing effort to use nature to generate fuels by the photosynthetic route.

Practical Experimental Schemes. After synthesizing glucose plants transform it into other substances, including cellulose. Hence it is possible to regard solar energy as being converted and stored in the form of cellulose. Crops can be treated to yield the cellulose which in turn can be the raw materials for chemical processes giving alcohols, other fuels, and useful raw materials. A good deal of study of such schemes, some in the sunny regions of Australia, seems to indicate that they could be exploited commercially in sunny climates and perhaps prove competitive if the difficulties of finding adequate supplies of coal and/or petroleum were to increase substantially. At present, however, no group claims the process is commercially viable. This is clearly evident from a reading of the proceedings of the UNESCO symposium in Adelaide of 1973 on Energy and How We Live. It is obvious that countries with clear sunny weather and the required supply of water are the most likely to benefit from such efforts to use solar energy, at least in the short term. The annual insolation on a horizontal surface does not vary much over the seasons in a country like Australia (see fig. 19). For Darwin in North Australia (latitude 12 °S) the mean daily insolation on an inclined surface in the best month is only 40 per cent more than the worst month (for Melbourne in the south it is 120 per cent). In the UK the summer and winter radiation can differ by a factor of about 6, which means of course that most schemes would need to solve a storage problem. Energy collected in the summer might have to be stored for the darker days of winter and it is here again that hydrogen storage could be vitally important.

We have mentioned the efficiency of photosynthesis, and a number of experimenters have made detailed study of such efficiencies in different circumstances. Before illustrating by reference to a particular measurement we should remind readers that solar energy depends on latitude but the total amount of energy received per year is generally the same within a factor of 2 over that part of the Earth which is densely populated. The variation between summer and winter is much more significant. Photosynthesis varies according to latitude but not so much as the data of fig. 19 would suggest. Work with well-watered crops has allowed estimates of potential net photosynthesis to be made and expressed as crop productivity in grams per sq.m per day. The values do not differ very widely for four-month periods, eight-month periods, and the year. The four-month period is important in regions of potentially high photosynthesis rates. Among these are North-West Canada, Southern

Scandinavia, and Iceland where values may be as high as 37 or 38 g per sq.m per day. At still higher latitudes the high productivity due to the long summer day is offset by the low temperature.

It has been shown that Sudan grass at Davis (University of California) could yield 51 g per sq.m per day and this was estimated as 49 per cent of the potential yield. In this instance the crop was putting into storage 3 per cent of the total solar radiation and 6·7 per cent of the visible radiation. Gates mentions that barley has been observed to convert as much as 14 per cent of incident visible light to carbon compounds. These values are unusually high and it has to be emphasised that while storage of solar energy may be about 3 per cent efficient in excellent conditions it is more likely to be about half this value, between 1 and 2 per cent. In fig. 33 the net productivity of seven environments is depicted; considerable differences are obvious.

A review article of Roger Lewin (*New Scientist*, May 1977, p. 474) is most instructive in assessing the wide range of views that are held about the value of biological solar energy conversion systems. Some biologists consider such bio-systems as one of the two most promising areas of future solar energy utilization but others are much more cautious. Almost all agree that by the year 2000 plant systems will be contributing significantly to the energy budget and probably expanding in a steady fashion. It is believed that such biological systems will assume a variety of forms. Lewin points out the important fact that most plants are wasteful of energy, effecting an energy conversion of rather less than 1 per cent on average with around 5 per cent as the maximum. At the same time he stresses that plants convert and store so that loss that would arise in providing for a separate storage process is avoided. Others are inclined to argue plants have achieved an extremely difficult chemical process, and their low conversion efficiency makes much greater success by man's efforts unlikely.

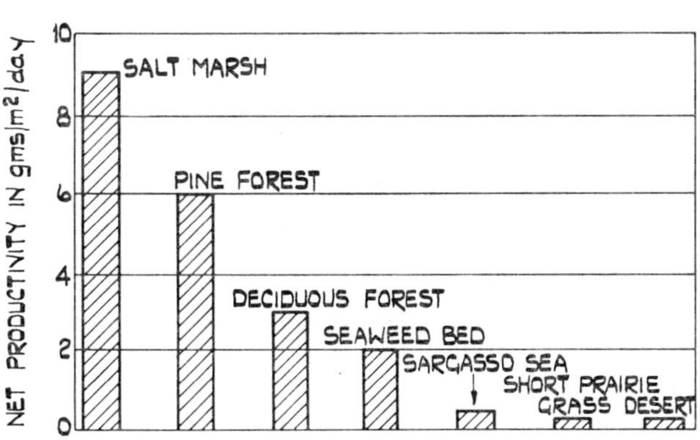

FIG. 33. Net productivity of various environments: salt marsh in Georgia, US; pine and deciduous forests in England; seaweed bed in Nova Scotia; the Sargasso Sea; prairie and desert in the US. [From D. M. Gates, *Scient. Am.* Sept. 1971, p. 97]

The total Earth surface receives 3×10^{24} joules of solar radiation per year and of this only about 0·2 per cent at most is trapped by plants and converted to chemicals, that is, stored. The energy content of food actually consumed is usually much less than the total used in the growth processes and in the UK for example it is only 2 per cent of the total. We can see that if we continued with present practices but met all our energy demand by plant production we would need very large areas of land compared with the amount of agricultural land required at this time for growing food stuffs. There are clearly ways of trying to overcome this problem of land utilization. Some researchers are seeking to grow crops where nothing is grown systematically. The Irish are trying to grow willow trees in worked-out peat bogs, and ocean kelp farming is being investigated in the US. More attention is being given to forestry and an energy multiplication of about 15 has been achieved by R. E. Inman in his cultivation of hardwoods. Problems facing the advocates here include that of transport, as wood is an awkward energy store to handle. Probably wood ceased to be one of the most important fuels because it is bulky and transport of adequate amounts is not only cumbersome but itself an energy-consuming process.

In the bio-energy field waste will prove a valuable resource. Thus most cane factories can generate their own electricity by burning the pulped fibre, and straw can be a valuable by-product. Microbiologists are examining the growing of algae for energy production. A remarkable example of success along these lines is found in the alga, Dunaliella Salina, which grows in the Dead Sea and produces glycerol as a defence against high osmotic pressure. The alga can, in the right conditions, manufacture glycerol to the extent of 85 per cent of its dry weight and this is the basis of a commercially successful venture. Algae feature also in some tentative schemes aimed at producing hydrogen. The productivity of algae in converting sewage to biomass seems promising (W. J. Oswald, *Solar Energy* **15**, 1973, p. 107).

All of the research programmes in bio-conversion of solar energy seem to offer encouraging prospects but it is somewhat early to predict what share of man's total energy budget will be provided in this way by the year 2000. Meanwhile in the countries which are primarily agricultural, wood, plant, and animal wastes continue to be a valuable form of energy resource, sometimes the main fuel.

Energy Flow and Agricultural Method. As we saw in the last section plant breeders and biochemists have a significant part to play in our attempts to solve our energy problems. A quite different approach to the biosphere, but nevertheless one which underlines the importance of their work, is concerned with investigations of energy flows within agricultural systems as a whole. Agriculture has been practised by man for many thousands of years and the raising of crops and husbanding of animals has been the most outstanding example of the exploitation of the energy that is stored in plants. Since the exploitation of the huge energy resources stored in the fossil fuels only began with the Industrial Revolution, this phase of human history is one of comparatively short duration and it is beneficial to look back into the past or to more primitive communities to see what lessons can be learned.

R. A. Rappaport has studied the flow of energy in an agricultural society (*Scient. Am.*, **225**, no. 3, 1971, p. 116) and he illustrates this vividly with the example of the primitive agricultural system of the Tsembaga farmers, a people who speak the Maring language and who live in the central highlands of the Australian Trust Territory of New Guinea about five degrees south of the Equator. Rappaport stresses that these farmers manage to pursue today a type of gardening that has existed for several thousands of years. The example of these farmers is particularly intriguing as they have only fire, muscle power, and the simplest of tools. It should be noted that a similar type of farming or gardening existed elsewhere, including medieval England where it was known as swiddening. The outstanding feature is the stability of the ecosystem that it secures, and in this it compares favourably with the ecological degradation that is introduced in modern farming with its ever-increasing demand for energy expenditure—expenditure in the manufacture of fertilizers, pesticides, and the use of agricultural machinery of all kinds.

Basically the method consists of creating a clearing in a heavily afforested region. The cuttings may be removed, burned, or allowed to rot according to needs. The garden is planted and harvested and then abandoned so that it soon becomes part of the forest again. A single planting is normal but occasionally the garden is used two or even three times before it is abandoned.

While a rain forest yields enough food to sustain about one human per square mile this is totally unsatisfactory from the point of view of man, and swiddening allows him to increase considerably the population density while maintaining the long-term stability of the total ecosystem. In the limit and in the case of fertile lands, the population density can approach that of industrialized societies, say several hundred per square mile. This figure is roughly that of Scotland but rather less than that of England. It is nevertheless a high number.

In practice the people have in cultivation about one quarter to one half of an acre per person, with some twenty times the area of cultivated land lying fallow. It is this high ratio of fallow land to cultivated land that ensures the stability of the total system and that the disturbance of the balance of nature is minimized. Horticulture provides about 99 per cent of the everyday diet of the people but additional protein of considerable value is obtained from pigs that are reared by allowing them to roam in the forests and supplying them with all the waste from the consumption of food by humans.

The examination of the energy flow for such a social system is less complicated than in an industrialized country, and twelve energy inputs were evaluated by Rappaport. These are interesting in themselves, illustrating as they do the various activities of moment in the community. They are clearing underbrush, clearing trees, fencing the garden, weeding and burning, placing soil and retainers, planting and weeding until the end of the harvest, general maintenance, the harvesting of sweet potato, taro, cassava and yam, and cartage. The human energy investment found by adding the twelve components is next compared with the energy yielded by the crops and the pigs, which consume part of the crops, and the surprising result is that the crop and pig energy yield is more than sixteen times the human energy expended. This factor is known as the energy gain. Swiddening is a highly efficient method of farming in countries which are naturally arboreous, and the instinct to save energy obviously results in an efficient farming or gardening system which has evolved over some thousands of years in such lands.

Many refinements to this method appear in practice and these include such aspects as the wise choice of crop for the different altitudes of the hilly country. The trees are regarded as the 'mother of gardens', indicating the insight of the people into the critical role of the trees in maintaining the stability of their total system. It is not always wise to use the most efficient plants, those that capture energy best. It can be more important to encourage plants that may be less efficient in trapping energy but which help to sustain other plants including those grown in the gardens for harvesting. Quite clearly, in this apparently contemporary example of a primitive agricultural system, the favourable ratio of the energy of the food crop to the human energy input is by no means the main consideration. The long-term stability of the system and its actual improvement in energy ratio (food energy to input energy) over long periods is equally important, if not of greater importance.

The so-called primitive peoples are in this sense remarkably able, and their cultivation and animal husbandry reaches the level of performance of an advanced technology, one that must be compared and contrasted with modern methods of agriculture practised today by industrial nations. In our society, fertilizers, pesticides, and other aids of modern agriculture were welcomed fairly uncritically as a means of increasing production to sustain rapidly expanding city populations, and possibilities of harmful long-term effects were ignored. Somewhat belatedly, serious criticism of modern practices is now advanced on the grounds that we disturb very radically the natural development towards balance of both plants and animals. The effect of pesticides on birds is one example, and the effect on insects and wild flowers as well as on weeds of some of these agents is another. The amount of energy consumed in their manufacture, and in the manufacture of fertilizers, also gives cause for concern (see p. 110).

It is perhaps very obvious that most people would be entirely unwilling to return to closed ecosystems such as practised by the Tsembaga people and to be dependent on human muscle power. We cannot readily contemplate being obliged to be farmers. In any case there is probably not enough fertile land to support us unless change occurred slowly with time to adjust populations downwards. But having agreed that the vast majority of people prefer to live in modern society, we have to question the basis of that society and sometimes reject so-called advances which are not truly advances in the longer term. Rappaport, in his penetrating study, thinks that the need for modern man to put more emphasis on the benefits of complexity and stability of the ecosystems which he controls, is a more pressing one than his need to overcome the threats of pollution, over-population, or even energy famine. As he points out, the solution is probably not readily reconciled with the values, goals, interests and political and economic institutions prevailing in industrialized and industrializing nations today. No modern nation seems interested in decreasing the population to the number that could be sustained by swiddening.

Energy Flow Diagrams. A good way of appreciating how differently energy is utilized in what are usually regarded as primitive societies, compared with the crude and somewhat primitive methods in most of the advanced industrial societies, is by means of energy flow diagrams. The papers by Earl Cook and others (see *Scient. Am.*, Sept. 1971) make use of this flow concept, a concept which is particularly appropriate when applied to energy which flows like the life-blood through our industrial, commercial, and domestic structure.

By comparing the present-day flow pattern of energy through the US economy with those of the past, Cook has provided a fascinating account of some extremely significant features in the development of the US to its present position as a world power. Likewise, comparing and contrasting the flow diagrams for energy as observed in a number of countries, the particular features of each emerge in a striking fashion. A most instructive way of presenting the broad features of historical changes in energy consumption with way of life is found in fig. 34. We shall further explore these changes in chapter 13.

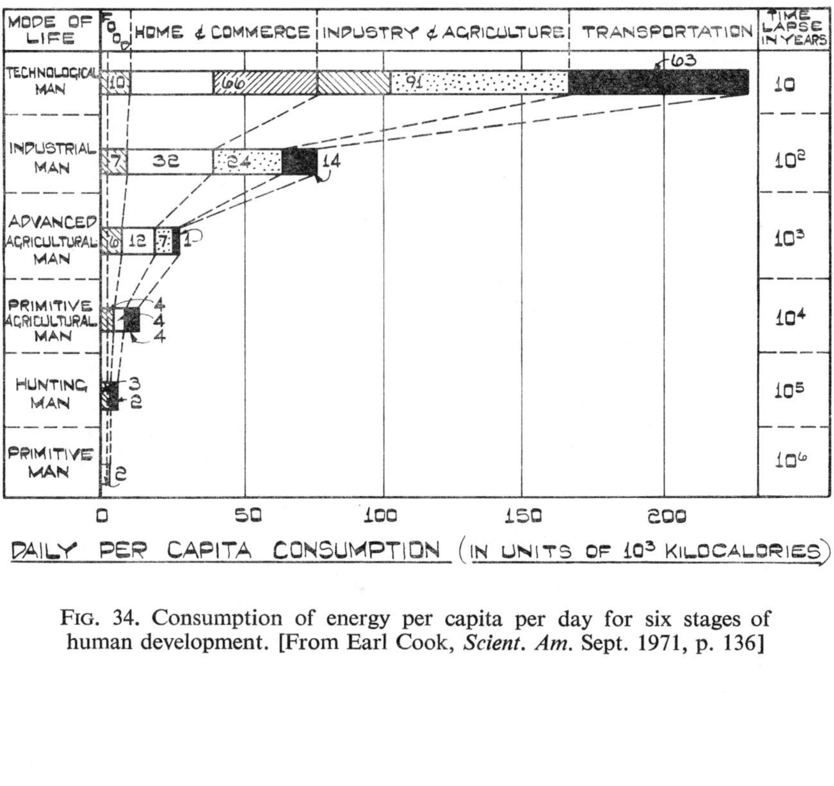

FIG. 34. Consumption of energy per capita per day for six stages of human development. [From Earl Cook, *Scient. Am.* Sept. 1971, p. 136]

The contrast between the strictly economical pattern of energy flow in the Tsembaga community with that in most 'advanced' nations is remarkable. We conclude quickly that industrial nations consume energy at a staggering rate. The rate of energy consumption per capita varies enormously and if we confine ourselves to considering fossil fuels, together with nuclear and hydro-electric sources, we find there is a factor of about 100 between the less developed, for instance Pakistan and Indonesia, and the advanced such as the US and Canada. The US and Canada exceed in consumption rate the majority of the industrialized countries such as the UK, West Germany, France, Australia, and even Sweden by a factor of 2. The average for the developed countries is roughly 3 times the world average, and this world average is roughly 7 times that for the underdeveloped countries. These are extremely large factors and Cook believes that some countries are much too wasteful of energy. It is interesting to note that he takes the view that the US is not a good model to follow; in particular he stresses the extremely wasteful use of petroleum in personal transport, that is, mainly in large and inefficient cars.

The form of the energy flow through the US economy is shown in fig. 35.

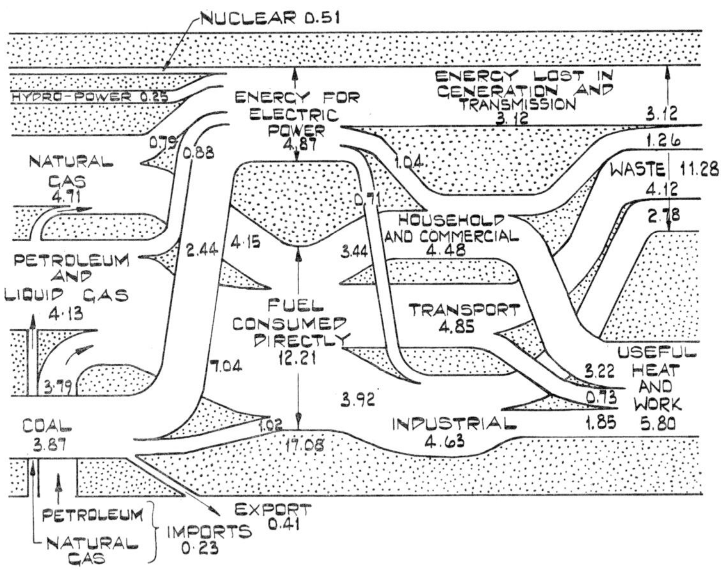

FIG. 35. Approximate flow of energy through the US economy, 1976. [From Earl Cook, *Energy Resources and the Environment*, Texas A & M University]

The width of the channels at the left hand or input side is a measure of the amount of each component of the input energy. Thus we see that nuclear at the top is relatively very small, hydro power is three times as large and natural gas ten or more times as large as hydro power. Petroleum and liquid natural gas roughly equal natural gas and, when imported petroleum is included, exceed it by nearly 30 per cent; coal is just under two-thirds of the other two main supplies. At the output end of the flow diagram (right-hand side) we find that useful heat and work are together about 43 per cent of the input and the waste is 57 per cent.

In most industrialized countries a somewhat similar situation obtains and only between one third and one half of the total input energy is usefully employed. This picture is in sad contrast to that deduced for the primitive agricultural society. Such flow diagrams for energy have been used a great deal by Cook and others, and they provide a vivid picture of how energy is produced, consumed, and applied in an industrial country. At the left side of the diagram we have fossil and other fuels streaming into the system and these fuels diverge as streams, sometimes breaking into several parts which are consumed in different ways. We find for example that coal may be used for the generation of electrical power, or consumed directly for household and other forms of heat, and it may also be exported. From left to right the total flow remains constant, and the input at the left equals the output at the right, making allowance for imports and exports shown at the bottom of the figure. As we have already noted, 'waste' may amount to nearly two-thirds of the output and this shows that in many societies, such as that in the US, energy has perhaps been too abundant and not wisely used. In various other countries the flow may be extremely differently represented and, for example, transport consumption may be a much less notable part, especially where the use of cars is much reduced and public transport predominates. Cook himself takes the view that with an energy consumption per capita under half that of the US, the flow diagram representing the UK scene is more satisfactory and possibly more representative of how industrial countries should behave. The ways in which energy is consumed in the UK are shown in fig. 36 together with those for the US. It should be realized the data change comparatively

20% INDUSTRIAL & COMMERCIAL HEATING	23% MANUFACTURE	25% RESIDENCES (HEATING & APPLIANCES)	21% TRANSPORT	10% IRON & STEEL	1% FARMING

(a) BREAKDOWN OF ENERGY CONSUMPTION FOR U.K.

37% INDUSTRY	32% TRANSPORT	17% RESIDENCES	14% COMMERCE

(b) BREAKDOWN FOR U.S ECONOMY, 1972

FIG. 36. Main areas of energy use in (a) UK (b) US. [Data for (a) taken from the *Observer*, 6th October 1974] rapidly by virtue of conservation measures, pricing policies, and the like. Climate plays a significant part in determining the pattern of consumption and the nature of the industrial activity is an important factor, for example heavy industries such as steel-making consume large amounts of energy.

The overall flow diagram again drives home a point that has been stressed previously. The Carnot cycle, which applies to all heat engines, and which was formulated in 1824, established the fact that heat engines cannot be highly efficient. At the time of Carnot's work the efficiency of machines was still around 1 per cent, although the work of James Watt and his successors had brought about a gradual increase from what had originally been extremely low values. Carnot established the theoretical limit to what could be achieved. Even today efficiencies are still comparatively low in absolute terms. For the internal combustion engine efficiencies fall below or around 25 per cent though a large diesel engine may reach almost 38 per cent. The overall efficiency can be approximately 40 per cent for a well-designed fossil-fuelled electricity generating station. These sample values show why the average amount of useful work is about 36 per cent of the equivalent energy consumed and why 64 per cent (nearly two-thirds) of the energy is degraded and is all too often simply 'waste' heat. While heat engines are intrinsically always short of 100 per cent efficient, and often very far short of 100 per cent, energy can of course be used very efficiently in certain industrial processes where heat is used directly such as glass-making or iron and steel making.

The tremendously steep rate of increase in the consumption of petroleum in the US was a consequence of the enormous growth in the automobile industry between, say, 1910 and the present day. Now there is a car per two people (including children) in the US. There had to be a correspondingly large increase in consumption and in fact the US had ample supplies and only recently needed to import petroleum. The rapidly increasing price of OPEC oil could lead to notable changes in the pattern of behaviour in the US. There large inefficient automobiles have been in fashion for half a century, but the future may see a reduction in the number of private cars and a decrease in the horsepower of the engines of automobiles. President Carter has shown that he would like to encourage such a change by fiscal measures similar in principle to the horsepower tax which prevailed in Britain for many years.

Major changes in customs can take place in a period of fifty years. Another illustration of this, and one which is evident on almost a world-wide basis, is in lighting. The dim light of the whale-oil lamp was largely replaced by the kerosene lamp just over one hundred years ago. The introduction of the Welsbach gas mantle then made a remarkable improvement in the way of life after dark. At the turn of the century a tentative start was made in the construction of electricity generating stations and lighting by electric lamps. Now cities and homes almost everywhere are brightly lit and a new degree of freedom has been provided. At the present moment we see the rapid introduction into our homes of energy-consuming machines such as water heaters, cookers, air-conditioners, refrigerators, freezers, irons, washing machines, clothes dryers, dish washers, electric blankets, fans, vacuum cleaners, toasters, and of course radio and television sets.

It is difficult to foresee any appreciable reduction in this area of demand. Unfortunately almost all of the needs listed and which lie in the domestic sector are met with the aid of electric power and for the most part this is relatively inefficiently generated from fossil fuels. On the other hand the power-operated devices in domestic use, even in the US, account for about 2·5 per cent only of the total consumed, while in commerce (offices, warehouses, shops, etc.) the corresponding figure is 5·4 per cent. Although worthwhile economies can certainly be made the sectors are not major ones and they need not cause alarm. In the industrial sector the same low overall inefficiency could not readily be tolerated. Convenience is important in the home but efficiency must be sought vigorously in industry. In recent years many large industrial firms have begun to pay close attention to the efficiency of their operations, and close study of energy budgets has been introduced. Direct burning of fuels persists in a large number of processes of manufacture because it results often in increased efficiency.

Energy Analysis. A great deal of thought has been devoted to this comparatively new field of study. The impetus for increased effort was provided by the steep rise of fuel prices, especially oil, and by the new awareness of the advent of an energy gap. Analysis can pinpoint areas where care has to be exercised. Thus it has been argued that photovoltaic conversion of solar energy is soundly based. The manufacture of the solar cells demands an energy expenditure but the life of the cells could be extremely long and the final energy return could thus greatly exceed the energy input. It has been argued that a few years of successful operation repays the initial energy investment and thereafter the cells bring about an energy gain. Similar considerations applied to other methods of energy production might lead to very different conclusions. Where large plant has to be constructed there is a large energy investment required (as well as a large capital investment). The plant must then operate for a certain calculable number of years to yield an energy dividend. By way of illustration it seems certain that wave energy will require large structures probably of steel. A typical 'duck' might measure 10 m × 10 m × 10 m. If the 'duck' does not survive corrosion well, or if it has to be replaced for some other reason, the energy investment in manufacture, in siting and in maintenance, could conceivably cost the equivalent in energy terms of too large a fraction of its available energy yield. Such considerations could be important in making the right decisions on future programmes of research and development.

Methods of energy analysis can be applied usefully in many fields of manufacture. Chemical and petro-chemical manufacturers have to examine carefully the energy cost per kg of most products, particularly those produced in bulk for sale to the public. All building materials have likewise to be examined in energy terms. Availability of energy at low cost resulted in some neglect of such matters for decades but the situation has improved rapidly.

Large industrial companies such as ICI study energy requirements in the manufacture of their products and those of competitive materials (see *The ICI View*, published by Plastics Division, ICI Limited, Aug. 1974).

The evaluation of the energy consumed in modern agricultural practice is not straightforward and the work of D. Pimentel and E. C. Tethune demonstrates this (*A. Rev. Energy*, **2**, 1977, p. 171). In the case of the US they calculate that the food system consumes 16·5 per cent of the US energy budget when all the indirect energy costs are included, excluding exported food. They observe that this value is very close to the percentage, 16·6, of the per capita disposable income spent on food. It is a figure in sharp contrast to the 80 per cent of personal income expended on food in many of the less developed countries, although admittedly many of these have hot climates so that the money spent on heating and housing is naturally less. At the same time, and it is a depressing fact to note, the energy in the food actually consumed in the US is decidedly less than one tenth of the amount required in making it available.

Much of the lavish expenditure of energy in modern food production was due to the availability of fossil fuels at low cost. This encouraged mechanization and the manufacture of fertilizers, pesticides, and herbicides, especially in the industrialized countries. Though the increase in productivity has been marked, when the energy content of the crops is compared with the energy required in the manufacture of the synthetic materials, the return in energy terms is found to be unsatisfactory. If the crops are used to feed animals, and the ainmals in turn are consumed as food, the energy value of food obtained in this fashion can frequently be less than the energy used in the overall food production process. Major innovations in the years between 1945 and 1975 resulted in the changes illustrated by way of example in table 11.1, which shows data for the cultivation of corn.

It is not easy to make sweeping conclusions from study of the table, or similar ones for other products. In the thirty years in question a 2·5 fold increase in protein yield was achieved at the cost of a 3·2 times increase in the energy input. This was accomplished at the same time as a reduction in the labour force by a factor of 3·4.

It is obvious that complex issues are raised by energy analysis of food systems, and energy ratios (kilocalorie return to kilocalorie input) do not in themselves provide sufficient information on which to base valid decisions. It has been pointed out by J. D. Nalewaja (*Proceedings North Central Weed Control Conference* **29**, 1974, p. 18) that the highest ratios are obtained with zero fossil energy input but the world could support far fewer people in such a situation. Ratios are a very useful additional piece of information in arriving at important decisions on issues of moment in agricultural policy but social and political aspects of the policy may be considered more important.

If it is difficult to resolve such questions of approach to crop production, it is clear that methods of bio-conversion of solar energy will also prove difficult to evaluate. It could be found that human labour will prove essential

TABLE 11.1

ANNUAL INPUTS PER HECTARE, US CORN PRODUCTION
(From D. Pimentel and E. C. Tethune, *Annual Review of Energy* **2**, 1977)

Input	1945 Quantity	1945 kcal	1975 Quantity	1975 kcal
Labour	57 hr	31,022	17 hr	9,250
Machinery	26 kg	539,000	55 kg	1,139,160
Fuel	140 lte	1,400,000	210 lte	2,100,000
Nitrogen	8 kg	121,440	128 kg	1,943,000
Phosphorus	8 kg	25,600	72 kg	230,000
Potassium	6 kg	13,200	80 kg	176,000
Seeds	10·5 kg	73,080	21 kg	146,160
Irrigation	—	103,740	—	780,000
Insecticides	0 kg	0	1·12 kg	81,312
Herbicides	0 kg	0	2·00 kg	145,200
Drying	—	42,320	—	375,000
Electricity	—	39,500	—	380,000
Transportation	—	49,400	—	180,000
Total		2,438,302		7,685,082
Corn yields	2,132 kg	7,419,360	5,394 kg	18,771,120
Protein yields	192 kg	—	485 kg	—
Energy ratio (kcal return / kcal input)	3·04		2·44	

in securing satisfactory results. In any case comprehensive mechanization may be found too expensive to be justified in energy terms or, even more importantly, to be justified in human terms, for there is always the problem of putting people out of work to be faced, especially where it involves 'a drift from the land'.

Part Two

FOSSIL FUELS AND THEIR DEPLETION

Chapter 12

Coal, Oil, and Natural Gas

Until a hundred years ago, wood was the most important energy source, and as late as the year 1900 it was still one of the major fuels consumed in the most advanced industrial country, the US. Between 1900 and 1925 more concentrated forms of fuel—coal, oil, and natural gas—rapidly became the chief energy sources and by 1950 wood was a comparatively minor element in the total supply (see fig. 37). This underlines the fact that near total dependence on the energy stored in the principal fossil fuels is almost entirely a phenomenon of the twentieth century. On the other hand wood had been, in many countries, the main fuel for some thousands of years. The contrast between the lengths of time that wood and fossil fuels have been dominant should drive home at least two lessons. The first is that the energy is stored in compact easily-used forms in the fossil fuels, and the second is that the rate of consumption of energy has risen extremely rapidly during the last hundred years. The second lesson is closely related to the first.

As we mentioned in chapter 11, all of the fossil fuels have been created through the agency of solar radiation. Leaves on plants capture a fraction of the solar radiation to which they are exposed and store it chemically by the process of photosynthesis. This store sustains the plant and indeed the animals living on the plants. Biologically stored energy is released in the process of oxidation at approximately the same rate as it is stored. However, if plant and animal matter is buried under anaerobic, or non-oxidising,

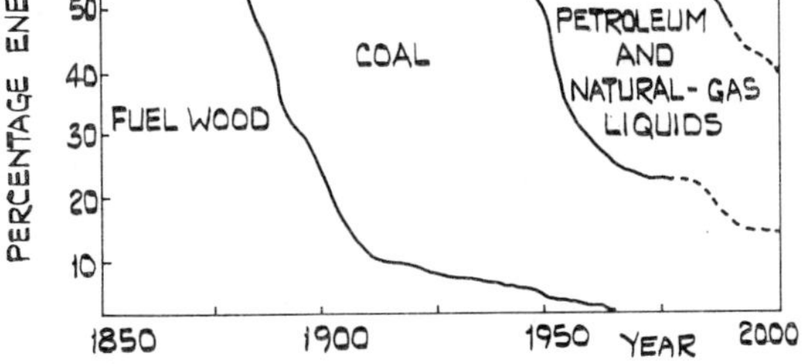

FIG. 37

Fossil fuel and other energy inputs into US economy as a function of time. [From Earl Cook, *Scient. Am.* Sept. 1971, p. 137] conditions then a gradual build-up of energy can and does occur. It is this energy of plants and animals, stored over the course of millions of years, that gives the fossil fuels their great value. In view of the long periods of time required to create fossil fuels, it is readily realized that we are consuming a non-renewable resource in using them as fuel. Moreover, coal and oil are chemically extremely complex and as a source of raw material for the chemical industries might prove irreplaceable in the future.

It is illuminating to compare the energy contents of the main fossil fuels. The use of a mixture of units, that is tons, barrels, and cubic metres, is historically well established and will only be relinquished slowly in practice. We convert the various units in table 12.1, adopting the joule as the reference unit connecting them together. We note that in energy content one ton of coal

TABLE 12.1

INTERCONNECTION OF QUANTITIES OF THE MAIN FOSSIL FUELS

Quantity	Energy content (joules)
Ton (of coal)	$30{\cdot}5\times10^{9}$
Barrel of oil ($\sim$35 imperial gallons)	$5{\cdot}92\times10^{9}$
Cubic metre (m^3) of natural gas	$3{\cdot}85\times10^{7}$

is roughly equivalent to five barrels of oil, and one barrel of oil to about 150 cu.m (at normal atmospheric pressure and temperature) of natural gas. In weight, 1 tonne of crude oil approximately equals 7·3 barrels.

A. COAL

Genesis. Swamps, bogs, and marshes provide ideal environments for the preservation of plant material, and peat develops. Coal is simply altered peat, produced by the higher temperature and pressure it has been subjected to when buried under thousands of feet of rocks. At various times and places during the Earth's history conditions were such that thousands of feet of alternating layers of peat, sand, and mud accumulated (frequently in subsiding tropical deltaic areas) to form 'coal measures'. The coal seams themselves are relatively thin layers which vary in kind (depending upon the heat and pressure the plant material has been subjected to). Coals of low rank are the brown coals or lignites; those of medium rank are known as bituminous coals; and the highest rank coals are anthracites.

Reserves. Studies throughout the world have shown the peak period of coal formation was the Carboniferous (*c.* 300 million years ago), but that significant amounts were also formed in various parts of the world throughout geological time since then. Estimates have been made on the geographical idstribution of coal reserves throughout the world, but in informed discussion of reserves of coal many complex factors have to be taken into account. These have been enumerated by G. Armstrong (*Phil. Trans. R. Soc.* **A276**, 1974, p. 439). He points out that geological factors, engineering innovations, and economic climate all influence the estimates of economically recoverable reserves, even though the geological environment and the gross total of reserves hardly change at all. To illustrate this important point we note that in October 1973 the assessments of Armstrong of gross reserves for the US were $2{\cdot}60 \times 10^{12}$ tonnes in seams to a depth of 1,830 m. However, of these huge totals only about one eighth, or 317×10^{9} tonnes, were in 'measured' deposits rather than 'inferred' deposits and within 305 m of the surface. Hence only these 317×10^{9} tonnes could be exploited by existing US mining techniques.

Similar qualifications apply to Great Britain where, in 1973, gross reserves were assessed at 163×10^{9} tonnes, of which 99×10^{9} tonnes were in 'known' deposits and in seams at least 60 cm thick within 1,220 m of the surface. The coal estimated to be economically recoverable in 1974 was only $3{\cdot}9 \times 10^{9}$ tonnes, or some $2\frac{1}{2}$ per cent of the possible gross reserves. Of these approximately 8 per cent was recoverable by surface mining, which is in general preferable on economic grounds although it does raise serious environmental problems.

World reserves of coal are subject to the same uncertainties that beset the calculations for the US and Great Britain. The World Energy Conference in 1968 stated that the total world reserves exceeded $8{\cdot}8 \times 10^{12}$ tonnes (of which $6{\cdot}7 \times 10^{12}$ were bituminous coal and anthracite and $2{\cdot}1 \times 10^{12}$ were brown coal and lignite). The figures included for the US and the UK were $1{\cdot}5 \times 10^{12}$ tonnes and $15{\cdot}5 \times 10^{9}$ tonnes respectively. As Armstrong remarks, these two totals considerably exceed the estimates of economically recoverable reserves in both cases and correspond approximately with those considered physically workable.

Making allowances for favourable changes in the economic situation (largely following sharp price increases for oil), Armstrong arrives at a total for the world of $4{\cdot}4 \times 10^{12}$ tonnes of economically recoverable coal. He stresses the huge size of this estimate. It amounts to 1,250 times the world output of 1975 given in *Mining Annual Review*, 1977, p. 87, as $3{\cdot}2 \times 10^{9}$ tonnes of coal. Even if the coal consumption increased by a factor of about 4 compared with 1975, the demand for such an amount (12×10^{9} tonnes per annum) could be met for more than three hundred years.

We deduce, therefore, that there is sufficient coal in the world to last a few hundred years at a much higher level of consumption than the present one. Is there, therefore, any reason to talk of energy shortage and fossil fuel depletion? This is the question which must be faced in examining the energy situation today but it is not a simple question to resolve.

In the first place the coal is not uniformly distributed among the nations that use it in quantity. A rough breakdown (table 12.2) by major geographical areas shows the variation. The figures highlight the important fact that there is extremely little coal in some highly populated areas of the world. This is also the case with other fossil fuels and with uranium, but the fact that coal

TABLE 12.2

WORLD DISTRIBUTION OF RECOVERABLE COAL ASSUMING TOTAL RESERVES = $4{\cdot}4 \times 10^{12}$ TONNES

(From G. Armstrong, *Phil. Trans. R. Soc.* **A276**, 1974, p. 450)

Area	Approximate % of total	Actual amount (10^{12} tonnes)
USSR	61·5	2·71
North America	17·5	0·77
China plus parts of Asia	17·0	0·75
Europe	2·5	0·11
Africa	1·5	0·07
South America	<0·5	<0·022
Australasia	<0·5	<0·022

NOTE: Paul Averitt estimates $10{\cdot}23 \times 10^{12}$ tonnes for the world total rather than the $4{\cdot}4 \times 10^{12}$ tonnes used above. The amount depends somewhat on thickness of seam and other factors. The geographical distribution differs and is favourable to the US ('Coal Resources of the United States, 1st January 1974', *U.S. Geol. Survey Bull.* **1412,** 1975, p. 91).

is not easily transported has been a prime cause of its decrease in popularity relative to oil in the last fifty years. Armstrong observes that out of a total world output of about 3×10^9 tonnes, only about 190×10^6 tonnes were exported from one country to another to correct local shortages and of this 190 million, almost 25 per cent, were exported from the US alone. This international transport of just over 6 per cent of world production contrasts markedly with the scale of export of other fossil fuels, oil and natural gas (in the form of liquefied natural gas): discounting the US, which is both a large producer and consumer of oil (in fact still the leading country in both), and setting aside the USSR, about 90 per cent of the production of world oil is exported from one country to another. If the US failed to export its 25 per cent contribution to the total coal exports, the basic imbalance in the geographical maldistribution of coal deposits would become still more serious.

This discussion of the difficulties of production and export leads to the conclusion that coal must be exploited with considerable determination if the depletion of oil is to be largely, or even partly, offset. Production involves massive investment of capital resources, and a time of about ten years between exploration and production at near planned output level. The US government has decided to increase US production from 550×10^6 tonnes in 1974 to $1{,}000 \times 10^6$ tonnes in 1985 and to $1{,}400 \times 10^6$ tonnes in the year 2000. The supporting comprehensive research programme will cost some £4,000 million. At the same time the USSR, producing slightly less coal than the US, will greatly increase production; expansion is planned in China, Poland, South Africa, and Australia also. The countries of the EEC are in a less assured position. In Britain developments are under way and these could lead to production in the 1980s of an additional 30×10^6 tonnes or so. This represents an increase of about one quarter of the present total and is in no sense a massive effort.

Most coal exploited to date is under land, but considerable reserves must exist under the sea. It is of course, like oil and natural gas, found only among the sedimentary rocks of the continental shelves and not in the true ocean bed, which consists of a comparatively thin layer of the igneous rock known as basalt. The technical problems of coal mining under deep waters have yet to be evaluated and resolved however.

Conversion Processes. Since coal is a less attractive fuel than oil a great deal of work is proceeding with the aim of making it a feedstock for conversion into other forms of energy. Coal conversion processes have attracted attention at various times during this century, not least in the war periods when petroleum supplies could be interrupted. Likewise the OPEC price changes since the shock increase of 1973 have resulted in increasing effort to convert coal to more versatile forms of product. A useful survey of conversion methods has been given by D. F. Williams, who also compares some aspects of the use of fossil fuels. His reserves and production figures, given in fig. 38a with some minor alterations of scale, show that oil constitutes about one eighth of the world's recoverable energy reserves (neglecting uranium) and natural gas only one sixteenth, but that these two fuels provide at present about two-thirds of world energy consumption. This pinpoints the dangers of the present (1977) supply and demand situation.

Williams points out that the EEC plans to produce much more nuclear electrical power and to import more natural gas but that, in the absence of major changes in existing technology, about 40 per cent of the total energy requirement would still have to be met by oil and gas because these are the readily transportable fuels. In view of the unfavourable reserves position for oil and gas and the continuing need for liquid and gaseous fuels, he contends that coal conversion technology is extremely important and deserves urgent development. The difficulty of transporting coal is another cogent argument in favour of its conversion. Nuclear plants provide electrical energy and the conversion of this to the equivalent hydrocarbon forms is not well advanced. The relatively slow programme of construction of nuclear stations adds force to this argument.

Williams compares coal with other hydrocarbons and his views are readily presented (see fig. 38b) in a form which emphasises its hydrogen deficiency. In general terms it can be said that the conversion of coal into clean liquid and gaseous hydrocarbon fuels requires the addition of hydrogen, which itself can be made from coal and water.

From the chemical point of view coal is complex and each molecule contains hundreds of atoms. In theory, however, both gasification and liquefaction can proceed readily along the routes outlined in the equations below:

A. Gasification

Gasification	$C+H_2O \rightarrow CO+H_2$
Combustion	$C+\frac{1}{2}O_2 \rightarrow CO$
	$CO+\frac{1}{2}O_2 \rightarrow CO_2$
Hydrogenation	$C+2H_2 \rightarrow CH_4$
Shift Reaction	$CO+H_2O \rightarrow CO_2+H_2$
Methanation	$CO+3H_2 \rightarrow CH_4+H_2O$.

B. Liquefaction

Synthesis: Coal $+O_2+H_2 \rightarrow CO+H_2O \rightarrow$ Hydrocarbons, etc.
Pyrolysis: Coal $\rightarrow$ Charcoal or Coke + tar + gas
Solvent Extraction: Charcoal + solvent $\rightarrow$ coal solution
Hydrogenation: Coal $+H_2 \rightarrow$ synthetic crude oil.

Many factors enter into the commercial exploitation of these reactions but it is certain that both gasification and liquefaction can be successfully achieved. Thus synthetic natural gas has been produced by the Lurgi process with promise of success in large-scale manufacture. Other developments that are going ahead include the so-called Cogas process. The schemes to achieve liquefaction on a large scale have not met insurmountable technical obstacles.

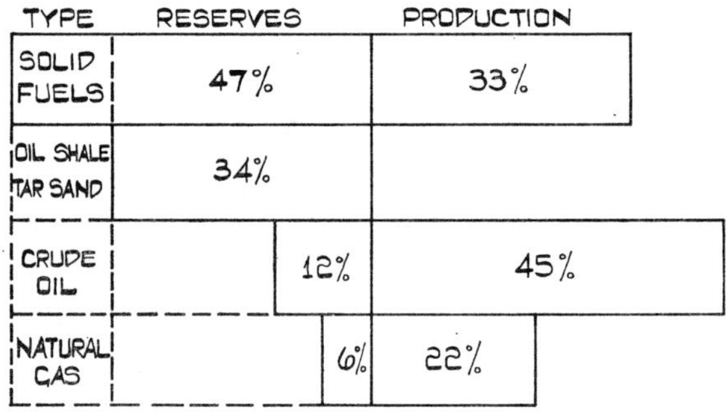

FIG. 38 (a)

World Reserves and production of fossil fuels

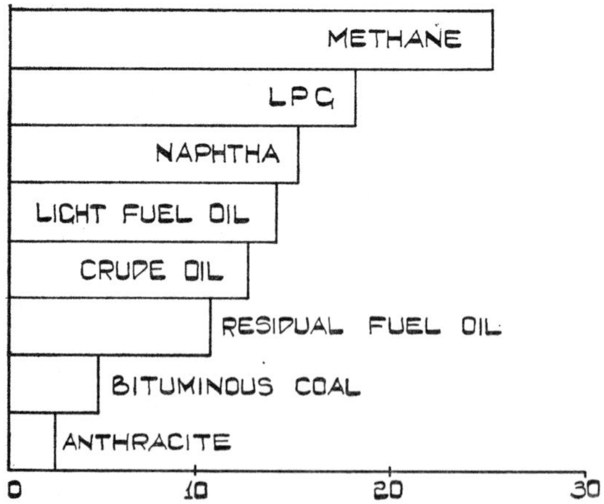

(b)

Percentage hydrogen content of various fuels. [Both from D. F. Williams, European Physical Society General Conference, Bucharest, 1975]

In the Lurgi process gasification of lump coal by reaction with steam and oxygen under high pressures proceeds in the reactor, and the coal is hydrogenated to methane as in the equations at A above. In the Cogas process, so named by the Cogas development company, decomposition of coal proceeds by pyrolysis to give volatiles and charcoal residues, which means that high quality gas and synthetic crude oil are delivered in accord with B above.

Much of the work on coal gasification is aimed at the utilization of gas with turbines capable of working at high inlet temperatures, say 1,200 °C or higher. Success in this would be a major step in overcoming the efficiency limitations of electricity generation.

Direct or underground gasification of coal is receiving renewed attention. Trials were arranged in the 1950s by the National Coal Board and the results appeared to suggest that with simpler coal seams of adequate size the technique might well prove successful (see P. N. Thompson *et al.*, *Underground Gasification of Coal*, National Coal Board, 1976).

All of these conversion studies indicate that coal can fill some of the functions of both oil and natural gas and, if appropriate programmes are pursued energetically, many of the drawbacks of coal can be overcome and some of the effects of the depletion of oil and natural gas offset, at least in part. Coal may, in short, prove a more attractive and more versatile fossil fuel.

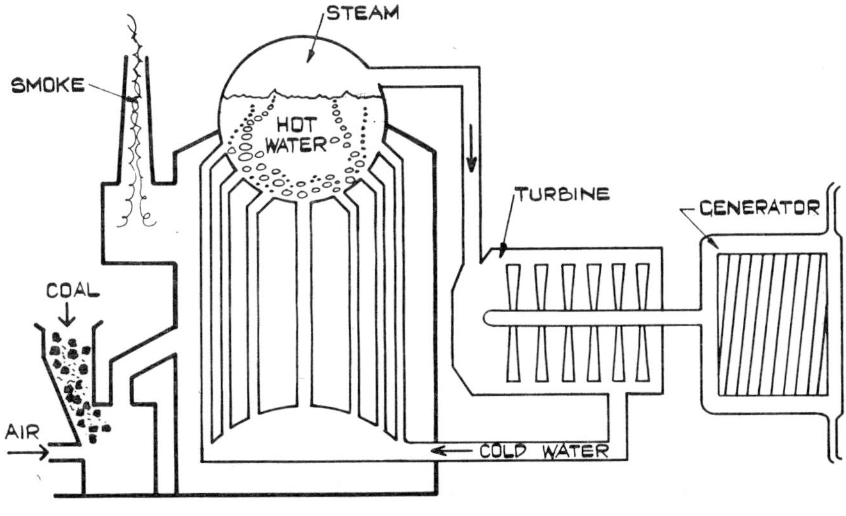

FIG. 39. Schematic diagram of typical coal-fired generating station

Technological Advances in Electricity Generation from Coal. The inefficiency with which coal burns has always been a matter for concern. The more or less standard form of coal-burning plant and its associated boiler, turbine, and generator is shown schematically in fig. 39. In practice there are many refinements in power-generating plants with considerable effort being taken to minimize wastage of energy. These are discussed in chapter 18.

Other substantial ventures may improve the performance of coal in the electricity generation. Among these are the introduction of fluidised bed combustion and of the magnetohydrodynamic generator. The success achieved to date with fluidised bed combustion represents a considerable step forward. If this type of combustion could be successfully introduced on an industrial scale, an increase in the overall efficiency of electrical power production would follow and, moreover, lower grade coals could be utilized. The programme has been described in a world context by J. Gibson (*Appl. Energy* **3**, 1977, p. 87) who believes the ultimate target should be the construction of large pressurised systems operating an air or combined cycle, and capable of generating electricity at efficiencies well above those now attained. The process should be economically attractive and capable of giving built-in and improved pollution control. The low grade coals might be of high-ash type and of comparatively high alkali and sulphur content. The engineering problems in the way of full success will prove troublesome but they should be within the reach of modern technology.

A schematic diagram of such a combustion system is shown in fig. 40. The combustor vessel is a refractory lined cylinder and fluidising air is fed to the bed through an array of bubble caps. Sometimes the bed is warmed initially to around 500 °C by gas burners and then brought to operating temperature. To

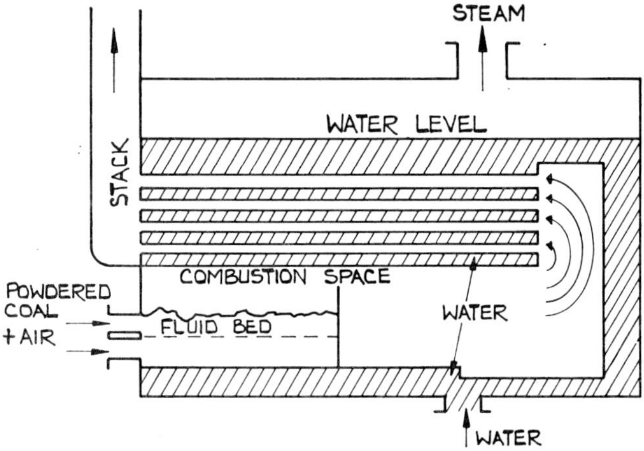

FIG. 40. Fluidised bed combustion for horizontal shell boiler

prevent clinkering extra water may be sprayed into the chambers. The suspension of particulate coal and impurity material behaves like a combustible gas and, burning at high temperature, the energy extraction from the fuel is more complete and the transfer of the thermal energy to the working fluid, usually water, is excellent. Tests with experimental and prototype rigs seem to prove that the fluidised-bed techniques will prove successful and greatly enhance the future prospects for coal-burning power plants. Orders for plant intended to operate in the 100 MW range were placed in 1977 by US companies.

The second method of enhancing the value of coal as a fuel for electricity generation is by means of the magnetohydrodynamic generator. In this arrangement, shown schematically in fig. 41, a hot gas from a combustion chamber is made electrically conducting by the addition of some seed material such as potassium carbonate. This material is injected into the flame of the fuel, and as the conducting gas streams from the chamber at supersonic speed it passes across a magnetic field. The interaction between the stream of ionized gas (plasma), which acts as a conductor, and the magnetic field resembles that in a conventional electrical generator and a current is caused to

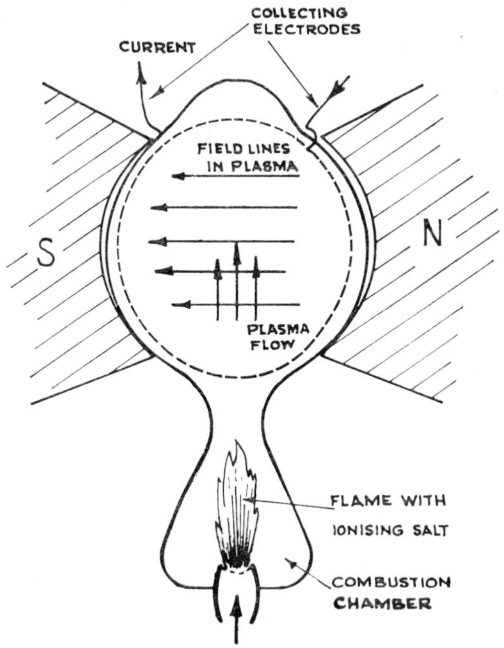

FIG. 41. Magnetohydrodynamic generator in which very high temperatures, over 2,000 °C, result in high conversion efficiency (to electrical energy). Impurities are chemically precipitated to lessen pollution, e.g. sulphur as alkali sulphates. The output electrodes, shown dashed, lie in planes parallel to the plane of the diagram.

flow between suitably arranged collecting electrodes. This device results in the direct production of electrical power from the heat of combustion. Fairly large experiments with apparatus achieving temperatures of 2,000 °C, and capable of yielding around 1 MW of power, seem to give conversion efficiencies of about 50 per cent. Impurities, e.g. sulphur and alkali sulphates, are chemically precipitated to lessen pollution. There are still a good many problems to be solved but the device demonstrates that there is still a lot of scope for advances in the use of coal, and other fossil fuels, in making electricity.

Grainger and others (see L. Grainger, *Phil. Trans. R. Soc.*, **A276**, 1974, p. 527) visualize the conversion of coal into other forms of energy in a number of unit processes which could be combined together in what is termed Coalplex. In such a multi-purpose complex the economies in capital cost, and the increases in efficiency in producing power and materials, can be substantial.

B. Oil

Resources. We noted in discussing coal reserves that coal was not distributed uniformly in the crust of the Earth. Oil is distributed even less uniformly than coal. While 66 per cent of the recoverable reserves are in OPEC countries, these countries occupy 8 per cent of the world land area and contain 6 per cent of the world population.

Few countries possess large amounts of both coal and oil. Among the fortunate ones are the USSR, the US, Canada, and China. Japan, on the other hand, has extremely limited reserves of either of these two fuels. This unfortunate maldistribution of fuel wealth has not been regarded as a really serious problem because oil has been transported easily in large oil tankers. Liquefied natural gas (LNG) is also being transported increasingly in specially designed ships, a modern achievement owing a great deal to the rapid development of cryogenic engineering, the science of low temperatures. The huge well-insulated vessels are not unlike oil tankers and they carry cargoes of liquefied natural gas in a similar fashion (plate 1 and fig. 42). The transportability of oil and natural gas in liquefied form is one of the main reasons for the dominance of these fuels as energy sources and the rapid rate of rise in consumption of both. Liquids and gas can be conveyed in pipes at high flow rates. There have been attempts to transport slurries of solids in pipes but no claims for success with large amounts over long distances have been made. Oil on the other hand has been piped over hundreds of miles at depths of hundreds of feet in deep waters like those of the North Sea.

Use. Not only is oil the most important fuel today, it is also the raw material of many major chemical industries. There are comparatively few industries in which coal is used in preference to oil other than in iron and steel

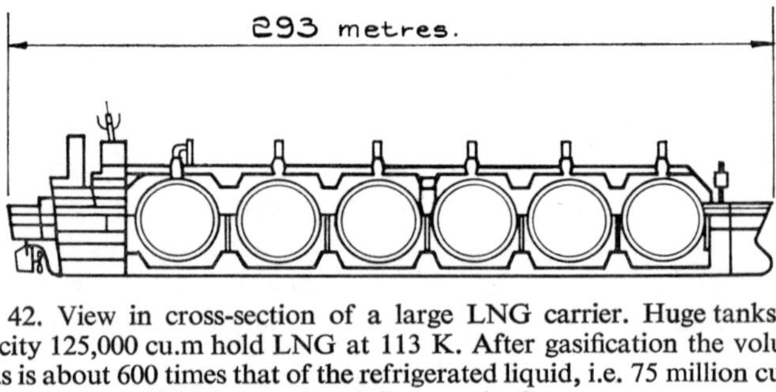

FIG. 42. View in cross-section of a large LNG carrier. Huge tanks of capacity 125,000 cu.m hold LNG at 113 K. After gasification the volume of gas is about 600 times that of the refrigerated liquid, i.e. 75 million cu.m. [Based on data of IU International Corporation]

making. Coal has also been replaced by oil in many domestic and commercial applications such as space and water heating and electricity generation (plate 2). Even in the iron and steel industry much of the need is for only high grade coals to provide coke of quality. In transportation coal has been displaced rapidly. Most modern locomotives are now of the diesel type and in cars, trucks, and aircraft the use of petroleum and the products of the petroleum refining industry is automatic.

The dependence on oil of so much of our modern industry is a matter of great concern when world resources are examined in some detail. It is not always easy to replace oil when it is used simply as a fuel in producing electricity or other energy forms but it may prove even more difficult to replace it in transportation, in the manufacture of chemicals of all kinds, and in the manufacture of many extremely important materials. It has been argued that it is so valuable it should not be used except when there is no alternative and we should conserve it for the many industrial applications for which it proves pre-eminently suitable, for instance as a feedstock for the chemical industries. It is in fact no exaggeration to say that the way of life in many of the advanced nations has come to be closely linked with ready availability of oil. This is a phenomenon of the twentieth century, as can be demonstrated by noting that in 1900 there were 8,000 automobiles in the US and in 1971 there were 94,000,000 in use. No such extraordinary change could be imagined without oil. It is difficult to picture 94,000,000 automobiles driven by steam from coal-fired boilers!

Genesis and Production. Oil was formed differently from coal and probably mainly from the remains of dead plankton and other micro-organisms accumulating in mud on the sea bed. Burial under other sediments took place before the organic material changed into oil by processes still as yet not completely understood. Subsequent compaction resulted in the oil being squeezed out from the muds into porous and permeable layers above. A variety of geological processes resulted in oil now being found mainly within porous sandstones and limestones of marine origin. An underground oil reservoir consists of a rock the pores of which are filled with brine, oil, and perhaps gas. Provided this reservoir rock has an impermeable cap rock above, an oil trap is formed. There are a great variety of possible oil traps with figs. 43 and 44 illustrating two of the simpler types. Oil is therefore looked for, and found, wherever there are thick sequences of sedimentary rocks, either on land

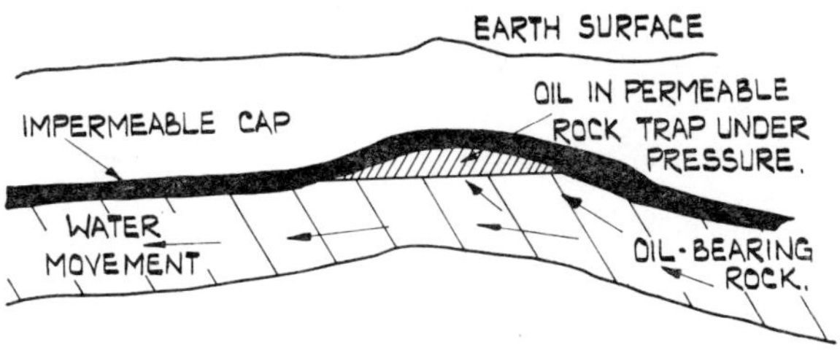

FIG. 43

Anticlinal trap for oil. The more buoyant oil is trapped as oil carried by water rises upwards and remains in the trap.

[Based on A. F. Fox, *The World of Ore*, Pergamon Press, 1964]

K

or beneath the seas covering the continental shelves of the world. The successful location and estimation of the size of oil traps is a matter of great skill. The oil is usually held under pressure and generally it contains one form of natural gas, a range of simple, gaseous hydrocarbons associated with oil and produced in a similar fashion to the oil itself. The gas helps to drive the oil to the surface and must often be flared off when it appears as a by-product of oil production. Because of the considerable depth of many oil wells, the oil is under great pressure.

The drilling of wells often results in the spontaneous release of oil and 'gushers' were a prominent feature of the early history of the industry. As techniques advanced the oil flow, as it occurred under the natural pressure, could be controlled with nearly 100 per cent success. Even today, however, the fraction of the total quantity of oil that can be extracted from a well without extraordinary difficulties seldom exceeds half. Research is continuing with the aim of increasing this fraction. The wells vary considerably in size and an increasingly large proportion of them is offshore. It may never be worthwhile attempting to use all the small deposits (one million barrels or less) and many under-sea deposits may prove virtually inaccessible. All of these features of oil distribution add to uncertainty in estimating the total amount available for consumption.

Production of oil in significant quantities started around 1880. Prior to that time James (Paraffin) Young had succeeded in launching an oil-shale industry in Scotland and his efforts coincided with the start in the US of a phenomenal growth industry. From 1890 to 1970 the average rate of increase of oil production was at 6·94 per cent per annum (doubling time = 10 years). Such a high rate of exponential growth cannot continue indefinitely. It is for this reason that during the last ten years world oil depletion has been an extremely frequent topic of debate. Prior to that the industry seemed to present a rosy picture though a few experts were expressing anxiety about the

FIG. 44

A fault trap formed by the slip of impervious rock above pervious oil-bearing rock

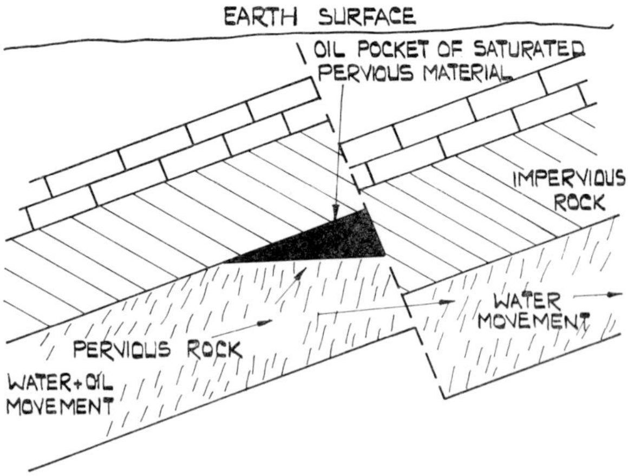

[Based on A. F. Fox, *The World of Ore*, Pergamon Press, 1964]

future. In 1968, of the total fossil fuel consumed in the US, 73 per cent was from petroleum and natural gas and 27 per cent from coal even though, as the reader will remember, coal is abundant in the US and relatively easily mined. How long can oil continue to be a fuel of such importance?

We have mentioned some of the main areas of difficulty in estimating world reserves of oil. There is nevertheless a good deal of experience in the exploration of oil deposits, and the proven yields of similar deposits can be used to estimate future yields. New fields of significance, again defined as a yield of one million barrels, are recorded in the US and from the data the difficulty of finding such wells can be deduced. Extrapolation of the data allows an estimate to be made of the probable total to be found and this is usually described as the estimated total resources. Other methods are in use and the reasonable agreement in the results gives some confidence in the answer. Similarly estimates of natural gas can be made. For oil, estimates of resources range from $1{,}350 \times 10^9$ barrels to $2{,}100 \times 10^9$ barrels, that is about three times the 'proven' oil reserves. The oil data include oil from offshore fields, oil already produced, and proved and probable reserves, together with future discoveries. Work since 1967 has not produced results that appreciably modify this presentation of the ultimate production of crude oil, though Sir Eric Drake (*Phil. Trans. R. Soc.*, **A276**, 1974, p. 453) favoured a somewhat lower value than that of Ryman (whose calculation was $2{,}100 \times 10^9$ barrels) and from some 16 expert estimates showed there could be grounds for adopting $1{,}900 \times 10^9$ barrels.

The Institute of Petroleum estimates proven world oil reserves in 1976 at $81{\cdot}8 \times 10^9$ tonnes, or 597×10^9 barrels (1 tonne = 7·3 barrels, and 1 barrel = 35 imperial gallons). The breakdown of their figures is given diagrammatically in fig. 45. World production in 1976 they put at $2{\cdot}95 \times 10^9$ tonnes, which is 59×10^6 barrels a day. Production has been steadily rising and Sir Eric Drake provides a forecast of world crude oil production between 1973 and 1985 (fig. 46), taking into account resources as distinct from proven resources of oil. As can be seen, the peaks in the main areas, excluding the North Sea, lie between 1980 and 1985. It would be wise to assume that the peak of production will be reached not later than sometime between 1990 and 2000. This does not leave much time to make good the deficiency of supply as against demand for such an important energy source. Urgent thought on closing the gap is clearly required.

Tar Sands and Oil Shales. Until recently it was believed tar sands and oil shales were an important resource but much less reliance is now placed upon them. Even so deposits could yield a substantial amount of oil by extraction and it has been established (see Harrison Brown, *A. Rev. Energy*, **1**, 1976, p. 26) that a total of about $3{,}000 \times 10^9$ barrels are available using shales with yields between 10 and 100 gallons per ton. The problems of extraction are formidable, the need for large quantities of water and the environmental disturbances caused by working the shales being among the more outstanding difficulties. With 1965 data, it was estimated about 190×10^9

barrels were reasonably easily recoverable. The price of the product would inevitably reflect the large capital investment involved and by 1977 no truly major venture had been launched. It is worth noting that if by any chance an additional total of, say, $2,000 \times 10^9$ barrels of oil were obtained from oil shales and tar sands, the grand total of world oil would be about doubled and an extra fifty or so years added to the time of exhaustion of supplies (on the assumption of 1977 consumption level). Like oil from oil wells the oil that can be extracted from shales and tar sands is finite in amount and they represent a non-renewable and extremely problematical source of supply. It could be wise, in view of environmental and ecological problems, to discount them as

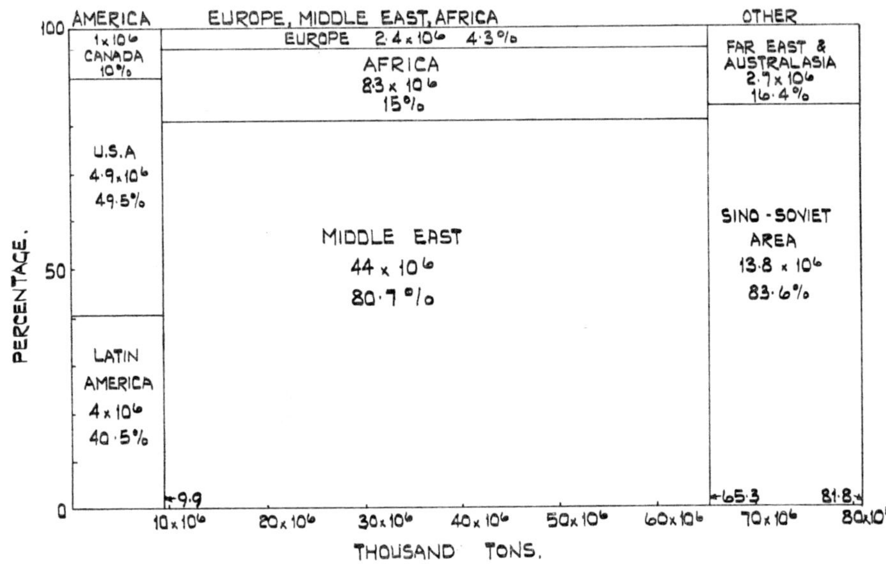

FIG. 45

World "proven" crude oil resources. [From *Know More About Oil. World Statistics*, Institute of Petroleum Information Services]

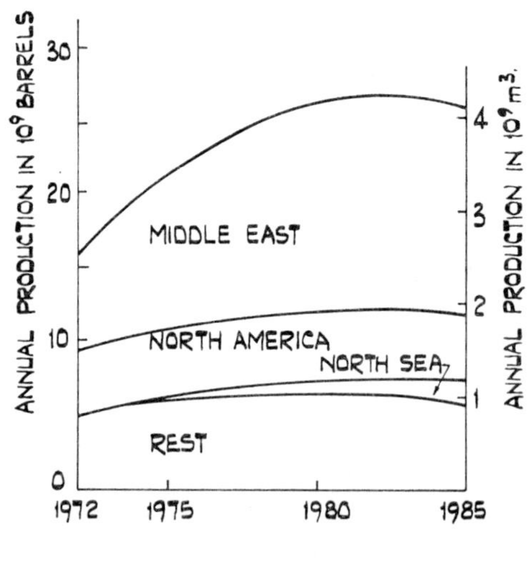

FIG. 46

Forecast of world crude oil production, excluding USSR, Eastern Europe and China. [From Sir Eric Drake, *Phil. Trans. R. Soc.* **A276**, 1974, p. 460] a viable source of oil in arriving at decisions that will affect the energy situation after 1990. In the longer term they may be utilized fairly widely for chemical industrial purposes rather than as a stop-gap in the fossil-fuel shortage. As chemical feedstock the extraction of oil could be much more prolonged, and this in itself would assist in helping to solve environmental difficulties.

Perhaps some elucidation of the environmental questions is required. Oil shale does not contain oil; it contains organic material referred to as kerogen and a high proportion of clayey material. Massive quantities of the shale (as well as overburden) have to be handled both in the mining and in the eventual destructive distillation process (e.g. a ton of shale may yield only 20 gallons of oil). The US has roughly two-thirds of the known recoverable resources (about 450 billion tonnes of oil) and in Colorado, Utah, and Wyoming the large amounts of water required in the extraction process are not easy to procure. Also the large outlay necessary to restore the land after processing the material could be a severe deterrent.

Tar sands consist of a mixture of sand, water, and highly viscous oil. Huge deposits are located in Canada (134 billion tonnes) and in Venezuela (575 billion tonnes). The best known, and probably the largest single source of all, is the Athabasca field in Alberta. The preliminary extraction of a mass of bitumen can involve handling nearly ten times as much mass of tar sand. In spite of the problems a surface-mining tar-sand project of Great Canadian Oil Sands Ltd has produced 2·5 million tonnes of oil per year since 1969 and a surface plant of Syncrude Canada is expected to yield 5 million tonnes of oil per year from 1979. A large number of pilot projects are also under way.

C. Natural Gas

Genesis and Production. The third of the major fossil fuels in use today is natural gas. This may be associated with oil and of common origin, or it may consist mainly of methane (CH_4). In many countries it is a highly valuable fuel in its own right. Methane used to be known as marsh gas and its occurrence in coal mines has to be monitored carefully as it presents a grave hazard. Among the more recent interesting occurrences of such gas is that in the rocks below the southern part of the North Sea, where its discovery raised interest in other parts of the continental shelf around Britain. Subsequent discoveries of oil and gas are, however, of completely different origin—the southern North Sea discoveries alone are of methane derived from devolatilisation of deeply buried coal seams. Natural gas occurring with oil (associated natural gas) may be found dissolved in the crude oil, when it is referred to as solution gas, or it may be found on top of underlying crude, when it is known as gas-cap gas. The volume of associated natural gas that can be recovered is dependent largely on the rate of exploitation of the oil itself. In very recent years it has become normal practice to pump gas back into the oil wells to maintain high pressure and increase the yield of oil from the well. Any improvement in this direction is extremely valuable.

In spite of considerable experience acquired over the last fifty years, in some of the industrialized countries, in the distribution of coal gas, natural gas has become an international commodity only since about 1950. In 1951 gas was exported by pipeline to the US from Canada. In 1964 Groningen gas was first exported from Holland and in the same year liquefied natural gas (LNG) was first brought from Algeria to Britain. Since that time both natural gas and LNG have become extremely important items in the international energy trade; for instance Japan imports a large fraction of its energy as LNG.

Reserves. The technology of large diameter pipes required for transport of gas in its natural state has been found difficult and only high quality cryogenic engineering paved the way to the transport of huge amounts of LNG. The result is that the reserves of natural gas are still very considerable today (1977) compared with total world consumption per annum. Possibly the fact that a gas is much less dense than a liquid leads one to expect that natural gas as an energy resource cannot be at all comparable with oil as a resource. In fact natural gas is closely comparable.

C. P. Coppack has pointed out the uncertainties in estimation of gas reserves but, while these are to be remembered, the figures are nevertheless most reassuring. His data are presented in figs. 47a and b which show known world gas reserves in 1972 and also the total anticipated from new discoveries. Added together they amount to 193×10^{12} cu.m which is, on a thermal basis, equivalent to a total of $1{\cdot}19 \times 10^{12}$ barrels of oil.

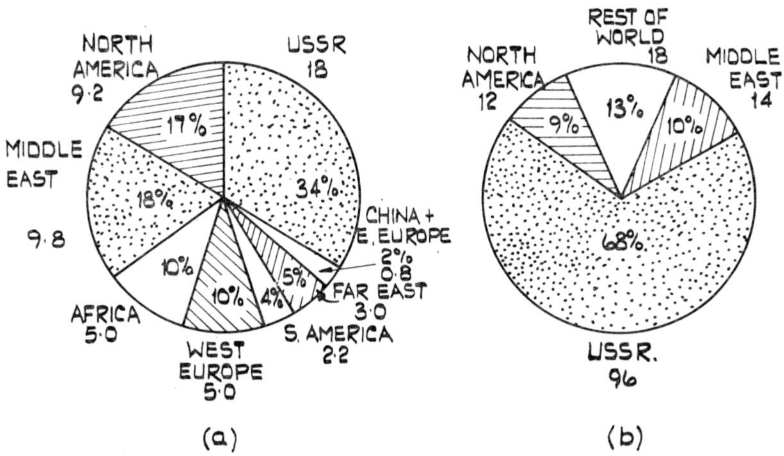

FIG. 47. Percentage distribution of proven and anticipated gas reserves in 10^{12} cubic metres.

(a) Proven world reserves in 1972. Total $= 53 \times 10^{12}$ cu.m, or $0{\cdot}33 \times 10^{12}$ barrels oil equivalent

(b) Anticipated reserves additional to those in (a). Amount $= 140 \times 10^{12}$ cu.m, or $0{\cdot}86 \times 10^{12}$ barrels oil equivalent.

(a) and (b) together give estimated grand total of 193×10^{12} cu.m, or $1{\cdot}19 \times 10^{12}$ barrels oil equivalent. [From C. P. Coppack, *Phil. Trans. R. Soc.* **A276**, 1974, p. 464]

If we compare this with the estimated world oil resources, which we took to be in the range $1{\cdot}35$ to $2{\cdot}10 \times 10^{12}$ barrels, we see that estimated gas reserves are indeed closely comparable, and even on the optimistic oil estimate equivalent to more than half of the estimated world oil resources. Gas is therefore one of the major fuels. In contrast to the near comparability in estimated reserves of the two fuels, the utilization of gas is markedly less than that of oil. Gas in 1972 supplied about 17 per cent of world energy but oil provided about 45 per cent. However, the disparity in these figures is not unexpected in view of the transportation problems with gas and the consequent delay in acquiring a world-wide market.

The geographical pattern of gas consumption is given by Coppack and this shows that the US is the predominant market (see fig. 48). The other major market areas are the USSR and Europe and both of these areas are rapidly expanding.

Among the interesting lessons to be learned from the data are, first, that known reserves can continue to provide gas at the 1972 level for

$$\frac{330 \times 10^9}{7 \times 10^9} \text{ years,}$$

that is for 47 years, or roughly 50 years since the data are very approximate. If we take the expectations of new discoveries into account we have a period of

$$\frac{(330+860) \times 10^9}{7 \times 10^9} \text{ years,}$$

that is, $\frac{1190}{7}$ years or 170 years.

Expressing this in another fashion, we appear to have $\frac{170}{3}$ years or almost 60 years of natural gas supply at three times the 1972 consumption level, provided all possible steps are taken to prove the potential reserves. In relative terms gas is a most promising source of energy, but with knowledge of the problems of discovery and delivery it is not likely to be able to replace oil within two to three decades. There are of course uses for oil which may never be met by supplies of natural gas, though many needs could be met by either fuel. On the other hand the demand for gas is rising relative to that for coal and oil, for gas offers advantages in many applications and not only in use

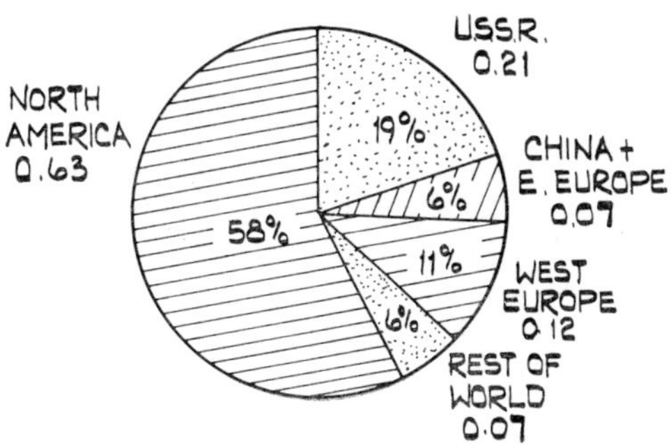

FIG. 48
Estimated natural gas consumption in 1972 in 10^9 cu.m. The total consumed $= 1{\cdot}1 \times 10^9$ cu.m., equivalent to 7×10^9 barrels of oil. [From C. P. Coppack, *Phil. Trans. R. Soc.* **A276**, 1974, p. 466]

as a heating fuel. It is a relatively clean fuel giving rise to less unfortunate environmental and social effects; for example the sulphur content of chimney effluent from burning gas is much lower than for coal or oil.

Consumption and Distribution. Japan is a major importer of fossil fuels due to its lack of indigenous supplies and its high rate of growth of energy consumption (around 11 per cent per annum). About 90 per cent of its total energy need is met by imported fuels which have had to satisfy increasingly strict criteria in order that the severe pollution problems in the densely populated industrial regions of the country can be alleviated. However, if Japan wishes to import gas on any scale she must construct fleets of liquefied natural gas ships.

The importance of natural gas in the US can be deduced from the fact that over a period of more than twenty years from 1950 the average growth rate of consumption of natural gas was 6·2 per cent per year, the highest rate of growth of the fossil fuels. In the event that natural gas supplies fail to match demand, there is the option of importing LNG or of manufacturing more synthetic natural gas (SNG) from coal or oil. We have discussed the technological problems involved in the case of gasification of coal and appreciated that a great deal has to be done to make commercial plants for the conversion process both efficient and economical. The environmental aspects of the massive open-cast mining expansion associated with a large expansion of coal conversion to SNG could well prove a stumbling block. The manufacture of SNG from oil is an established process but here the problem is getting enough feedstock for a large expansion of the gas supply. The feedstocks are naphtha-type fractions and gas liquids, and these are in strong demand for the petrochemical industries. It is clear that there are considerable complexities to be faced in the gas industry and that these are more likely to be resolved by expanding the natural gas supply than by large programmes for conversion of coal or oil to synthetic natural gas.

The geographical location of some of the world's most abundant reserves will make extraction and transportation to the customers a great problem. Thus the US is likely to remain a major consumer for decades and therefore the biggest potential importer. The USSR with possibly two-thirds of the potential world reserves has perhaps half of this in Western Siberia. The problems of drilling and piping natural gas in the severe climatic conditions are likely to be formidable, and pipeline systems themselves raise considerable technical and investment problems.

The geographical distribution of natural gas reserves seems to leave the US and Japan with the most severe problems, if we consider only the highly industrialized countries. The Middle East has the capability to export, and likewise the USSR in the longer term. The situation in Europe is reasonably good with fields tolerably well spread and Algerian reserves close enough to be imported with ease. The expansion of consumption in Europe continues at a high rate but differences in the position of the various countries are revealed on close examination. In some cases the transportation of the gas to the customer cannot be done economically because of the lack of an adequately extensive network of modern pipelines. Vigorous expansion of the network in Europe continues and this should help to smooth out supplies. In the US a well-developed network of natural gas pipelines already exists. Almost a quarter of a million miles of high pressure pipelines, and over a half million miles of lower pressure lines, ensure the delivery of the gas from the production areas to consumers in most parts of the country. No other part of the world is so well provided with a distribution network. In Europe much of the distribution network that carried manufactured gas has been converted to convey natural gas. Pipeline problems, for instance in making natural gas from Middle East fields available in Western Europe, may be overcome by large diameter pipeline networks on an international scale.

The conversion of natural gas to methanol (methyl alcohol) has attractions for transport by ship. The methanol can be reconverted to yield good quality natural gas, or it may itself serve as a clean fuel in industrial applications. In the re-gasification process there is substantial absorption of heat and the intriguing idea of producing 'cold' has led to schemes in which such 'cold' would be utilized.

Gas Technology. As the reader will have gathered, the increasing importance of natural gas has been reflected in the advances made in gas technology in recent years, advances that have had relevance to work on conversion of coal and oil to gas. A number of publications demonstrate the increased scale on which gasification is studied and representative of these is the paper by B. Robson (*Energy Res.* **1**, 1977, p. 157) and that by W. W. Bodle, K. C. Vyas, and N. Talwalker (*Clean Fuels from Coal*, Symposium II, Chicago, 1975, p. 53). The flexibility introduced by the ability to convert a solid, liquid, or gaseous fuel to either of the other two forms may prove important as networks of pipelines are brought into operation. Ryozo Hiwada (*Chem. Econ. Eng. Rev.*, Oct. 1976, p. 7) has emphasised this point, and he mentions the possibility of nuclear power being used to produce hydrogen which could, of course, be delivered in gaseous form. This brings in a topic of great general importance not closely linked to natural gas and sometimes described in terms such as 'hydrogen economy'. We defer consideration of that concept until chapter 20, merely noting here that hydrogen does not exist in nature in the free state and it is not an energy resource as such. On the other hand it could well be the end-product of the conversion of the primary energy of other sources such as the fossil fuels, nuclear fuels, and solar energy itself.

D. Fossil Fuels in Modern Society

Overall Patterns of Consumption. Before leaving the discussion of fossil fuels as such, it is essential to convey to readers a full appreciation of the outstanding position they occupy in a modern industrial society. Chauncey Starr has demonstrated, in the manner of fig. 49, that fossil fuels accounted in 1977 for more than 90 per cent of the total energy input into the US economy. As we have said before, the highest average rate of increase is for natural gas. The same trend is evident in a number of other countries. The depletion that is forecast for oil, and later for natural gas, will clearly present an immense problem when regarded in the light of the dependence implied in fig. 49. Coal can again be made capable of filling a greater role, and nuclear energy likewise, but the gap left by depletion of oil and natural gas will be an enormous one. This leads us into the subject of our next chapter.

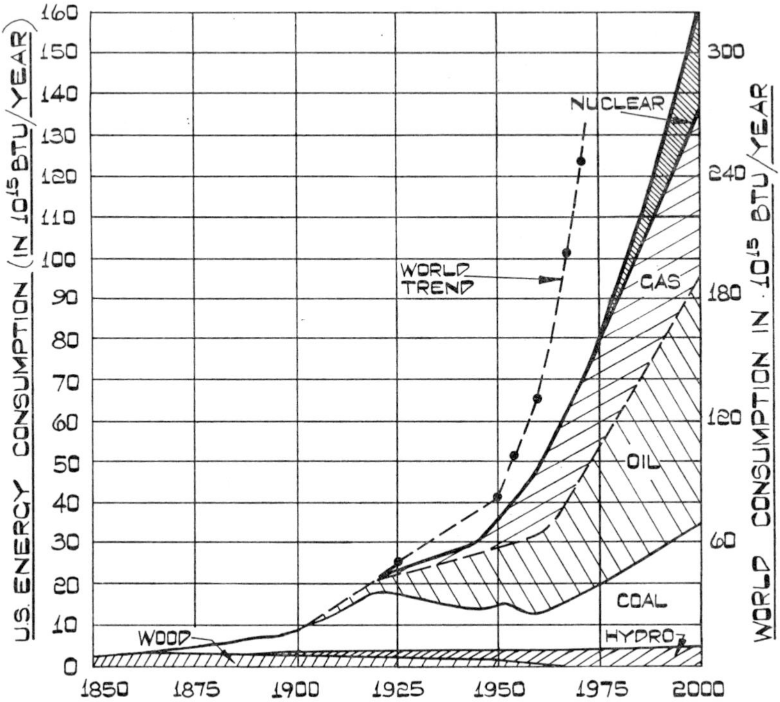

FIG. 49. US energy consumption with time, extrapolated up to 2000. The scale on the right is for world total figures. Note that 50×10^6 btu = 1 kW-yr. [From Chauncey Starr, *Scient. Am.* Sept. 1971, p. 39]

Chapter 13

The Energy Gap

We have already noted several times the very considerable changes that occurred in the way of life and the consumption of energy that took place from the beginning of the Industrial Revolution onwards.

The contrast between the energy pattern in, for example, the US economy today and that prevailing in the past, is vividly portrayed by the method of presentation of daily energy consumption used by Cook in fig. 34. From the time of primitive man, about one million years ago, when he managed without fire and depended entirely on the food he acquired, we pass to hunting man of about 100,000 years ago. He had wood for heating and for cooking his more plentiful food. Primitive agricultural man grew crops and settled in reasonably fertile regions. He was in fact living in a fashion somewhat similar to that followed by the Tsembaga people (see chapter 11). Advanced agricultural man, like those occupying north-west Europe around the year 1400, had some coal for heating, together with water and wind power and animal transport. The industrial man of the nineteenth century had the steam engine and other aids, while modern man has all the technological advances allied to the exploitation of fossil fuels and electricity.

Today people in the US use energy to an extent unknown in human history, and the components indicated in the figure add up to a total of 230,000 kilocalories per person per day. This total is perhaps more easily appreciated if we convert it to an equivalent horsepower. Making the calculation with suitable conversions, we find that in 1970 a man in the US when working an 8-hour day was provided with the work capacity equivalent to that of an engine of 45 horsepower. This is a fantastically high power and should be contrasted with that of the Tsembaga primitive agricultural man who manages to use the Sun to increase his own muscle power by a factor of 16. From fig. 34 his capacity is $\frac{16}{230}$ of 1970 technological man or $\frac{16}{230}$ of 45 horsepower, that is, approximately 3 horsepower. The choice of the 8-hour working day boosts both of these figures artificially. If we consider the energy is used uniformly over a 24-hour day the results are 15 horsepower and 1 horsepower. The fifteen-fold factor raises several far-reaching issues. Is such a high level of energy expenditure justified or necessary? Assuming it is, should the energy come mainly from virtually irreplaceable fossil fuels or should these be used conservatively in the interests of future generations? Is man's ingenuity adequate to continue to find energy in a universe richly endowed with it? These are all issues giving rise to continuing debate today and to which we must find the answers.

The great increase in the power output of basic machines was the springboard of the Industrial Revolution and the basic machines themselves depended on the exploitation of fossil energy resources. These points are well illustrated by Chauncey Starr. The data, plotted as shown in fig. 50, establish the fact that after the start of the Industrial Revolution, between 1700 and 1750, the maximum power output of certain types of machine increased extremely rapidly and an improvement by a factor of about a million has ensued. The rate of increase with time in the power output of the internal combustion engine has been very steep indeed. It is worth stressing the comparison made by Starr between modern times and earlier times. The early stages of the Industrial Revolution depended on the application of the waterwheel and the windmill as prime movers. These were fixed devices in the geographic sense and this lack of mobility determined the location of industries. With the introduction of compact sources of energy such as coal, the steam engine became attractive as a means of overcoming this and other limitations of water power. It was possible to transport fuel to the machine, and the increasing capacity and efficiency of the chosen machine made this more and more practicable. Thus industry was freed from a number of constraints and expanded enormously. The maximum power of man, ox, or horse is puny compared with the 1,000 MW output of a large steam turbine.

Most of the large growth in the power output of machines or energy-conversion devices has happened in the past hundred years since the past

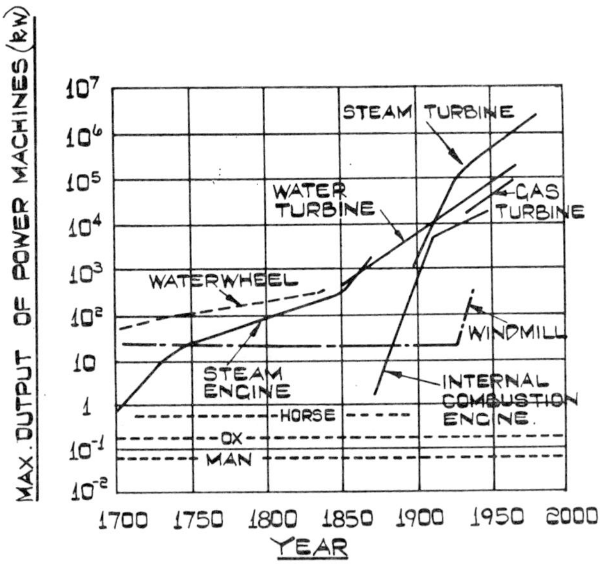

FIG. 50. Power output of basic machines. Note division on output scale represents a factor of 10. Modern steam turbines can deliver one million times the power of the earliest steam engine (under 1 kW equivalent). [From Chauncey Starr, *Scient. Am.* Sept. 1971, p. 38]

century has been one of highly accelerated energy consumption; in fact total energy utilization has risen exponentially (see p. 140 below). The curves of fig. 49 show how, for the US, consumption of different forms of energy has grown. It is clear that only wood has effectively disappeared as a viable fuel in industrialized countries, such as the US, whereas in 1850 it supplied over 90 per cent of all energy needs. In 1900 coal had become pre-eminent as the energy source, but from about 1925 oil and natural gas began to grow extremely rapidly and the demand for coal abated. Hydro power has made a steadily increasing contribution, and nuclear is now emerging as a new energy source. The same kind of overall pattern applies to a good number of industrialized societies as well as the US, but with differences depending mainly on the availability of the different fossil fuels in each case. For a large number of countries the energy consumption has shown an approximately exponential rate of increase of consumption.

In view of the fact that the fossil fuels have held a dominant position in the energy supply to the advanced nations during the present century, we shall look at some of the published data on consumption in a world-wide context and on a regional basis. The figure for world-wide energy consumption in 1972 given by J. Darmstadter and S. H. Schurr is 250×10^{18} J. Of this grand total no less than 46 per cent was supplied as oil. In the decade 1962–72 the world growth rate of energy consumption was $5\frac{1}{2}$ per cent and, taking 2 per cent as the population growth rate per year, the growth rate of per capita consumption equals $3\frac{1}{2}$ per cent. The data for representative years from 1925 to 1972 are given in table 13.1.

TABLE 13.1

WORLD ENERGY CONSUMPTION AND POPULATION, SELECTED YEARS, 1925–72

(From J. Darmstadter and S. H. Schurr, *Phil. Trans. R. Soc.* **A276**, 1974, p. 414)

Year	Total energy consumption in 10^{18} J	Population in billions (10^9)	Energy consumption per capita in 10^9 J
1925	47	1·890	24·6
1950	81	2·505	32·4
1955	105	2·726	38·6
1960	131	2·990	43·8
1965	169	3·281	51·6
1968	200	3·484	57·5
1970	226	3·609	62·6
1971	235	3·678	64·2
1972	250	3·747	66·8

The average annual percentage rates of change in total world energy consumption and in population can be shown, from the kind of data of this table, to be reasonably uniform, especially since about 1950. This feature is brought out in the values entered in table 13.2.

TABLE 13.2

AVERAGE ANNUAL PERCENTAGE RATES OF CHANGE
(From J. Darmstadter and S. H. Schurr, *Phil. Trans. R. Soc.* **A276,** 1974, p. 414)

Period over which averaged	Energy consumption	Population	Energy consumed per capita
1925–50	2·2	1·1	1·1
1950–55	5·3	1·7	3·6
1955–60	4·5	1·9	2·5
1960–65	5·3	1·9	3·4
1965–70	5·9	1·9	3·9
1970–72	5·2	1·9	3·2

The total growth data obscure rather notable differences in the rates for different geographical regions. For example in the USSR the fraction of total world energy consumed in 1925 was a mere 2 per cent but in 1970 it was 15 per cent. In contrast with this the fraction of the total energy consumed in the US was 45 per cent in 1950 and 33 per cent in 1970.

It may be useful to digress here for a moment to explain the concept of *exponential* growth or decay, and illustrate it with a familiar example. Most

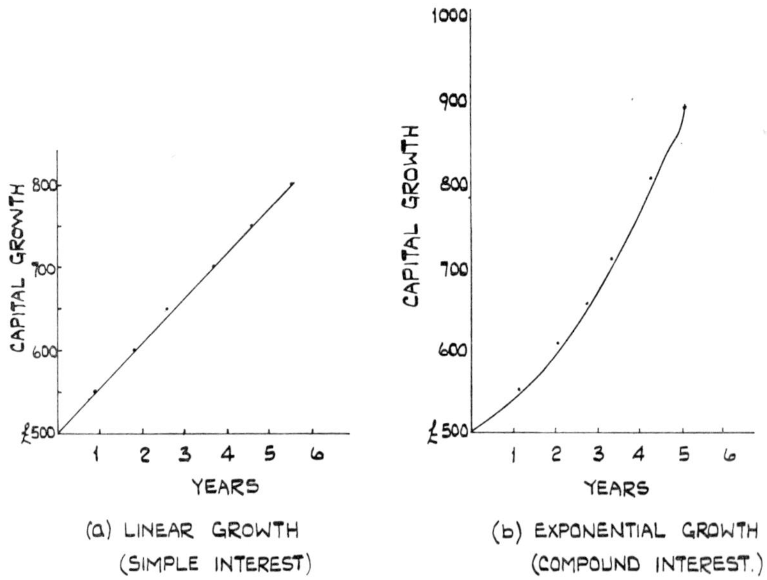

FIG. 51. (a) Linear growth (simple interest). (b) Exponential growth (compound interest)

of us appreciate the difference between simple and compound interest. In the case of simple interest, a certain percentage of the original sum of money is paid out on that sum each year. If we have £500 which we lend at 10 per cent p.a. our capital will grow by increments of £50, and a graph of the increase will be represented by a straight line (fig. 51a).

In the case of compound interest, however, the borrower not only continues to borrow the £500 but also borrows, at the same rate of interest, the amount by which our capital has increased. He is then paying a fixed percentage of interest (as before) but on an ever-increasing amount of borrowed money. Our capital naturally grows much more quickly and a graph of the increase is represented by an exponential curve (fig. 51b).

Exponential growth, or decay, is thus a percentage change on an original number which itself alters according to that same percentage. The change at any one time can be calculated mathematically by means of the formula $N=N_0e^{\lambda t}$ where, for the present example, $N_0=500$ and $\lambda=0{\cdot}1$ (representing 10 per cent) and t is the time in years. The value of N is the total capital after t years.

Many natural processes proceed in exponential fashion. For example, we can express the decay of a number of radioactive atoms by the equation

$$N=N_0e^{-\lambda t}$$

where N is the number of atoms remaining unchanged out of an original number N_0 after a time t. The value of the constant e is 2·72 and it is the base of natural logarithms. The value of λ is known as the decay constant. In decay processes the time to half-value, or the half-life, is an important concept. It is found by putting

$$N=\tfrac{1}{2}N_0 \quad \text{so}$$
$$\tfrac{1}{2}N_0=N_0e^{-\lambda t} \quad \text{or} \quad e^{\lambda t}=2.$$

Hence taking logarithms to base e

$$\lambda t_{\frac{1}{2}} \log_e e=\log_e 2=0{\cdot}693$$
$$\text{so} \quad t_{\frac{1}{2}}=\frac{0{\cdot}693}{\lambda}$$

In growth phenomena where N increases from the value N_0 with time t we have

$$N=N_0e^{\lambda t}$$

and the doubling time t_D is found from

$$2N_0=N_0e^{\lambda t}D$$
$$\text{i.e.} \quad \lambda t_D=\log_e 2=0{\cdot}693$$
$$\text{so} \quad t_D=\frac{0{\cdot}693}{\lambda} \text{ as in decay.}$$

If we assume that λ is the growth constant and use $\lambda=0{\cdot}02$ which means a 2 per cent increase in a value each year, we find

$$t_D=\frac{0{\cdot}693}{0{\cdot}02}=34{\cdot}65 \text{ yr.}$$

Thus with an exponential rate of growth of world energy consumption of 2 per cent per year the energy consumption doubles in a period of 34·6 years.

In his paper 'On Strategies and Fate' (*Physics Technol.* **8**, no. 4, 1977, p. 157, Energy Special Issue of the Institute of Physics) C. Marchetti shows that world energy consumption, including the contribution from wood and farm waste, has grown exponentially over a period exceeding one hundred years and that it exhibits a 2 per cent annual increase. Thus while the points in fig. 49 for the world trend may not appear to establish sufficiently clearly that world energy consumption grows exponentially, fig. 52 from the work of Marchetti establishes this fact quite decidedly and establishes a doubling time t_D of approximately thirty-five years (2 per cent per year growth). The value obtained here is apparently less than that indicated in table 13·2, but it is assessed over a much longer time-scale and may well be a better indicator of a lasting trend than values over a period of twenty to thirty years as given in the table. In both cases energy consumption per capita is involved. The attention of the reader is drawn to the fact that when the scale on the ordinate, or y-axis, of a graph for the exponentially varying quantity is logarithmic the plot gives a straight line. If the reader transforms, for example, fig. 51b so that the ordinate scale is logarithmic, the curve will be transformed to a straight line.

Many explanations can be offered for differing rates of growth of con-

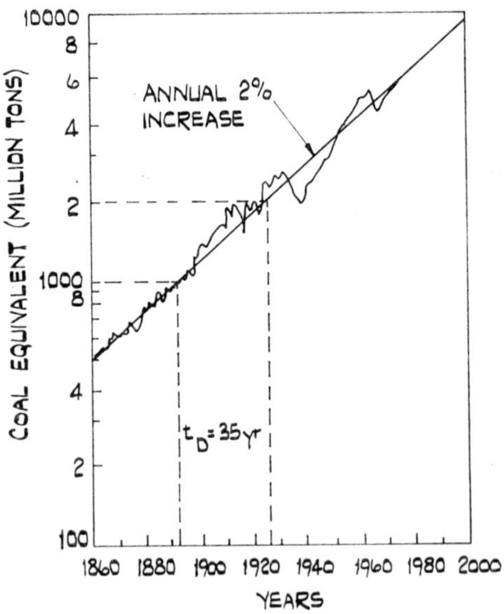

FIG. 52

World energy consumption. [From C. Marchetti, *Physics Technol.* **8**, no. 4, 1977] sumption of energy in different countries, among them differing efficiencies with which the energy is used, a factor now influencing downwards the growth rate in the advanced industrial countries. Also attempts to do without energy (as opposed to learning to make the best use of it) exert a fairly marked downwards influence on consumption. One of the most important factors is the price of energy. Cheapness for several decades undoubtedly raised the levels of demand.

Among the most striking of all the features demonstrated by growth data is the simple fact that throughout the world, and for all the major regions, there is a positive per capita energy consumption growth rate. This is found to be true even when there are relatively large percentage growth rates in population. Energy consumption everywhere is rising more rapidly than population. We give some data illustrating these conclusions in table 13.3.

TABLE 13.3

AVERAGE ANNUAL PERCENTAGE RATES OF CHANGE IN ENERGY CONSUMPTION (EC), IN POPULATION (P), AND IN ENERGY CONSUMPTION PER CAPITA (EC/P), IN WORLD REGIONS FOR 1960–70
(From J. Darmstadter and S. H. Schurr, *Phil. Trans. R. Soc.* **A276**, 1974, p. 415)

Region	EC	P	EC/P
US	4·2	1·3	2·9
Canada	6·1	1·8	4·2
Western Europe	6·3	0·9	5·3
Oceania	5·8	2·2	3·5
Latin America	6·3	2·9	3·4
Non-Communist Asia (excluding Japan)	7·7	2·5	5·0
Japan	11·9	1·0	10·7
Africa	5·4	2·4	3·0
USSR	6·0	1·2	4·7
Communist East Europe	4·7	0·7	4·0
Communist Asia	5·6	1·6	3·9
World	5·6	1·9	3·6

The picture is very different from that of the exceptional case of the Tsembaga people, described in chapter 11. Almost everywhere there is a more or less exponential rate of growth of per capita energy consumption. No doubt the ways in which this increased amount of energy is used in different countries differ now, and they will continue to differ in the future, but the common factors in utilization are important. The vast majority of nations express the desire to become more prosperous and provide at the same time more work for their peoples. In terms of energy this usually implies increasing consumption of power in improving the average standard of living and in manufacturing more material goods. Transportation, as an element in the total consumption of fuel, increases in step with these changes. The movement of materials and goods grows continuously, and people expect greater mobility for personal and other reasons. Perhaps these points can be expressed simply in the statement that throughout the world there is an apparent effort to increase the per capita value of the gross national product (GNP). The connection between energy expenditure and GNP has been examined by a good many experts and presented in a variety of ways. Most conclude that there is a rough correlation between energy and gross national product, though there may be considerable differences of opinion about the exact conclusions to be drawn from the apparent correlation. The data used by Earl Cook are chosen here in fig. 53. The unique position of the US is obvious in the figure, and if we consider that a line is drawn between the point representing the US and the origin there is a spread of points above and below the line. As we would expect a broad band is more appropriate, the breadth of the band depending on the emphasis in a given country on heavy industry, on services, on manufacture of specialized goods and materials, and of course

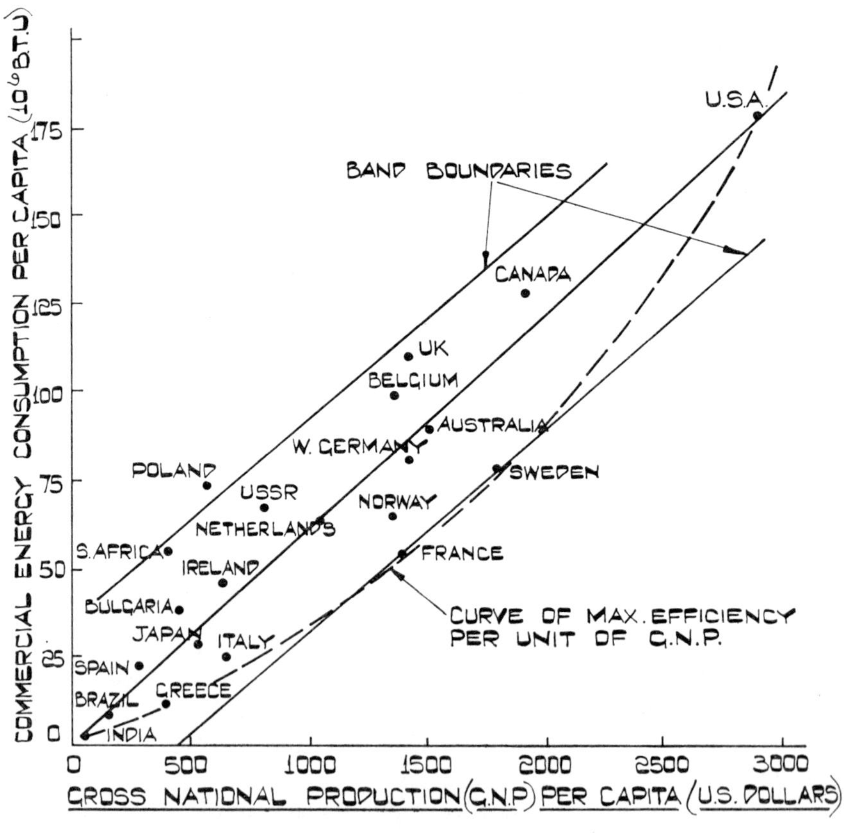

FIG. 53. Relationship between energy consumption and GNP. [From Earl Cook, *Scient. Am.* Sept. 1971, p. 135]

on the efficiency of energy-consuming machinery and processes in all sectors including agriculture. Climate itself plays an appreciable part in determining the level of consumption of both commercial energy and domestic energy.

It is clear that the aspirations of peoples in most countries of the world will tend to move points on the figure from near the origin towards the US. No one argues that the energy consumption level in the US can be, or should be, reached on a world-wide scale. The number of private cars in the US is a luxury and an energy-expensive one which would appear to be denied almost everywhere else in the future. On the other hand the use of energy in industry and commerce is more efficient in the US than in many other countries. This point is made by Beckmann (*The Health Hazards of Not Going Nuclear*, *op. cit.*, p. 159) in which he seems to use 1972 data which differ somewhat from Cook's data, but this does not apparently affect the conclusion which he draws—it could equally be drawn from the Cook values. In essence Beckmann argues that the dotted curve in fig. 53 indicates the curve on which lie those countries where the least amount of energy is consumed in relation to GNP. Beckmann contends that all countries above the curve are less efficient than those on the curve, among which is the US. While clearly this argument has some validity, it fails to allow for a wide range of other factors that clearly influence the efficiency of energy utilization and that can be beyond the control of a single country in many cases. There are differences among countries that are difficult to explain and which may be partly due to variations in the calculations of GNP which are not too consistently or precisely defined. Thus the very different consumption levels of France and the UK cannot be easily resolved.

Although a unique position is held by the US in leading the world in per capita GNP and in per capita energy expenditure, as has been noted the rate of growth of energy consumption on a world basis exceeds at present the rate of growth in the US. In fig. 49 the data for world consumption was plotted, and the curve is seen to rise more steeply than that for the US total since about 1925.

The curves spell out extremely clearly the essence of the problem facing us all. How can the continuous growth of energy consumption, at the high rate maintained during the past fifty years, be sustained? It is a question presenting grave problems to politicians, to scientists and engineers, and indeed to everyone. The momentum in society towards continuation of major trends is tremendously strong. Industries depend critically on ready availability of energy; the standard of material well-being and the style of life, even life expectancy itself, are also to some extent determined by the energy supply.

For such reasons it seemed wise to survey in this volume the potential capability of new sources of energy, and this has been one of the principal themes up to this point. We have seen that there are new sources to be exploited given adequate dedication to the pursuit of research and development in a number of fields, some of which have not been supported sufficiently enthusiastically in the past. However, if we look again at the US energy consumption (fig. 49) we can appreciate the scale of the problem now before us. The share of the energy market filled by oil, gas, and coal is very large. If these fossil fuels cannot be produced in adequate amounts the gap between demand and supply will become critically large and it could do so in a short period, in ten or twenty years. Production from US oil wells is not rising, and the amount of oil imported into the US is presently increasing by a factor of four in a decade. The situation in the US is not entirely unrepresentative of a large number of advanced nations, and that in Japan is even more grave.

Demand and Total Resources. A good many careful studies of the possibility of an energy gap have been made, and most experts agree that an energy gap will arise within two or three decades. The agreement of the participants in the Workshop on Alternative Energy Strategies, as given in *Energy: Global Prospects 1985–2000* (McGraw-Hill, 1977) would appear to justify fully such a statement. The gap is forecast mainly on the grounds of depletion of petroleum and gas supplies. Even in countries where there seem to be abundant reserves of coal there is no feeling of security for the future. As we noted in the last chapter, the sector of the market occupied by coal has in most instances declined steadily even where it can still be mined successfully. Oil is versatile and convenient and it was a comparatively cheap fuel until 1973, when OPEC policy changed. Too much reliance for too long has been placed on oil, and wishful thinking on the part of consumers has resulted in over-ready acceptance of the assurances of producers who were themselves apt to underestimate the gravity of the problems of using tar sands and oil shales. However, it is not too late to recover if the problems are tackled vigorously. Perhaps the complacency so manifest in most quarters until around 1970 was partly founded on the growth of the nuclear energy industry, which seemed to offer a guarantee of almost unlimited supplies. Now there are serious questions raised as to the part nuclear power can play. We shall examine this matter in some detail in chapters 15, 16, and 26–28.

In 1971 Chauncey Starr probably spoke for the majority of scientists and engineers, and possibly for the public at large, when he said 'there is no question that nuclear power is a saving technical development for the energy prospects of mankind'. He was stating this belief in the light of US energy supplies and US capacity to meet all demands. He gave the estimates shown in table 13.4. (The reader will find the differences between thermal and breeder reactors described on pp. 158–172.) Starr defined the economically recoverable fuel supply as the quantity available at no more than twice 1971 costs (presumably in real terms). The estimates showed that the fossil fuel reserves when defined in this way were barely equivalent, in the case of the US, to twice the cumulative demand to the year 2000. He argued that the case for the breeder reactor was self-evident on this basis. Alternative options had to be considered closely and Starr did provide his estimates of the contribution renewable sources might be able to provide by the year 2000. He included in his list wood, farm waste, photosynthesis fuel, hydro power, tidal power, wind power, geothermal power, and the direct conversion of solar radiation

TABLE 13.4

FUEL SUPPLY AND DEMAND 1960–2000
(From C. Starr, *Scient. Am.*, Sept. 1971, p. 43)

Depletable supply 10^{12} watt-years	World	US
Coal	670–1,000	160–230
Petroleum	100– 200	20– 35
Gas	70– 170	20– 35
Subtotal	840–1,370	200–300
Thermal	~3,000	~300
(Nuclear)		
Breeder	~300,000	~30,000
Cumulative demand 1960–70	350– 700	100–140

and solar space heating, omitting from the list wave power. He concluded that by the year 2000 the contribution from all of these for the US fell far short of meeting the gap likely to develop between demand and supply at that date. In the total world situation he thought the position was more favourable, largely because it could be assumed that a substantial contribution could be made by wood and hydro power in the less developed countries.

Starr's estimates of alternative supplies are given in table 13.5 but it

TABLE 13.5

RENEWABLE ENERGY SUPPLIES AND ESTIMATED DEMAND
(From C. Starr, *Scient. Am.*, Sept. 1971, p. 43)

Continuous supply in 10^{12} watts	World Maximum	World Possible by 2000	US Maximum	US Possible by 2000
Solar radiation	28,000		1,600	
Wood	3	1·3	0·1	0·05
Farm waste	2	0·6	0·2	0·00
Photosynthesis fuel	8	0·01	0·5	0·001
Hydro power	3	1·0	0·3	0·1
Wind power	0·1	0·01	0·01	0·001
Direct conversion	?	0·01	?	0·001
Space heating	0·6	0·006	0·01	0·001
Non-solar				
Tidal	1·0	0·06	0·1	0·06
Geothermal	0·06	0·006	0·01	0·006
Total	18+	3	1·2	0·2
Annual demand in year 2000 10^{12} watts	~15		~5 to 6	

should be stressed they are based on certain views of the rate of development of new sources and certain ideas of the economic costs of fuel. In the world situation coal could in theory delay for a considerable time the serious shortage of supply as against demand, but there are problems in that answer as we shall see presently.

With nuclear energy replacing fossil fuels there has to be a substantial change in the pattern of fuel use, since nuclear energy can only be used to generate electricity. However, even with a quite determined change-over to electric power, Starr calculates that by the year 2000 no less than 40 per cent of the US energy needs will still have to be met with fossil fuels.

Experts in most of the advanced countries realize time is running short to avoid a fossil fuel crisis, and most see nuclear energy as the only practical alternative. Some believe much remains to be done by energy conservation but, though this is true, most experts believe not enough can be done to compensate for depletion of oil and natural gas. Others take the view that the life style of people must be radically altered, if necessary by enforced regulations. It is difficult to believe that legislation to ensure that nuclear energy will not be used at least to help to fill the gap, and possibly completely fill the gap, will attract political backing.

In the UK the recent discovery of considerable resources of oil and natural gas in the North Sea might be regarded as adequate insurance against fossil fuel depletion. The oil fields have been developed rapidly since 1970,

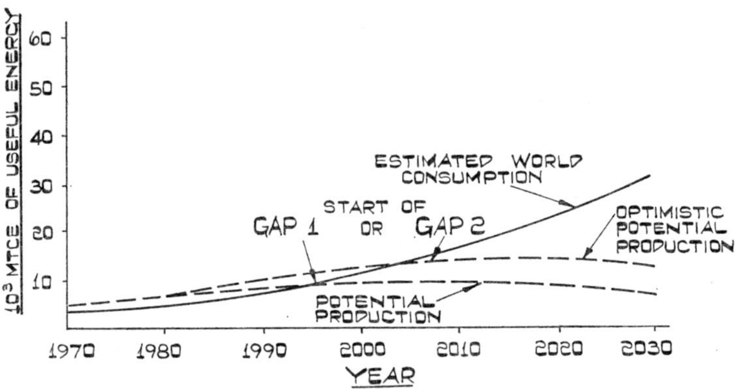

FIG. 54. Predicted energy gap in millions of tons coal equivalent (MTCE). Note that the useful energy is taken to be 45% of the total fuel energy, and that tar sands and oil shales are not included. A possible greatly increased level of coal production is excluded from the calculations. [From L. G. Brookes, *Atom* **234**, 1976, p. 95]

and by 1980 will be providing an amount approximately equivalent to the UK total consumption of oil in 1977. The UK population is about $1{\cdot}5 \times 10^{-2}$ of the world population, and the British oil fields will be able to provide about 1979 this fraction of the estimated total world oil production. It follows that the UK can hope to provide itself with a per capita petroleum supply greater than the world average, but this is of course low because so little oil is used in under-developed countries. The UK has moreover better than average per capita coal resources. It has been estimated by L. G. Brookes, however, that by the year 2000 the UK will be faced with a gap between supply and demand of fossil fuels. He argues persuasively (*Atom* **234**, April 1976, p. 95) that unless the advanced countries, including Britain, move consistently and continuously forward with a nuclear reactor construction programme there will be a severe failure to meet energy demand with supply. His estimates are given in fig. 54 and they show that a gap will appear between 1997 and 2007, that is twenty to thirty years from now.

Brookes advocates the use of uranium and thorium in thermal reactors, together with a combined thermal plus breeder programme. He takes the view that the time-scale is so short this plan is the only one that can offer a realistic means of preventing a gap. He provides a good deal of information on uranium resources and the requirements for uranium according to different choices of reactor systems. Much of his argument would apply to other industrialized countries. He finds that, in the case of the UK, 0·4 M tons is the total uranium requirement, assuming a thermal plus breeder reactor system is used. This amount he regards as consistent with a UK share of a world total of 20 million tons (20 million tons is his best estimate of proven and probable reserves of uranium from sources which can be economically worked). This share is 2 per cent of the world total and well above the share that would be available to the UK on a population basis, a point which perhaps illustrates the gravity of the energy crisis that could hit the advanced industrialized countries. Moreover, the uranium need would be very much increased, by much more than a factor of 5, if the breeder is not used. According to Brookes, if the UK adopted an all-thermal programme of reactor construction, 10 per cent of world production of uranium would be needed. It is difficult to see that this would be forthcoming. There is of course a great deal of uranium in very low concentration and also there is a very large amount in the oceans but the difficulties of extraction from these sources are substantial.

The question of a world energy gap is crucial in policy decisions for the future. President Carter has repeatedly spoken of the gravity of the position as seen from the US point of view. At the risk of some duplication of data and the appearance of a certain amount of divergence of opinion, we should note the ideas of T. M. Fry on revisions of published data. These are included in the figures shown here in table 13.6.

The tar-sands in the table refer to Alberta deposits only and these are reinforced by large reserves elsewhere (in Venezuela, in Madagascar, and elsewhere) which could bring the total up by a factor of about 3. The breeder reactor total could be as high as $2{,}000 \times 10^{21}$ J if it is assumed 1 per cent of the uranium in the oceans is recovered and, if this recovery is left aside as questionable, the alternative figure of about 200 is to be preferred. This figure

TABLE 13.6

ENERGY CONTENTS OF WORLD INITIAL RESERVES OF FUEL
(From T. M. Fry, *Phil. Trans. R. Soc.* **A276,** 1974, p. 606)

Fuel	Quantity recoverable in kg (save natural gas)	Conversion to energy factors in joules per kg and m³ for gas	Energy content in 10^{21} J
Coal and lignite	$7{\cdot}6\times10^{15}$	$2{\cdot}6\times10^{7}$	200
Oil			
petroleum liquids	$2{\cdot}7\times10^{14}$	$4{\cdot}3\times10^{7}$	10
tar-sand oil	$4{\cdot}1\times10^{13}$	$4{\cdot}4\times10^{7}$	2
shale oil, min.	$2{\cdot}6\times10^{13}$	$4{\cdot}6\times10^{7}$	1
shale oil, max.	$(4{\cdot}5\times10^{14})$	$4{\cdot}6\times10^{7}$	(20)
Natural gas	$2{\cdot}8\times10^{14}$ m³	$3{\cdot}75\times10^{7}$ j/m³	10
Uranium		$8{\cdot}3\times10^{13}$ for $U^{235}+U^{238}$	
Thermal reactors, once-through cycle	$3{\cdot}5\times10^{9}$		1
Thermal reactors, plutonium recycle	(6×10^{9})		(10)
Fast breeder reactors	4×10^{10}		200
Thorium	?		(100)
Lithium (fusion) land	10^{11}		500
oceanic	$2{\cdot}5\times10^{12}$		12,500
Cumulative consumption to 1990			7

assumes a reactor efficiency of about 33 per cent and a doubling only of the relatively readily recoverable reserves of $3{\cdot}5\times10^{9}$ kg. All of the U^{235} and U^{238} is assumed to be used up, releasing $8{\cdot}3\times10^{13}$ J/kg.

The last entry in the table assumes that a 5 per cent growth rate in total world energy consumption continues during the 1970s and 1980s. This is in accord with a 1972 world consumption of $0{\cdot}25\times10^{21}$ J per year and the 5 per cent overall figure does not allow for a world population growth of close to 2 per cent per year, so in reality the per capita energy increase is at the comparatively modest rate of 3 per cent per year.

It does seem clear that by 1990 the petroleum and natural gas reserves will begin to show signs of serious depletion. If oil-shale and tar-sand sources are discounted for environmental or any other reasons, coal and the uranium-breeder supplies are the only two ample sources. It would appear from these figures, and on the stated assumptions, that an energy gap arises from the notable lack of success in locating in recent years sufficient additional reserves of oil and natural gas. This last point has been stressed in talks in 1977 about the US position made by President Carter, but the US is not in a unique position.

We have made mention of other advanced industrial countries. There is no shortage of concrete evidence of their fear of a possible gap. In both West Germany and France reactor construction programmes are continuing, often in spite of demonstrations. France is pressing ahead with breeder reactor plans and Super-Phoenix is the first of a range of commercial breeders. The US has undertaken to build about 100 reactors of the PWR type in countries overseas, including Japan. These contracts are now under critical review after design deficiencies of the PWR cooling system were revealed in the Harrisburg incident in 1979, when the flow of coolant was interrupted by gas bubble formation. Despite concern for the energy supply for the future, there is considerable opposition to the introduction of nuclear power on a large scale and we study the reasons for the clash of opinions and beliefs in the chapters 26–28.

Depletion of the fossil fuel resources presents the most serious challenge of this technological age. The right answer must be found, or extremely grave problems affecting world peace and stability will engulf us all at the turn of the century.

Part Three

THE NUCLEAR OPTION

Chapter 14

Fission and Fusion Phenomena

Fission and fusion were described in outline on pp. 20 to 23. We are all aware that the understanding of these phenomena has led to the creation of weapons which have changed, and are changing, the course of history.

The original atom bombs exploded over Hiroshima and Nagasaki in 1945 were fission devices containing uranium (principally as the light isotope U^{235}_{92}) and plutonium (Pu^{239}_{94}). These atoms, in sufficient or greater amount than that known as the critical mass, have a very unusual property—they can take part in a self-sustaining reaction. In the bomb the mass is rapidly made super-critical. This can be accomplished, for example, by rapid powerful compression of a mass of such quantity and shape that when resting it is below critical but after compression is super-critical. In the dense super-critical state the mass undergoes extremely rapid disintegration by fission and in this way yields explosively an enormous amount of power. In the super-critical state the injection of a few neutrons brings about the break-up of the uranium (or plutonium) nuclei with the release of more neutrons and much energy. The new neutrons, being fast and energetic, trigger the collapse of further nuclei and the process repeats extremely rapidly and spreads throughout the mass of the bomb core. The total fission stage of the energy release can be over in an extremely short time.

The moderation or slowing-up of neutrons was described in pp. 21 and 22 and some of the properties of slow neutrons were explained. They readily interact with many nuclei to make radioactive isotopes of many elements and they can cause fission in uranium and certain other heavy elements. However, slow neutrons are not useful in an explosive device as the yield of a weapon depends markedly on the rate at which the energy is released. For this reason the isotopes U^{235} and Pu^{239} are particularly important in weapon manufacture; both undergo fission comparatively readily when bombarded with fast neutrons. It is an ironical fact that U^{235} and Pu^{239} are also the isotopes that prove to be the most important fuels for energy production in reactors.

Later, the development of the hydrogen bomb exploited the process of nuclear fusion. When the temperature is raised in a mass of highly compressed hydrogen, in the form of deuterium and tritium, nuclear reactions occur in which the nuclei join together rather than disintegrate. At the same time energetic neutrons are released in the process of fusion. In appropriate conditions the reaction proceeds extremely rapidly through the mass and it is of course accompanied with the release of huge quantities of energy, much of it in the form of kinetic energy of product nuclei. In this fusion device, as in the fission one, the total mass of the products is slightly less than that of the reactants. Among the important products are fast neutrons. Quite literally part of the original matter disappears and a reduced amount of matter of different form plus a great amount of radiation appears. The radiation is the energy equivalent of the reduction in mass, and bearing in mind $E=mc^2$ (p. 15) the reader will appreciate the vast amounts that are liberated.

For the purposes of controlled production of nuclear energy on an industrial scale we appear to have open to us the possibility of working on two cases only—firstly, the release in fission reactors of the energy stored in the heaviest nuclei, such as uranium and thorium, and, secondly, the release of the energy stored in hydrogen in the heavier of the two naturally occurring forms of hydrogen, deuterium, by the process of fusion. Only these two processes seem remotely practicable at present but by successfully exploiting these two we could obtain vast amounts of energy to help to meet our power needs.

Even if we were completely successful in the use of these processes we would of course be very far from success in releasing all the stored nuclear energy. Theoretically much more complete success would involve the use of many of our light elements as fusion materials to produce, say, an element like iron; or the break-up of most of our heavy elements to finish again with elements like iron or copper or barium. Indeed if the process of nuclear combustion after the 'big bang' had proceeded much further, our universe would consist predominantly of radiation energy and matter in the form of elements like iron. The activation or triggering energy required to bring about the nuclear reactions that yield medium-weight nuclei, starting with light or heavy elements, is so great the reactions are virtually impossible to achieve. In the hot stars the situation is very different—for example, the core of our Sun consists mainly of protons which burn fiercely with immense energy release, forming helium in the reactions—but nevertheless it would appear rare that the nuclear 'fires' in our stellar systems are in the condition to proceed to the extreme limit, the extreme resulting in the burning to near-complete depletion of the fuel and the end products being mainly medium-weight elements close to iron in atomic weight. It is an interesting point that the variety of the natural elements in our Universe and, of course, in that particular part we know best, our own Earth, arises from the failure of the 'big bang' to proceed to the ultimate limit of the burning process. A considerable fraction of the energy of our Universe remains locked in the existing stable elements or, more precisely, in the nuclei of these elements.

Development of Research Techniques. As we have stressed, nuclear reactions usually require a considerable amount of activation energy, and for this reason nuclear physicists use 'accelerators' to provide beams of particles of high energy. In accelerators, charged particles such as protons or deuterons are given a very great velocity by passing them through high-voltage electric fields. The particles are then used as 'bullets' with which to cause transmutation of the target nuclei. Prior to the introduction of accelerators to effect nuclear transmutations, nuclear physicists used the energetic particles provided by nature itself in pursuing their early researches. The only energetic bombarding agents available to them were alpha particles from small amounts of certain natural radioactive elements and cosmic rays. Rutherford found, as early as 1917, that he could transmute nuclei with alpha particles as energetic 'agents provocateurs'. The frequency of such transmutation was extremely low. Many bombarding agents, including alpha particles, can now be produced in much greater intensity and with extremely high energy by means of accelerators such as the Van de Graaf generator or the cyclotron. The interested reader is referred to the many excellent textbooks dealing with the techniques, for example *Atomic Physics* by J. Willmott (John Wiley, 1978). Today not only can the experimenter use man-made alpha particles, but he can in fact, by various ingenious means, provide himself with intense beams of very high energy protons or neutrons, or deuterons (nuclei of deuterium), or tritons (nuclei of tritium), or electrons or gamma rays. More recently scientists have successfully created beams of charged atoms of considerably higher mass, mainly for use in more specialised experiments. In some of the most recent work beams of neutrinos have been generated. In some experiments in fusion research energetic beams of neutral atoms have been produced. In practice beams of energetic charged particles such as deuterons are passed through thin sheets of matter or thin layers of gas, and the charge on the particles is neutralised so that energetic neutral atoms emerge in the form of a beam.

The neutron was not very widely applied in transmutation studies until the discovery of fission in 1939. It is true that, Chadwick having discovered the neutron in 1932, the introduction of accelerators soon afterwards enabled physicists to bring about the release of neutrons in considerable numbers. At the same time Joliot-Curie and others showed that with special mixtures of radioactive elements and certain substances useful intensities of neutrons could be obtained. Fermi and his team of co-workers proved also that by slowing down neutrons they could make use of the very greatly increased power to penetrate nuclei and readily study capture and various types of disintegration of nuclei. But it was the discovery of fission, or rather the realization that fission could be brought about by relatively simple means, which really resulted in a revolution in the thinking of most physicists. In the fission reactor neutrons are generated at intensities which can be comparatively easily controlled. It provides an intense flux of neutrons in a large volume and is actually a powerful tool for transmuting many nuclei. Material enclosed within the reactor assembly is subject to intense neutron bombardment. Fluxes between 10^{12} and 10^{17} neutrons per cm^2 per second can be produced. Transmutation of elements placed in such fluxes can proceed on a scale greatly exceeding in total the transmutation achieved with particles from accelerators. New elements could be made in kilogramme quantities.

With the discovery of fission and the successful operation of the first thermal fission reactor in 1942, it was obvious that for the first time in human history a really basic form of energy release had come within the control of man. The hydrogen bomb was another step in the same direction but the fusion process has not yet been controlled in the same way as fission, as we shall see in chapter 17.

The Oklo Phenomenon. We have described briefly the ways in which the techniques of fission and fusion were first devised by scientists but it was discovered in 1972 that a fission reactor had operated in natural conditions about 1·8 billion years ago. The Oklo river is a small stream passing near Mounana, Gabon, and in the same neighbourhood is a large open-pit uraninite mine. French scientists of the Commissariat à l'Energie Atomique discovered the open mine was actually at the site of an ancient spontaneous chain reactor. At the time the natural reactor started to operate the uranium 235 isotope would have been at a concentration of approximately 3 per cent, a good deal higher than the present day value of 0·72 per cent. (The concentration of U^{235} is increased artificially to about 3 per cent in the modern pressurized water reactors to enable them to perform satisfactorily.)

Measurements made at the site and in various laboratories proved the reactor operated well and provided an integrated output of 15,000 MW-years, which is the value for a modest modern power-generating nuclear reactor giving a thermal energy output of 1,000 MW for some fifteen years. The performance level seems surprisingly high and Alvin Weinberg (*Nature* **266**, 1977, p. 206) regards the discovery as a most exciting one. He uses the discovery to confirm his view that some of the hazards of the nuclear energy industry are imagined or exaggerated. He contends that the finding of the natural reactor site helps to provide answers to many problems such as the waste disposal one. About 5 tons of fission products and 2 or so tons of plutonium were formed in the Oklo reactor and most of the fission products and plutonium remained in place after the reactor ceased operation. Weinburg considers the actinides (elements with atomic numbers from 89 to 103) and strontium 90 decayed essentially *in situ* though one of the more troublesome byproducts, caesium 137, which is soluble, did migrate. Weinburg believes that the Oklo findings show that a thermal fission reactor which operated entirely naturally almost 2 billion years ago neither produced any catastrophe nor brought long-standing problems of disposal of radioactive waste. In the view of D. A. Gray the findings are promising enough for further research on the burial of radioactive waste to be worthwhile ('Geological Aspects of Uranium in the Environment', *Geol. Soc. Miscellaneous Papers*, **7**, 1978, p. 67).

The Oklo observations tend to prove that thermal reactors are rather unsophisticated machines which can operate well for years without showing a tendency to explode. With some reassurance obtained from this natural reactor we shall examine in the next chapter the design features of some of the man-made reactors now in operation.

As regards fusion, we have no knowledge of fusion occurring on Earth and indeed it is very unlikely to have happened. There would never have been temperatures and pressures sufficiently high to contain the required amount of extremely hot reacting hydrogen in a small volume.

Chapter 15

Fission Reactors

The design of reactors has been the subject of a great deal of research and development since the first simple machine was brought into operation in 1942. Between then and 1948 all the reactors, such as those at Hanford in the US, were designed for the production of plutonium as a nuclear explosive and not for electricity production. The first reactor to make electrical energy was a fast breeder and it went into operation at Idaho Falls. Lord Hinton has given an interesting account of this early period in *The Breeder Reactor* (ed. J. S. Forrest, Scottish Academic Press, 1977).

Nuclear reactors all operate on more or less the same theoretical principles. Uranium, being potentially fissile, is usually packed into rods (grouped as the core of the reactor), and the reaction proceeds when the amount of uranium exceeds the critical mass. Control rods, usually of cadmium or boron, are present in the core and are arranged so that they absorb the surplus neutrons if the reactor power begins to exceed the required value. The heat produced by the fission process is carried away by a coolant which circulates at high temperature between the core and a boiler. From this boiler steam is taken to turbines coupled to generators where the energy acquired in the core is converted into electricity. The core is sometimes surrounded by a so-called 'blanket' of uranium which absorbs surplus neutrons. Many different reactor designs have been produced and tested in several countries in the past thirty years (table 15.1). There are advantages and disadvantages to the various solutions already established and it may take a long time for a generally preferred design, or a very few preferred designs, to emerge and gain wide acceptance.

Among the varieties of reactors the difference between thermal reactors of many types on the one hand, and fast breeder reactors on the other, is probably the most basic. Here the outstanding difference lies in the energy spectrum of the neutrons which are effective in sustaining the reaction. In comparing reactor designs the record of performance over the years, the inherent safety of operation, the ease of maintenance, the cost of construction and the price of a unit of power matter greatly. Thus length of experience, based on the history of development and commercial use, counts for a great deal. It is reassuring to note that none of the power reactors of any type has proved a failure in practice.

Thermal Reactors. Though the physics of all reactors is in principle simple the design of a power reactor intended for reliable operation over a period of

TABLE 15.1

PRINCIPAL TYPES OF REACTOR

Reactor type		Fuel	Moderator	Coolant
Natural Uranium Reactors	Magnox	Uranium metal in magnesium alloy sheath (120 tons)	Pure graphite 1,150 tons	Carbon dioxide at high pressure
	CANDU	Natural uranium metal held in stainless steel tubes	Pure heavy water surrounding fuel tubes (also partly heavy water coolant)	Heavy water passed at pressure through tubes holding the fuel
	Steam-generating heavy water SGHWR	Uranium oxide in zirconium case (enrichment: zero to say 1·5% of U^{235})	Heavy water (100 or more tons)	Light water
Enriched Reactors	Advanced gas cooled AGR	Uranium oxide enriched to 2·5% U^{235} in zirconium alloy sheath	Pure graphite	Carbon dioxide at high pressure
	Pressurised water, PWR	Uranium metal encased in stainless steel and enriched to between 2 and 5% of U^{235}	Light water at high pressure (about 160 bars)	Light water
	Boiling water, BWR	Uranium oxide of 1·5% U^{235} or more. About 50 to 60 tons	Light water (about 70 bars pressure)	Light water
Breeders	Dounreay fast reactor (DFR)	Uranium metal enriched to 45·5% U^{235}, clad in niobium		Mixture of molten sodium and potassium
	Prototype fast reactor (PFR)	Uranium and plutonium oxide		Molten sodium

twenty to thirty years is much more complex. Uranium consists of two main isotopes—the abundant U^{238} (99·3 per cent approximately of all natural uranium) and the lighter and much less abundant U^{235} (only about 0·7 per cent of the total in natural uranium). An almost trivial amount of U^{234} is also present. The behaviour of the two principal isotopes when bombarded with neutrons, which we outlined in chapter 3, depends very much on the speed or energy of the neutrons. Thus U^{235} is relatively readily transmuted in the process of fission with fast or slow (thermal) neutrons, while U^{238}, though much less likely to undergo fission with fast or slow neutrons, complicates reactor design by readily absorbing neutrons at particular values of the neutron energy, known as resonant energies, rather above thermal. The use of natural uranium as fuel in thermal reactors involves the designer therefore in the careful incorporation of a considerable amount of moderator to reduce the energy of the neutrons in large steps. The average energy of the fission neutrons in well-designed thermal reactors is rapidly reduced to low value and it becomes comparatively unlikely that potentially useful neutrons, that is, neutrons that are required to cause fission in U^{235} will have energies at which they will be simply absorbed in U^{238} without bringing about energy release in a fission process.

In constructing reactors on the industrial scale only three materials have been used as moderator, namely 'light' or ordinary water, 'heavy' water or some other oxide of deuterium, and pure carbon. Collisions of neutrons with hydrogen atoms roughly halve the neutron energy in each collision so hydrogen is excellent in this respect as moderator, but since the gas is not dense enough to act as a moderator on its own, materials containing hydrogen are substituted, particularly water. Of the moderators mentioned light water, in which the hydrogen is present in its common isotopic form with an atomic weight of one, is in some ways the least attractive, for hydrogen also captures neutrons with appreciable probability and this causes a loss of useful neutrons in the reactor. Deuterium captures neutrons less readily and is thus a better moderator, though much more expensive. Because of the great convenience of light water, and the long experience with steam as a prime mover for industrial machines, it is often used as the coolant; it can even be tolerated as a moderator in spite of the neutron absorption loss that its presence entails.

In the earlier years of the development of reactor technology the tendency was to use natural uranium fuel elements in the way described here, and both British and Canadian practice illustrate this feature. The British Magnox reactors such as the ones at Hunterston, constructed during the 1950s and 1960s, used natural uranium with pure graphite as the moderator and carbon dioxide gas as coolant (fig. 55). More than twenty years of successful operation have been achieved. On the basis of this performance it is generally assumed such gas-cooled reactors can have operating lives of twenty-five years or more. In Canada much progress was made with natural uranium as fuel and with heavy water as moderator and coolant; such reactors are known as CANDU type (Canadian deuterium, uranium reactor). In these British and Canadian designs the economical use of the fission neutrons is an important feature. A variant of the CANDU form should be mentioned. If light water is substituted for heavy water as coolant, mainly to economise in heavy water, the reactor may be described as a steam-generating heavy water reactor (SGHWR). This variant has been intensely studied but it is not presently favoured.

Isotope Separation. The reactor designer is freed from many difficulties if the fuel is enriched in the U^{235} component by methods described below. The core size can be very considerably reduced, and in this and other ways undesirable loss of neutrons through capture in the moderator, coolant, and construction materials can be avoided. Much the same result is achieved by using plutonium (Pu^{239}) as a component of the reactor fuel (see p. 164 below) but this technique is more often regarded as applying in breeder reactors. In the case of the fuel or core, consisting entirely of uranium—and to date this is the normal situation in thermal reactors—the enrichment brings the percentage of U^{235} well above its natural value of 0·7 per cent. A value of 3 per cent is often used. Such a value is sufficiently high to facilitate the task of the designer, and readily achieved when full-scale plant is available to separate the two isotopes.

The separation of uranium isotopes was exploited originally for weapon manufacture and only later used to provide enriched fuel for reactors. The first serious attempts at isotope separation on a kilogramme scale started around 1942. The early successful methods were based on electromagnetic separation and on gaseous diffusion, the latter method still being practised on a massive scale. In the first method beams of ions of uranium are produced in evacuated vessels placed within a magnetic field. The ions U^{238} and U^{235} follow approximately semi-circular but, being of different mass, slightly different paths, making possible the collection of the atoms of the isotopes in separate pockets. In this way uranium can be enriched by a factor of over 100, and in fact uranium with an abundance of U^{235} in excess of 90 per cent, as against the usual value of 0·7 per cent, can be made available. In the gaseous

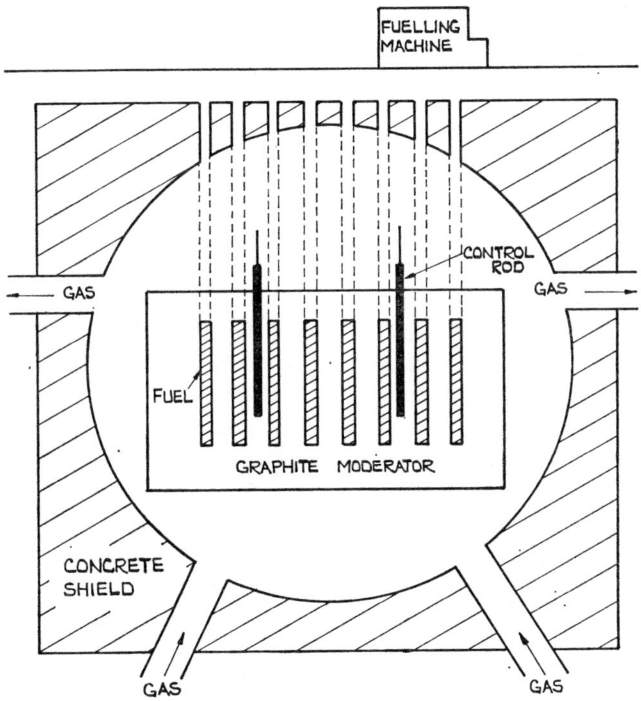

FIG. 55. Schematic diagram of Magnox reactor assembly

diffusion method, a gas or vapour of uranium passes through a series of membranes and the lighter isotope is gradually separated from the heavier by virtue of the slightly higher rate of diffusion which it possesses. Although very high enrichment factors are not readily achieved by this means it is easily possible to raise the normal abundance of 0·7 per cent to 5 per cent or 10 per cent.

Thus by operating isotope enrichment machines the proportion of U^{235} in the uranium made available for reactors can be altered substantially, and a whole range of designs of modern reactors depends on the availability of such enriched fuel. It must be stressed nevertheless that the enrichment involves a considerable amount of extra effort, including the expenditure of a good deal of electrical energy, as both methods of isotope separation discussed here are expensive in energy terms. Work continues with a view to improving the methods of isotope separation and among those we should note the growing importance of separation in centrifuges.

Enriched Fuel Reactors. A very considerable range of designs of enriched fuel reactors exists and a number of them have been used successfully in helping to meet national energy needs. The most developed at present are the light water moderated reactors (LWR) in which modest enrichment to between 2 per cent and 5 per cent is used, mainly to compensate for the neutron loss that occurs in the hydrogen of the light water and in some of the construction materials. The reactors in this class in which the water is used as a moderator and coolant are of two main types. In the boiling water reactors (BWR), shown in fig. 56, the water is allowed to boil within the reactor vessel; while in the pressurised water reactors (PWR) the pressure inside the reactor vessel and cooling circuit is high enough to prevent the water boiling up to a temperature of 350 °C, or 623 K (fig. 57). This means the PWR type is maintained at a pressure of about 160 atmospheres. The high-pressure water is passed to a heat exchanger to deliver heat to a secondary water system at low pressure; boiling proceeds in the heat exchanger and the steam so produced drives a turbine-generator system as in a conventional coal-fired or oil-fired generating station. In the BWR the pressure is roughly half that used on the PWR, about 70 atmospheres, so the boiling occurs around 300 °C (573 K). The steam from the reactor core is passed at high temperature and pressure directly to the turbines. In a fairly typical example of such a BWR the core contains 57·5 tons of uranium oxide enriched to 1·5 per cent of U^{235}. In a comparable PWR (Yankee) the core weighs 21 tons and the uranium rods of the core are enriched to 3·4 per cent of U^{235}. Control rods of 80 per cent silver, 15 per cent boron and 5 per cent cadmium, all able to absorb slow neutrons readily, are employed.

In 1978 the number of LWRs in operation exceeded 200. Most were manufactured in the US and some of them were exported; increasingly, similar reactors, particularly of PWR type, are being manufactured in other countries, especially the German Federal Republic. It is thought that they will have a useful working life, comparable to Magnox, of twenty-five years or more. The natural successor to Magnox, but using enriched fuel, is the Advanced Gas-cooled Reactor (AGR), such as those at Hunterston (Ayrshire) and Hinkley Point (Somerset).

Mention has been made of the steam generating heavy water reactor

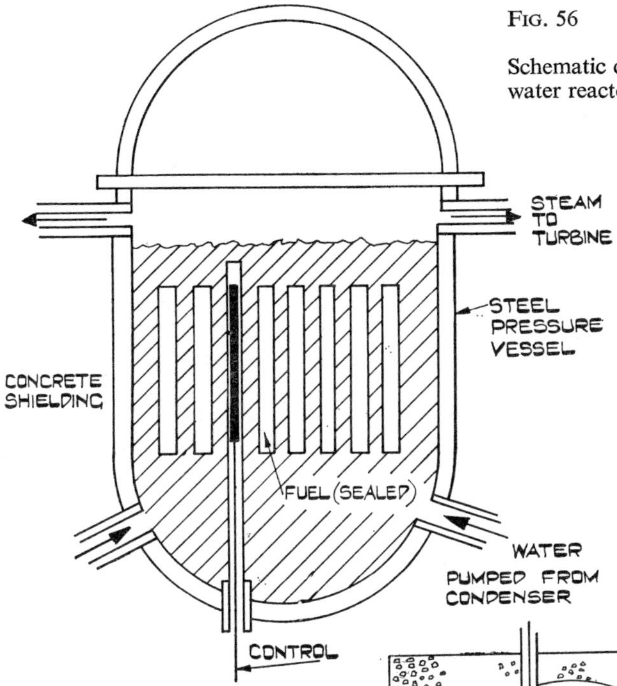

FIG. 56

Schematic diagram of typical boiling water reactor

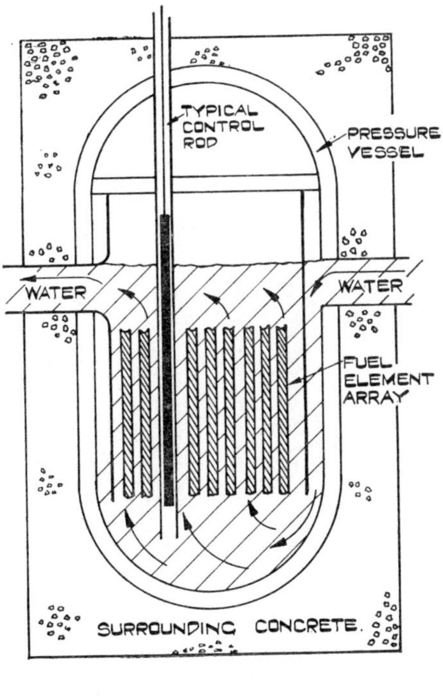

FIG. 57

Schematic diagram of pressurised water reactor

(SGHWR) illustrated in fig. 58. In a prototype the uranium oxide fuel elements are enclosed in zirconium alloy and immersed in some 120 tons of heavy water at atmospheric pressure. Light water is used as the coolant and, as such, it captures neutrons, but to a much lesser extent than when it is used as a moderator. In fact if the reactor is designed so that the capture of neutrons in the coolant is kept to a minimum, the SGHWR can be constructed to use natural uranium fuel elements rather than slightly enriched uranium.

Plutonium Production. The production of plutonium permits a very different approach. This route eliminates entirely the need to separate isotopes, and in terms of energy consumption and capital investment is more attractive. Some of the neutrons present in huge numbers in a reactor are captured by the U^{238}. The reaction results in the formation of uranium U^{239}. This unstable isotope decays by beta emission to neptunium which, in turn, decays to give plutonium. The decay chain can be shown symbolically:

$$U^{238}_{92} + n^1_0 \rightarrow U^{239}_{92} \rightarrow Np^{239}_{93} + e_- \rightarrow Pu^{239}_{94} + e_-.$$

Now plutonium is itself radioactive and the isotope Pu^{239} has a relatively long half-life for decay by emission of alpha particles—only half of it changes in 24,000 years. It is, however, a fissile material and can be used instead of U^{235}

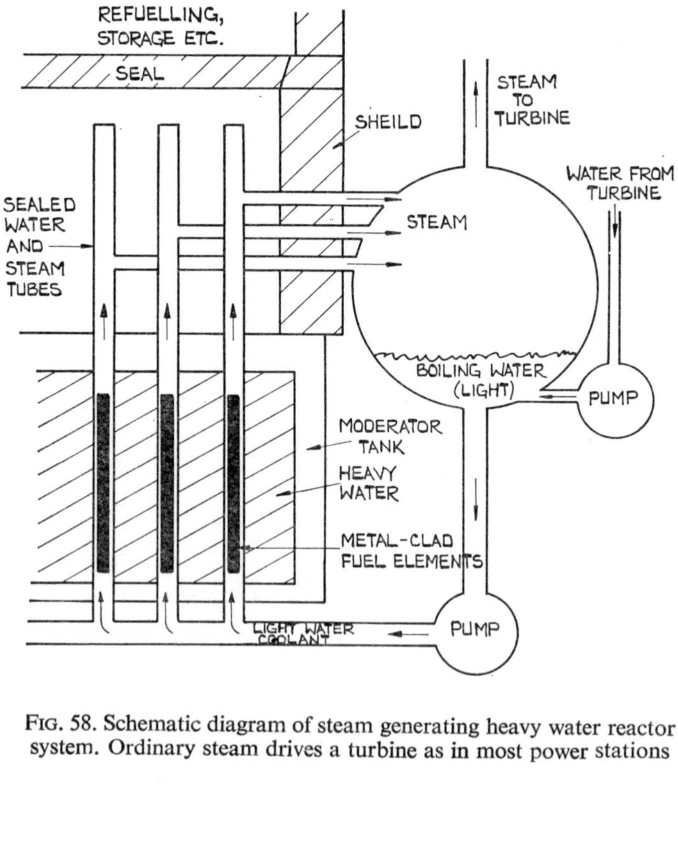

FIG. 58. Schematic diagram of steam generating heavy water reactor system. Ordinary steam drives a turbine as in most power stations

PLATE 1. Liquid natural gas ships of IU International.
[Photograph from the Gotaas-Larsen Shipping Corporation.]

PLATE 2

The CEGB's 2,000-MW oil-fired Fawley power station, Southampton. One of the features is the combined circular control and administration building. The station receives its oil fuel supplies direct by pipeline from the nearby Esso Refinery. The main station building consists of boiler house, deaerator bay, and electrical annexe, and turbine hall. There is one 650-ft multi-flue chimney and a 400-kV indoor substation.

Reclamation work commenced in 1962 and the station was officially opened on 6th May 1971. [Photograph from the CEGB]

[on facing page

(b) The advanced gas-cooled reactor at Windscale, Cumbria. Building started in 1958 and the reactor went into operation four years later. [Photographs from the United Kingdom Atomic Energy Authority.]

PLATE 3 (a) The prototype fast reactor at Dounreay, which produced its first power in 1974.

Plate 4

Fully bladed shaft being lowered into position during the assembly of a Parsons 30-MW single-cylinder industrial turbine. Steam at 900 lb/sq.in, 900 °F is supplied to the turbine which drives a direct coupled, air-cooled, generator at 3,000 rev/min. [Photograph from Parsons] as a nuclear fuel. The manufacture of plutonium as Pu^{239} may be an incidental or a deliberate feature of reactor design. When the design is such as to maximize plutonium yield the reactor is known as the 'breeder' type, which we mentioned above (p. 158). A breeder reactor is a machine of a most unusual nature. In it the working fuel embedded in various other materials is used to provide us with energy, and during its consumption it transmutes some of the accompanying materials into fresh fuel, greater in amount than that which has been consumed.

Capture of Thermal Neutrons. We discussed in chapter 3 and p. 160 the moderation of neutrons to thermal energies. A very important property of thermal neutrons can be calculated by applying the theory of the great French physicist, Prince Louis de Broglie. He shows that particles are like 'wave packets'—matter and waves being simply complementary forms of the same thing. A particle of mass m moving with velocity v, and having therefore momentum mv, can be represented by a de Broglie wave of wavelength λ given by the equation

$$\lambda = \frac{h}{mv}$$

where h is the Planck constant ($6{\cdot}63 \times 10^{-34}$ J sec). If then a neutron is thermal and moving at a speed of, say, 10^3 m/sec we find

$$\lambda = \frac{6{\cdot}63 \times 10^{-34}}{1837 \times 9 \times 10^{-31} \times 10^3} \cong 3 \times 10^{-10} \text{ m.}$$

This example shows a thermal neutron has an associated wavelength roughly equal to the diameter of a large atom and much larger than the diameter of a typical nucleus (around 10^{-14} m). In fact a fast neutron and most small nuclei would have a particle diameter of only a few times 10^{-15} m. We can appreciate by this calculation that the influence of a slow particle, in particular a slow neutron, may extend over a range some 10,000 times larger than its own diameter (when it is regarded simply as a particle). This kind of interaction at a distance helps to explain the remarkably high probabilities of neutron capture in certain nuclei when the neutrons have thermal, or certain resonant energies, and underlines the need to examine carefully the whole range of processes which could affect the total neutron economy of a reactor. For example, the increasing probability of capture with neutron thermalisation affects the choice of all materials that are exposed to neutrons. Moreover, the presence of impurities in core material, moderator, or coolant can be extremely troublesome.

Neutron Economy in Reactors. The precise average value of the number of neutrons released in fission is clearly one of the critical parameters in evaluating the practicability of any reactor design. We start with the equation

$$\text{U} + \text{n} \rightarrow \text{A} + \text{B} + \text{x neutrons}$$

where U indicates a uranium isotope, for most practical purposes U^{235} and n is an incident neutron. The symbols A and B represent product nuclei formed as the result of fission, and these may vary considerably (typical examples are molybdenum and barium) from one fission process to another. The number of neutrons released, denoted by x, is usually between 2 and 3 for U^{235} where n is a neutron of average fission energy. The thermal energy of the reaction is derived for the most part from the high kinetic energies given to the fragments A and B at the moment of fission, though a good fraction of it appears subsequent to the reaction as A and B decay radioactively.

We have to examine closely the neutron balance, and this balance is obviously dependent on the particular type and design of the system. If we assume we are concerned with natural uranium fuel with graphite as moderator and, say, carbon dioxide gas as coolant (the Magnox type) we can assume the following figures as approximately representative of the fractions of the enormous flux of neutrons which participate in the main processes that occur in a reactor:

Neutrons from U^{235} per fission	2·26
Neutron to maintain reaction	1·00
Excess	1·26
Absorbed by capture in moderator	0·25
Absorbed in other materials	0·05
Escape from core to shield etc.	0·05
Absorbed in U^{235} (capture)	0·15
Total loss =	0·50
Neutrons available for other uses such as plutonium breeding	0·76

A small percentage, which is not included in the calculation, might also be lost in the coolant, and a small fraction has to be left for control purposes. Nevertheless in a reactor giving such a balance we have plenty of neutrons to spare and they can be used in various ways.

Regeneration. The figures given above can be taken as reasonably representative of the situation in a simple power-generating reactor containing natural uranium. The availability of neutrons in excess of those required to sustain power and make good the losses proves attractive, and it can certainly be turned to substantial economic advantage in what is known as the regenerative mode of operation of a reactor. In designing for regeneration we arrange to make as much potential fuel as we can while the reactor is used for power production. In the example above we could possibly arrange for capture of almost all of the excess neutrons in the abundant nuclei of the uranium U^{238} itself, which in any case is the major part of the fuel. Additionally we may so construct the reactor that U^{238} in a uranium blanket system absorbs the excess. Assuming this is done we see that for each 100 neutrons used to bring about fission in a sustained reacting condition there would be up to 76 neutrons in excess. If these are captured in U^{238} we get 76 or more atoms of U^{239} formed, which subsequently transmute to Pu^{239}. As a fissile material Pu^{239} is similar in many ways to U^{235}, so taking plutonium as equivalent to U^{235} we have in effect 76 gm or more of Pu^{239} returned to us as usable fuel for each 100 gm U^{235} consumed. In a sense we have managed to get 100 per cent energy output for less than 30 per cent consumption of the fuel. By repeating the cycle we can arrange to get some 3 to 4 times as much power as the amount of U^{235} in natural uranium would provide. We are assuming in saying this that we can actually replace U^{235} in a reactor with Pu^{239} but some design changes would of course be required. The early reactors were used in this fashion, for they were in effect machines built to manufacture plutonium for nuclear weapons. The hazards of handling plutonium are considerably greater than those encountered with natural uranium or enriched uranium, and this makes the use of plutonium a more complicated matter.

It is obviously tempting to try to construct a reactor able to convert all of a charge of U^{238} to plutonium, which could then be consumed totally by fission in reactors. To be able to do this would tremendously increase the value of uranium as a nuclear fuel, as almost all of the available natural uranium could, if required, be used in due course for power production. When the production of plutonium is the principal aim we become involved in the design of the fast breeder reactor (FBR).

Breeder Reactors. In principle moderated enriched fuel reactors would not seem to be useful for breeding, but in practice they can be operated so that spare neutrons are available for the conversion of U^{238} in natural uranium or U^{238} in depleted uranium (natural uranium from which much of the U^{235} has been removed for other purposes) to Pu^{239}. In practice, with a combined system of thermal and breeder reactors, where the thermal produces the plutonium fuel for the early breeder reactors, probably about 10 per cent of the total uranium stock can be usefully consumed. Some of the factors controlling uranium utilization have been examined by S. E. Hunt (*The Breeder Reactor*, ed. J. S. Forrest, *op. cit.*). The most important parameter is the breeder ratio, that is the atoms of Pu^{239} made in the U^{238} of the core and the blanket per atom of core undergoing fission. The percentage of total uranium that can be usefully burned rises steeply to 70 per cent or 80 per cent as the ratio reaches unity and exceeds it. The variation is shown in fig. 59.

Thorium. The spare neutrons generated by moderated fuel reactors could also be used to convert thorium to U^{233}, though this has yet to be carried out on a commercial scale. Thorium is a relatively abundant element in nature, though accurate estimates are not yet available, and breeding of fissile material from thorium could prove important if nuclear energy is required to meet a large fraction of the total world energy needs for some hundreds of years. Thorium used as a blanket of material around the core can capture neutrons thus:

$$Th^{232}_{90} + n^1_0 \rightarrow Th^{233}_{90} \rightarrow Pa^{233}_{91} + e_- \rightarrow U^{233}_{92} + e_-.$$

The uranium isotope U^{233} artificially produced in this way is separated chemically from the thorium in which it is produced. It has properties somewhat similar to those of U^{235}, and can be used in much the same way as U^{235} in reactors. It could possibly prove superior to U^{235} as a fuel in fast reactors. It has been argued that a well-designed thermal reactor of the SGHWR type could provide sufficient spare neutrons to enable the breeding of plutonium and U^{233} to be carried to the point where a large fraction of all natural uranium and thorium could be consumed in reactors. Although most nuclear work for two or so decades has been concentrated on uranium and much less experience with thorium has been acquired, in theory we have in such use of thorium a considerable increase in our energy resources.

Harrison Brown notes that . . . 'about 30 per cent of uranium and thorium in an average granite is situated in such a way that it can be washed out of pulverised rock by using dilute acid and . . . the energy "profit" from one ton of average granite is equivalent to about 15 tons of coal'. Since huge outcrops of granites are found in various parts of the world that contain considerably more than average concentrations of uranium and thorium 'the energy potentially available from such rocks in the US is thousands of times larger than that of all oil and shales combined'. It is clear, as Harrison Brown says, 'the deployment of breeder reactors will have a profound effect upon the uranium supply problem'.

Fred Hoyle has argued the case for using thorium, but with some unjustifiable assumptions. The production of U^{233} from thorium is the preferred method and his table, shown below, makes a strong point. Cross-sections give

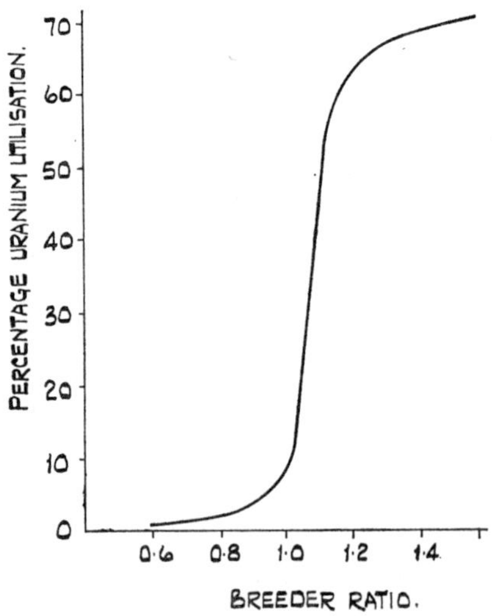

FIG. 59

Uranium utilization versus breeder ratio, with recycling. [From S. E. Hunt in *The Breeder Reactor*, ed. J. S. Forrest, p. 54, Scottish Academic Press, 1977]

TABLE 15.2

THERMAL NEUTRON CROSS-SECTIONS (10^{-24} CM)
(From F. Hoyle, *Energy or Extinction*, Heinemann, 1977, p. 50)

Atom	For neutron capture	For fission
Th^{232}	7·3 (useful for breeding)	—
U^{233}	53 (unwanted)	525
U^{234}	90 (unwanted)	—
U^{235}	106 (unwanted)	577
U^{238}	2·7 (for breeding)	—
Pu^{239}	286 (unwanted)	742
Pu^{240}	450 (unwanted)	—

a convenient measure of the probability of nuclear interactions.

It is argued that breeding U^{233} from Th^{232} is much more practicable in thermal reactors than is breeding of Pu^{239} from U^{238} as in the former case there is much less self-poisoning. From this angle Hoyle argues for 5 per cent burn-up of the fuels, thorium and uranium as a mixture, in a thermal reactor of CANDU type. The argument from that point is concerned with the economics of electricity generation. He contends that the relatively low price of electricity will permit a high uranium and thorium purchase price and make much of these fuels available as a result. All of this argument is speculative and it assumes successful design of operationally trouble-free and reliable reactors of a form not yet entirely proven. It is difficult to see such a route being successfully exploited in the relatively short time available before the energy gap becomes acute. It is an idea based on some proposals that have been made before but in the total energy context it overlooks the tremendous care that has to be taken in introducing a different type of reactor and a new fuel economy. It is all too easy to ignore the crucial point that industrial uranium reactors have been successfully operated for more than twenty years but only now is the nuclear contribution to total energy supply becoming a worthwhile component.

The comparison of uranium and thorium as fissile materials, and the neutron economy in thermal and breeder reactors, is treated with insight by Leslie Shepherd in his paper in *Nature* (**265**, 1977, p. 394).

Fast Breeders in Operation. A reactor fairly representative of some of those reactors now in operation in several countries is the first one built in the UK, the DFR (Dounreay Fast Reactor), operated safely at Dounreay from 1959 to 1977. After this long period of use as a tool for scientific and engineering studies of fast breeder operations, it was closed down to release staff for work on a Prototype Fast Reactor (PFR) at the same site. The core of DFR consisted of uranium enriched to 45·5 per cent U^{235} and clad in niobium, a rare metal which has a low neutron absorption. The coolant was a molten mixture of sodium and potassium. About 20 tons of depleted uranium formed the blanket. It proved satisfactory in operation and gave valuable information on which to base the design of the larger PFR.

This second reactor was put into service in 1977 and can produce 600 MW of heat and 250 MW of electricity. Theoretically about 1·2 atoms of plutonium Pu^{239} will be available on average from absorption in the blanket for each atom of U^{235} consumed in the core. The PFR has operated successfully, despite the anticipated engineering troubles, and shown that the main problems concerned with welding and other manufacturing techniques have been overcome. It is intended to be the forerunner of the commercial fast reactors (CFR), each of more than 1000 MW electrical output, which it is hoped will be built in the future as part of a combined thermal and fast reactor programme. Each of the CFRs would operate at a power level such as to provide enough plutonium to fuel another reactor within twelve or fifteen years. Work on fast reactors using helium gas as the coolant indicates the increase secured in the flux of neutrons for making plutonium would possibly reduce this time to about eight years.

Designs of FBRs vary widely depending on the nature of the core and the coolant. A schematic diagram is shown in fig. 60. A slightly more elaborate diagram of an actual plant, the PFR at Dounreay, appears in fig. 61 and a photograph on plate 3a.

A number of excellent prototype breeder reactors, operating at around 100 to 250 MW electrical output against 1,000 MW in the commercial version, have performed very well indeed. In the USSR and in France, as well as in the UK, the examination of the behaviour of prototype power-generating fast

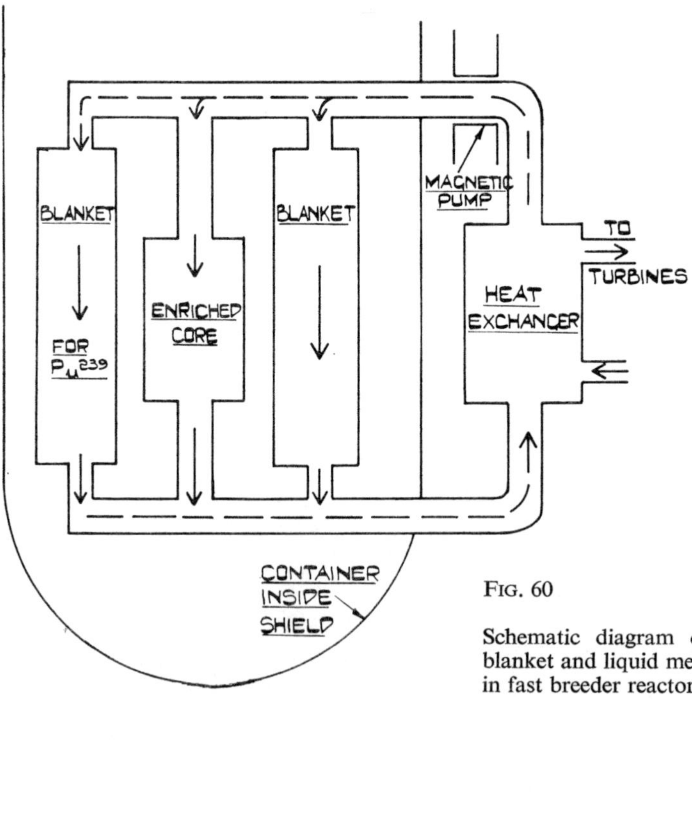

FIG. 60

Schematic diagram of core, breeding blanket and liquid metal heat extraction in fast breeder reactor

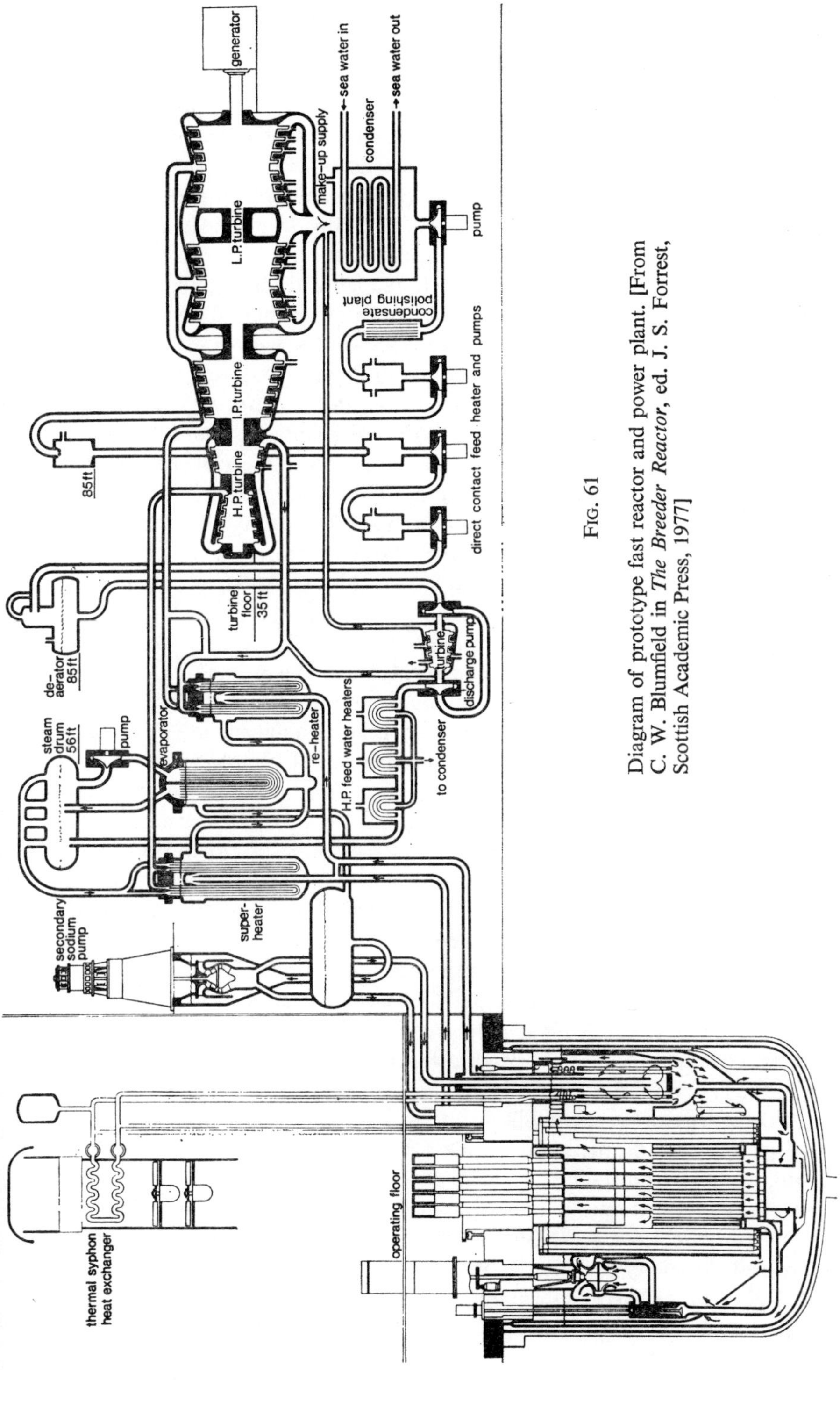

FIG. 61

Diagram of prototype fast reactor and power plant. [From C. W. Blumfield in *The Breeder Reactor*, ed. J. S. Forrest, Scottish Academic Press, 1977] reactors has been carried to an advanced stage, and all the experimental and prototype machines have produced useful results without serious setbacks arising. Scientists have explored the safety aspects and most expert opinion is reassuring on this subject. In the case of the DFR at Dounreay some very searching experiments to establish the safety of the machine have been successfully completed. Some of these are described in *The Breeder Reactor* (ed. J. S. Forrest, *op. cit.*) and the paper by C. W. Blumfield gives considerable detail of British work. The paper by H. C. Simpson provides a most helpful review of work on fast reactors in other countries.

Structure and Size of Reactors. The reader will have gathered that there are so many varieties of reactors built on an experimental as well as a commercial scale it is impossible to generalise about their structure or size. If the fuel is highly enriched, say 90 per cent or more of U^{235}, a reactor can be small and compact, perhaps a metre in height by half a metre in diameter if constructed as a cylinder containing a solution of a suitable uranium salt such as uranyl sulphate. On the other hand many of the commercial reactors contain thousands of fuel pins about 1 cm to 2·4 cm in diameter and several metres long. The core itself measures several metres in height, width, and depth, and it is usually enclosed in a comparatively large reactor vessel. The vessel itself is often contained in a massive radiation shielding structure of dense concrete some 1 m thick. Such containers assume various shapes but are commonly some 10 m or so in height, width, and depth.

Achievements of the Nuclear Industry. It would be wrong to finish this brief review of reactor designs without saying something about the overall achievements of the last thirty years or so. The story of nuclear power is one of the most fascinating in the history of science and technology. It reaches its culmination in the fast breeder. In such a machine hundreds of megawatts of heat can be produced in a volume of around 1 cu.m, and successfully used to produce electricity. Moreover, the energy can be released continuously, and after about ten or more years of useful operation enough fuel created in the blanket of natural or depleted uranium to make a fresh core and also provide some of the fuel for a second new core.

The core within the breeder is the nearest approach to the philosophers' stone that science has yet provided. If we equate energy with gold—and some would insist that energy is a more realistic form of wealth than gold—the breeder core provides energy or wealth for years. By transmuting the elements U^{238} and Th^{232} to more readily fissile elements it multiplies its capacity to provide energy or wealth while it operates.

In 1978, only France and the USSR had breeders in commercial service though others were being constructed by the UK, the US, Japan, and the German Federal Republic in conjunction with Belgium and the Netherlands. It is from the longer established thermal reactor that the world gets its increased electricity supplies. In many countries including the US, the USSR, the UK, France, West Germany, Japan, and Sweden, the contribution of nuclear generated electrical power to the total electricity supply industry is already comparatively large. In the UK it represents some 15 per cent of the total electricity supply and, taking Scotland on its own, it supplies about 30 per cent. In the US it averages about 11 per cent. In most of these countries it would have been substantially higher at the time of writing (1978) but for two main impediments. Large nuclear stations are expensive to build and obtaining the capital for the investment often provides an obstacle, although electricity from nuclear plant may be comparatively cheap when compared with that generated by coal- or oil-fired plants; the economics of the nuclear industry are considered in the next chapter. The second impediment to more rapid progress is public concern about the safety of nuclear plant, particularly plant associated with breeders, and protests in many countries such as the UK, the US, France, Japan, Sweden, and West Germany have delayed nuclear construction programmes and reduced them in scale. Questions of safety are dealt with in chapters 26–28.

An additional complication which has contributed to the continuing uncertainty about the rate of expansion of nuclear power is the difficulty of choosing between competitive systems. The expansion of the world nuclear industry has principally depended on the success of the light water reactor using enriched fuel, especially in those cases where reactors, including fuel, have been exported from the US. The two hundred reactors of PWR type which we mentioned earlier do not include those used in defence vessels such as submarines where the reactor has a much smaller output, of the order of 50 MW. However, the UK investment in gas-cooled reactors of the AGR type has been substantial, though not successful in export terms. The comparison of the advantages and disadvantages of these two very distinct types of machine is not at all straightforward. There are many aspects of operation and maintenance that are not all easy to quantify. Thus generic or in-built safety is important; ease of access for maintenance requires careful consideration; and the conversion efficiency of thermal to electrical energy is clearly a parameter of major significance. When these and other features, including those associated with coupling to the turbine, are assessed there is no entirely clear-cut overall advantage in adopting one of these two systems rather than the other. There is also a case to be made for CANDU based on the inherent advantages of heavy water as moderator and even as coolant. The CANDU system claims advantage over the other two in the degree of overall consumption of natural uranium that can be achieved without breeding. The final preference could depend on whether breeding is adopted or abandoned, a decision linked to the problem of fuel-reprocessing. Meanwhile it can be said with confidence that at least these three types of thermal reactor perform very satisfactorily and any of the three could serve well if a choice had to be made. Greater international collaboration on thermal reactors would have been helpful in the past and fortunately there are signs that this is being appreciated in the development of the commercial fast breeder reactors.

Chapter 16

The Economics of Nuclear Power

The arguments for and against nuclear power are complicated but the most important factors can be set forth in a straightforward fashion, as we hope to show in this chapter. Underlying all the other arguments is the question of supplies. As we have stressed time and again, on the present pattern of consumption oil will last only some fifty or so years and coal some hundreds of years. In the uranium reactor, especially the 'breeder' type, we have an energy-producing device which could have a very long future limited only by the total amount of uranium and thorium that could be made available. Although a few experts express a certain pessimism about the extent of the recoverable uranium reserves they are certainly considerable, and most experts believe there is sufficient to meet a large programme lasting some hundreds of years at least.

The Cost of Producing Electricity by Nuclear Means. Scientists, engineers, and politicians are all concerned about the cost of energy. Where a customer, industrial or domestic, requires heat rather than electrical power it is usually simpler and more economical for him to use one of the common fossil fuels, since it is not possible to use uranium in a straightforward way as a primary fuel, for obvious reasons. However, if electrical power is required, the position is quite different. Until 1973 it was generally assumed, except among experts, that fossil fuels would continue to be comparatively abundant and that the cost of a unit of electrical energy produced in coal- or oil-burning stations would remain relatively low. The cost of a unit did not depend closely upon the cost of extraction and transportation of the fuels and, moreover, the capital cost of the generating stations was not high. Admittedly turbo-generator plant was sophisticated, but well proven and of long life. Boilers were well understood and no longer regarded as very sophisticated pieces of engineering. On the other hand, generating stations which incorporated uranium reactors tended to have considerably higher capital costs per unit of power output, because the nuclear reactor was a novel and rather massive structure designed for maximum safety and incorporating advanced engineering features. Its maintenance also involved complicated techniques. Therefore, although uranium was much cheaper than fossil fuels when the costs of the quantity of fuel needed to produce one unit of electrical energy were compared, the scale of the initial investments meant that the working life of the nuclear-fuelled plant had to be long if the price of a unit of electrical energy from it was to compete with the price of the unit from more orthodox plants.

Opinion altered substantially in favour of nuclear power once it was firmly established that nuclear plants could be reliable for periods of twenty years or more. The full importance of their long life was appreciated only when the price of oil and coal rose steeply after 1973 (fig. 62). The price of electricity from a reactor station is much less dependent on the price of the material in the core than that from a fossil fuel station is upon the cost of coal or oil.

The relationship between construction costs, relative fuel prices, length of

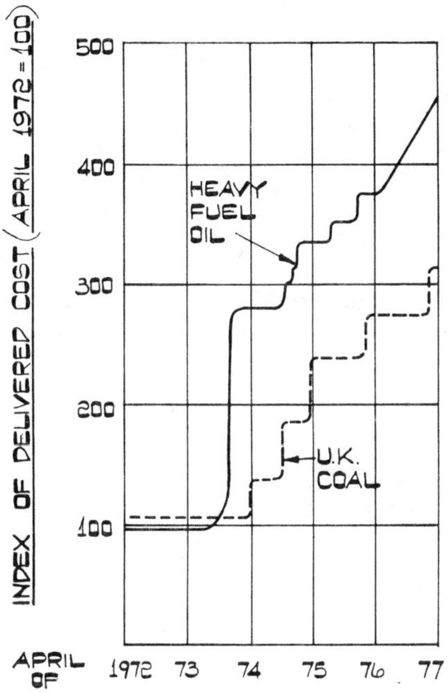

FIG. 62

Approximate delivered cost of heavy fuel oil (including duty) and of UK coal from April 1972 to March 1977. [Based on data of CEGB Annual Report and Accounts of 1976/77.] The effect of OPEC decisions on pricing is clearly evident

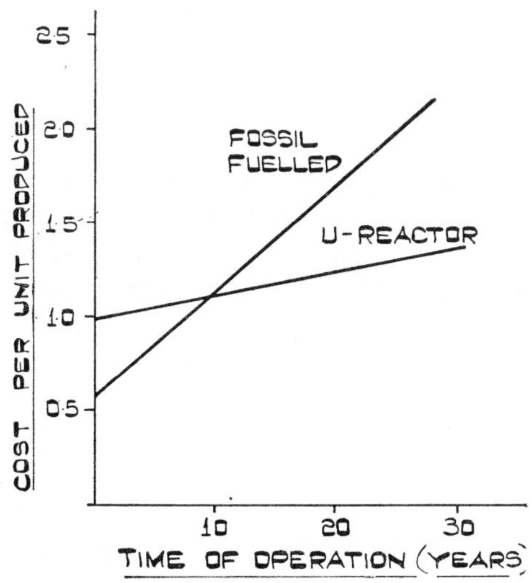

FIG. 63

The relationship between time of operation and cost per unit of power produced by fossil-fuelled and reactor stations station life and their influences on the cost of a unit of electricity are illustrated in fig. 63. The prices of fuels are subject to rapid change and the diagrams are illustrative and not intended for literal interpretation. It appears certain that the trend of rising cost of fossil fuels will continue and that, in consequence, the price of nuclear power will become more and more attractive. The larger investment required initially for the reactor power station has to be reflected in the price to the customer but, assuming a station life of ten or so years, the cost per unit is less than the fossil-produced power, and after twenty years there is a real advantage of nuclear over fossil-fuelled stations. According to the Chairman of the United Kingdom Atomic Energy Authority, Sir John Hill, the unit of electricity provided by nuclear stations, built at the same time as oil-fired and coal-fired stations, was respectively 40 per cent and 30 per cent cheaper than from these two sources (UKAEA Annual Report 1976/77). The price doubling time for fossil fuel is very conservatively put at ten years. In the past, when coal and oil were cheaper, the price of different fuels was probably given too much weight in arriving at policy decisions, decisions which were often against nuclear installations. In addition, and admittedly in hindsight, probably it is unwise for any country to depend very heavily on imported fuel for, as we mentioned often before, the price of oil can be fixed almost arbitrarily by the producing countries to the disadvantage of the consumers.

As far as the construction of modern nuclear plant is concerned, it is still the case that the high costs are occasioned not so much by the construction of the reactor or reprocessing plants themselves, but by the creation of safety systems to guard against major accidents and to protect the people who operate the plant from day to day. Among them are the provision of a massive concrete radiation shield for the reactor; automatic shut-down systems to prevent the reactor running out of control; a vessel strong enough to contain the molten mass if the reactor should 'run away'; and, in every type of plant, sophisticated remote-handling devices. Safety factors may account for as much as 50 per cent of the total cost of nuclear plant and consequently much of the debate about the price of nuclear power centres on their sufficiency or oversufficiency.

Efficiency of Reactors. Our discussion of costs must not be construed as meaning that nuclear plants are necessarily more efficient than other types. As with all other heat engines, nuclear reactors have only limited efficiency. Carnot proved that the ultimate efficiency of any heat engine equalled

$$\frac{T_1-T_2}{T_1}$$

where T_1 and T_2 are the output and input temperatures expressed in degrees Kelvin. In most commercial reactors in operation today the efficiency is between 30 per cent and 40 per cent, and this moderate efficiency is almost inevitable because all are primarily heat sources. For example, the Magnox reactors (fig. 55) which are gas cooled, have an output temperature of say

370 °C or 634 K, an input temperature of 120 °C or 393 K, and an efficiency which is approximately 38 per cent. When a number of other energy losses are taken into account, it is not surprising that the overall efficiency of conversion of thermal to electrical energy is between 30 and 31 per cent for Magnox. The highest efficiency is obtained with the high temperature reactor, HTR. In one form of this reactor, the fuel consists of small pellets of enriched uranium oxide encased in carbides and carbon, and the coolant is helium gas. With an outlet working temperature of 800 °C or more the efficiency can exceed 40 per cent. No commercial scale HTR has been put into operation by 1978 but, in the longer term, when considerable research and development problems have been overcome, scientists see the ideal reactor as an HTR type in which the heat is used directly to bring about water dissociation. The hydrogen thus produced contains as a fuel a large fraction of the heat energy of the reactor. This exciting, if somewhat speculative, prospect has been explored principally with machines such as the HTR in the forefront. It is described in more detail in chapter 20.

The most widely used reactors are light water reactors of the BWR (boiling water) or PWR (pressurised water) types, and the operating temperature in each case is usually in the range 250 °C to 300 °C. The resulting efficiencies are lower than for good fossil-fuelled plants, which normally operate in the range 530 °C to 550 °C, and hence more waste heat is released to the air or dissipated in water per unit of useful energy produced. However, the AGR (advanced gas-cooled) of British design, currently in operation, has shown that it performs satisfactorily at 560 °C, and the operating efficiency has been measured at 39·4 per cent, higher than for most typical fossil-burning stations. Taken as a whole, then, there is no sound reason for believing that reactor plants cause more 'thermal pollution' than fossil fuel ones. Normally the remaining 60 per cent or so of the energy released from all power plants is simply wasted thermal energy. The waste heat, which is diffused later into the atmosphere or into the large mass of water from which the coolant is drawn, can give rise to problems discussed in the next section.

District Heating. Many schemes for 'district heating', that is, schemes to use the discarded energy as heat for a nearby town, have been proposed. Economies in fuel utilization can be guaranteed but an extensive township has to exist, or be built, in the vicinity of the station to use the waste heat to near full advantage, and it is seldom that such a large combined programme can be financed. Moreover, in the case of nuclear plant, it is contrary to the safety philosophy which has resulted in the past in the choice of isolated sites for large nuclear stations; since there is virtually no experience of the effects of accidents at nuclear sites, reactors are sited in accordance with rules relating to the population living within certain distances of the site. Although the rules may be conservatively formulated, most experts would be unwilling to neglect or discard them. Only time can give assurance on the scale of the effects of nuclear incidents. An interesting review of the position reached with schemes for combining district heating and power generation has been given by G. Hammond (*Physics Technol.*, **8**, 1977, p. 163). He notes that significant savings in fuel are possible but that the capital expenditure required is high.

Where district heating is regarded as impracticable or undesirable for cities, it should be possible to use the waste heat from reactors, or any other type of generating plant, to improve the yield of horticultural or agricultural ventures. It is not at all difficult, in a comprehensive programme of development, to ensure that some reactor generating stations are sited where extensive use of the heat to improve crop productivity is practicable. In some instances a manufacturing plant situated relatively close to a power station can use advantageously heat, as well as electrical energy, in a combined heat and electricity scheme, and a few such power plants are in operation.

Thermal Pollution. The overall effect of discharging much heat into the air and water of the environment has been much investigated but is still imperfectly understood. The natural balance is, of course, disturbed when any large power plant withdraws large volumes of water and discharges it at somewhat higher temperatures into rivers or lakes or seas. The amount of oxygen retained in the water is reduced and the resulting oxygen-deficient water is less able to sustain many common varieties of fish, while at the same time it provides a suitable habitat for various undesirable forms of algae and fungi. Nevertheless desirable changes arising from the increased temperature of water near power plants have been observed; for example oysters grow more rapidly.

All of the energy consumed by a population is inevitably degraded to heat. Though the proportion of this heat discharged into water, and thus widely dispersed, is by no means negligible, most of it goes directly into the atmosphere. This means that within, and close to, densely populated cities there is a heated column of air, which must give rise to localized weather effects. Urban heat production arises from the fuels used in combustion processes, and in powering machines, and from electrical energy used in heating and lighting and electrical motors. The temperature of some cities may remain overnight about 5 °C above that of the surrounding countryside, because of the release of day-time heat stored in the materials of all kinds of buildings, and the reflection of outgoing radiation in the atmospheric pollution layer. Apart from increasing discomfort in summer (possibly counterbalanced by comfort in winter) it is not obvious that the effects of the released heat are harmful. More study of this subject, which has received increasing attention in recent years, could prove reassuring.

In *Atmosphere, Weather and Climate* (R. G. Barry and R. J. Chorley, Methven, 1971), the influence of urban life on local climate is examined. The authors study not only heat production, but also the modification of atmospheric composition and alteration of surface configuration and 'roughness' (a term used to describe the modified nature of the terrain). The increases in the temperature of cities as compared with the surrounding countryside have been measured and found to vary a good deal according to topography. It is greatest for sheltered sites with light winds, with heat loss from buildings by radiation usually the most important single factor. Among the most notable increases in mean annual temperature is that of Osaka, about 2·6 °C over one hundred years. This increase is not negligible, but it is scarcely alarming. The wind and rainfall differences between city and countryside may be more significant than temperature.

Clearly the phenomena are complex and other detailed studies are required, but unless there is a very major increase in energy consumption there seems to be no good reason to be unduly worried about thermal pollution. We must remember that solar energy is enormous compared with the amount of equivalent thermal energy released by man-made devices, exceeding it by a factor of well over a thousand. Man-made waste heat may be unevenly distributed, and more concentrated in large cities than it ought to be, but the total is almost negligible compared with the total thermal flux, and if necessary the distribution can be corrected in the future. The concentration of buildings in many of the largest cities of the world has been much too great, and city parks serve many useful purposes including that of heat dilution.

Chapter 17

Energy from Fusion

Schemes to release useful energy by nuclear fusion rather than fission are still an idea rather than a reality.

The introduction of accelerators in the 1930s saw the beginning of a decade of truly exciting discoveries in physics, and knowledge of nuclear interactions increased rapidly. In the early experiments of the 1930s, mentioned briefly in chapter 14, deuterons proved to be extremely effective bombarding particles, and accelerators, operating at a few hundred thousand volts, brought about a large range of disintegrations in many elements. Soon it was realized that the astonishing ease with which deuterons reacted with some elements was a troublesome feature. Deuterons falling on a target soon form a thin layer of absorbed deuterium and this in turn fuses with the incident deuterons to form the lighter isotope of helium and emit neutrons. The reaction is—

$$H_1^2 + H_1^2 \rightarrow He_2^3 + n_0^1 + 3{\cdot}2 \text{ MeV} \qquad (1)$$

The ease with which this reaction proceeds led some physicists to consider nuclear power as a potential energy resource even prior to the discovery of fission in February 1939. There seemed, however, to be formidable barriers to constructing a useful energy machine even though the fusion reaction was decidedly exothermic, that is, it generated heat. The total kinetic energy of the two resulting particles, the light helium nucleus He_2^3 and the neutron, is expressed as 3·2 MeV (million electron-volts). The electron-volt (defined on p. xiii) is an energy unit extremely suitable for evaluation of such energy yields. A readily measurable output of neutrons can be obtained when the bombarding deuteron had an energy of, for example, only 32 KeV (one thousand eV). Since by expending this energy we release 3·2 MeV, the ratio of output energy to energy supplied per reaction is

$$\frac{3{\cdot}2 \times 10^6 \text{ eV}}{32 \times 10^3 \text{ eV}} = 10^2.$$

This hundredfold increase is extremely attractive. Unfortunately, however, the reaction occurs with a low probability, even if more frequently than most reactions of this type, for during bombardment the incident fast deuterons can lose their energy in many other ways. Collisions with electrons cause excitation and ionization of target atoms, and it is relatively seldom that the fast deuteron collides effectively with a deuterium nucleus. It is only about one deuteron in every million that brings about the desired nuclear fusion. This was the riddle that faced physicists for many years and seemed to make it extremely improbable that a net energy gain would be obtained by such exothermic nuclear reactions. In this relatively favourable case almost all the particles expend their energy uselessly and there is an adverse energy balance, adverse by a factor of around 10^4.

The situation improved somewhat when it was discovered that the reaction between deuterons and tritium proceeded more readily than deuteron-deuterium reactions for a given deuteron bombarding energy. In early work it had been found that tritium was sometimes the product of a second reaction that took place between deuterons and deuterium, distinct from the reaction of equation (1) above, namely

$$H_1^2+H_1^2 \rightarrow H_1^3+H_1^1+4{\cdot}0 \text{ MeV} \qquad (2)$$

Tritium as such could not be made in useful amounts in this way. However, when an abundant source of neutrons became available, and the first fission reactors were well suited for that purpose, another means of producing tritium in quantity was to hand. The lithium isotopes Li_3^6 and Li_3^7 yield tritium when bombarded with neutrons, the reactions being

$$Li_3^6+n_0^1 \rightarrow He_2^4+H_1^3 \qquad (3)$$
$$Li_3^7+n_0^1 \rightarrow He_2^4+H_1^3+n_0^1 \qquad (4)$$

Using tritium produced in quantity by these means, the study of the reaction between deuterons and tritium showed clearly that the reaction was not only easily produced but it was also extremely attractive by virtue of the high energy released in the process. We know that the particular reaction

$$H_1^3+H_1^2 \rightarrow He_2^4+n_0^1+17{\cdot}6 \text{ MeV} \qquad (5)$$

proceeds at even lower bombarding voltage than the two deuteron-deuterium ones of equations (1) and (2) above. For this reason it could be important that tritium can be made as in the deuterium-deuterium reaction (2) discussed above, because tritium is obviously an excellent fusion reactor gas. However, since in two of the reactions, namely (1) and (5), fast neutrons are released, these fast neutrons can be used to produce tritium in lithium as in (3) and (4) thus providing a second method of making tritium. If a lithium blanket (or mass of material containing lithium) is put around a fusion reactor it acts as a tritium source. The way is open therefore to use deuterium and tritium in any suitable mixture when our first aim is to produce energy by fusion, and the ease with which reaction (5) proceeds makes it the clear favourite.

Several encouraging features emerge when we examine the fusion process. First, as we have already noted, the reactions proceed well at what seem to be comparatively modest energies, at any rate on the scale familiar to the nuclear physicist. Thus a bombarding voltage of several KeV suffices to bring about the deuteron-triton reaction. In terms of temperature this finding is rather less reassuring. Particles of energy as low as 3 KeV are able to react, but even an energy as low as this corresponds to the average energy of particles in plasma (very hot highly ionized gas) at 30×10^6 K. We realize that in our Sun and most of the observable stars in our Universe hydrogen is fused to helium, liberating immense amounts of energy. In the hydrogen bomb man copied this process, achieving a sufficiently high temperature although the time it is maintained is very short. We can appreciate the scale on which energy is released by fusion if we consider that the bomb releases energy equivalent to that yielded by millions of tons of a typical chemical explosive such as TNT. Thus in a very small fraction of a second we can release from such a device the energy equivalent of several million tons of coal.

We thus have one of the most challenging situations in modern physics. We have a tremendous potential energy resource very similar to that which maintains the burning processes in many stars and we have succeeded in releasing energy in somewhat similar fashion but at an explosive rate only. We would like to control the rate at which such fusion of hydrogen can be achieved. It is worth reminding ourselves that the conditions prevailing in the stars are completely different from any we can simulate on the laboratory scale or in a power station.

Our experiments are always carried out in plasmas to ensure that the bombarding charged particles do not uselessly expend their kinetic energy in exciting and ionizing bound electrons of atoms in the target material. Not only is it difficult to raise the temperature of the plasmas to that obtaining in stars but in most stars like our Sun the mass is so great that the gravitational force is at a value very much higher than any conceivable on Earth and very few of the inter-acting particles can escape. One of the principal difficulties in Earth-based machines is, as we shall see, the attempt to create within an experimental apparatus, and with the very limited means at our disposal, the equivalent of the enormous gravitational force which prevents the plasma from escaping.

Another great difficulty in controlled thermonuclear work is the initial ignition. In the case of explosive release of fusion energy, as in the hydrogen bomb, the temperature necessary to maintain the fusion process is achieved apparently by the use of an atomic trigger not unlike the early atomic bombs, which had an energy release equivalent to thousands of tons of TNT. Such a method is certainly not one that can be considered feasible for power production for peaceful applications.

Work on fusion is interesting for its own sake and the achievements in the understanding of the physics of plasma have been notable indeed. However, from the start of the work it has been fully appreciated that, if fusion could be successfully initiated and controlled, a most attractive way of providing an enormous energy resource would be established. Deuterium is present in our oceans as one part in every 6,000 parts of water and this means we have 100,000 times as much energy resource in deuterium as in all the uranium reserves. If tritium should prove essential there is a possible restriction to the scale of development, considering the reserves of lithium from which the tritium might have to be produced. Table 13.6 of chapter 13 shows lithium reserves reduce the total energy available in fusion to the same order as that from a breeder type of fission programme. Nick Walton and Ed Spooner have examined this matter (*Nature* **261**, 1976, p. 533) and concluded that the demand for lithium, apart from fusion requirements, is likely to be greater than or equal to the fusion demand. The annual demand has risen by 10 per cent for more than fifty years and might continue steeply upwards. They concluded that the figures for cumulative demand and presently known resources are similar, and deduced that there is no cause for complacency in recent assessments such as those of King Hubbert (UN and IAEA, *Proceedings of Fourth International Conference*, Geneva, 7, 1972, p. 467). There is nevertheless no lack of encouragement to solve fusion problems although many large-scale experiments in a number of countries including the UK, the US, the USSR, West Germany, and France have shown that they are indeed formidable.

History of Research into Fusion. Following somewhat speculative thinking about controlled fusion reactions, experimental work in the UK was begun in the laboratories of Sir George Thomson at Imperial College, London, in 1947 and independently at the Clarendon Laboratory, Oxford, by P. C. Thonemann. Both investigations were intended to create plasma in a toroidal, or doughnut-shaped, vessel. Such a shape was obviously appealing since the usual high rate of loss of energetic particles that occurs at electrodes mounted in straight discharge vessels would conceivably be avoided. The torus was arranged so that it was virtually a single-turn secondary coil of a transformer (fig. 64). It was anticipated also that the 'self-pinching', or compression

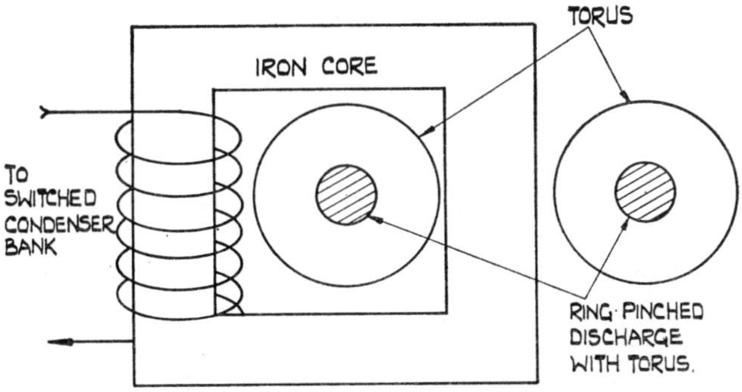

FIG. 64. Arrangement for producing pinched toroidal discharge

towards the axis, that necessarily occurred when a current passed through the ionized gas produced within the vessel, would lead to concentration of the ionized atoms or molecules near the axis of the discharge vessel and hence there would be much less loss of 'heated' gas to the container walls than might otherwise occur. More important, the self-pinching was regarded as a very effective method of heating the ionized gas to higher temperature. The pressure of the gas had to be fairly low, about one millionth of one atmosphere (10^{-3} torr), in order that very high values of current, often in the range of 1 to 10 million amperes, could be sustained during the pulse produced by releasing the energy stored in a large bank of suitably charged condensers. The results obtained from study and measurement of the discharges produced in the earliest experiments were indeed encouraging. Shortly after this promising start a much expanded programme was introduced at the Atomic Energy Research Establishment (AERE), Harwell, and Thonemann was responsible for the supervision of a substantial experimental attack on the problem. In parallel it was arranged that the work of the Imperial College team would be largely transferred to the Associated Electrical Industries Laboratories at Aldermaston Court. At this stage the British Government classsified all of the work as secret.

Later it became clear that work on the fusion problem was being pursued in the US and in the USSR. The results of many of the early studies in all three countries were apparently confidential in view of defence and economic implications, but I. Kurchatov described, in a notable lecture at AERE, Harwell, in 1956, results of some early work done in the USSR with very high current discharges between electrodes mounted in cylindrical vessels. Similar studies at the Atomic Weapons Research Establishment and elsewhere were being pursued at the same time. Soon the work in all the participating centres became declassified. In 1957 representatives of all the involved nations began to meet at international conferences and publish accounts of their work; plasma research was made free of all official restrictions, and much of the progress in the field up to 1957 was outlined in the papers presented at the Third International Conference on Ionization Phenomena in Gases, Venice (1957). The Second Geneva Conference in 1958 heralded the publication of a great deal of work on fusion. Since that time it has become a very impressive part of the programme of international co-operation in science. Extremely sophisticated projects are continuing in a number of countries and steady progress towards the final goal of sustained fusion reaction with net energy release is made. At present there is no guarantee all the problems can be overcome but, vice versa, no scientist can prove that the problems are insoluble. The possible energy dividend is so great the work must surely be pursued on a large scale in those countries with experience and the resources, and international co-operation should be encouraged in every possible way.

The apparently successful early experiments on the Zeta (zero energy thermonuclear apparatus) machine at Harwell were found in 1957 to be misleading, for it had not been thoroughly appreciated that the observation techniques required for study of hot plasma presented very special problems. In addition instabilities of the plasma could result in spurious results: the emission of neutrons did not necessarily mean the deuteron-deuteron or deuteron-triton reactions were proceeding in thermonuclear fashion, for acceleration of a fraction of the particles could occur and neutrons could be produced in ways not unlike those obtaining in an accelerator. These neutrons are sometimes described as 'trivial' neutrons. Such acceleration was recognized as remote from the desired goal which was of course the self-sustaining process of combustion of deuterium or of deuterium-tritium mixtures.

Quantification of the Conditions Necessary for Self-sustaining Fusion. As we have already indicated, thermonuclear burning only takes place when certain physical conditions are fulfilled. Let us illustrate the point by analogy with a more familiar type of combustion. In a coal fire we provide ourselves with a container or grate in which we build up the fuel and apply a light in some suitable fashion. In doing so we hope to get the temperature of at least part of the fire high enough to ensure that the main bulk of the fuel will ignite and burn, thus maintaining the fire. We have to start by putting in the grate enough fuel to generate adequate heat to burn itself up thoroughly, in spite of heat loss from the fire by radiation etc. The grate itself must be made in such a fashion that as burning proceeds the hotter embers do not too readily fall out of the grate and so bring about extinction. Very similar requirements exist in a thermonuclear device. We have to provide enough fuel, and raise it to a sufficiently high temperature, for self-sustaining reactions to proceed in spite of loss of hot material to the walls of the vessel. We can quantify one of these requirements by saying the density of the plasma in our fusion device should be at least n atoms per cubic centimetre. The material at this density n, when raised to the reacting temperature, must not escape too rapidly, as in the analogous case of fuel falling from the fire, and this condition is expressed by saying the containment time of the hot plasma should be at least t. Obviously n and t are inter-related and the value of nt is the important quantity. The ignition of the fuel and its 'burning' also implies that a certain minimum temperature T has to be sustained in the material in some way. The important criteria for fusion are:

(1) T not less than 30×10^6 K and preferably at 10^8 K in the most favourable case of deuterium and tritium mixtures;
(2) nt not less than 10^{14} atoms-sec per cu.cm with t, the energy containment time, about 1 sec.

The second condition, given here in a simplified form, is known as the Lawson criterion. For a fuller account of it we refer interested readers to an excellent discussion paper by R. S. Pease (*British Association Proceedings*, September 1976, and *Physics Technol.*, **8**, 1977, p. 144).

By injecting special beams of energetic particles into the vessel we can satisfy condition (1) above, but we may still be far from satisfying condition (2). To satisfy both (1) and (2) simultaneously is a very difficult task, not least because of the inter-connections between n, t and T_i, where T_i denotes the ion temperature.

The Present State of Research. Many elegant experiments have been launched in the last twenty years in efforts to achieve the conditions for fusion outlined in the previous section. Among them are experiments in which the energy of injected beams of atoms is multiplied though self-sustainment of the reactions is not achieved. Success with such special work as in the colliding beam torus is regarded as interesting, but the primary goal is of course sustained reaction. In this general survey it is not appropriate to discuss the many other studies in detail. It is sufficient to say that steady progress continues to be made in various ways and T_i, nt and t are all approaching, in a somewhat step-wise fashion, the desired goal. The steps in the advance are mainly due to the changes in scale of the experiments. The machines take some years to build and they are designed to include the good features of earlier equipment while seeking better performance in the light of fuller knowledge. The progress is made very evident in the curves of fig. 65 which is based on the report of R. S. Pease.

Among the other aspects of fusion being intensively studied are instabilities in hot plasma and the means to eliminate or reduce their effect.

The confinement of hot plasma in vessels by the application of electrical and magnetic fields is another field of research. The theory of the various types of 'magnetic bottle' that have been devised is complicated, and the experiments that have led to certain changes and innovations in design have themselves presented considerable difficulties in interpretation.

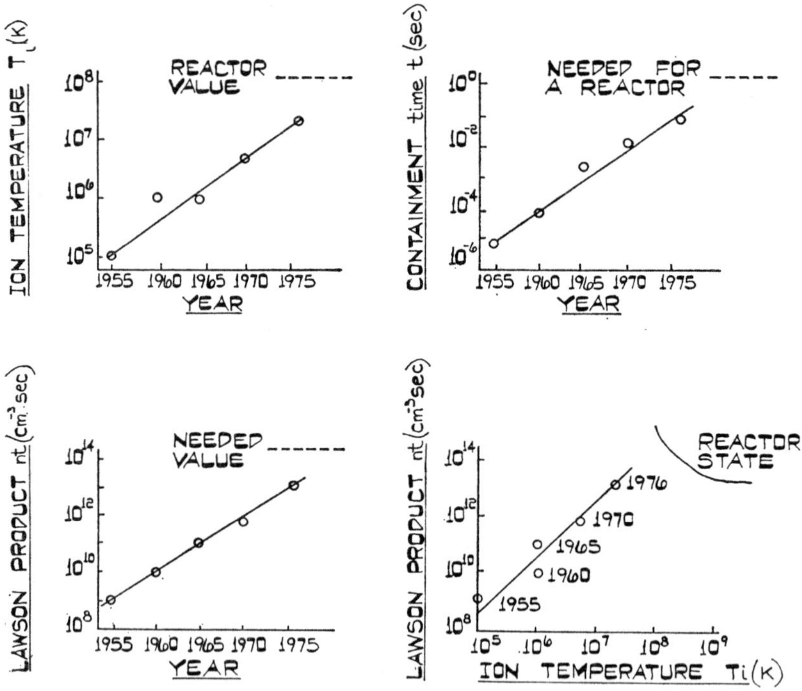

FIG. 65. Progress towards a reactor state. The graphs show experimental results for ion temperature T_i, containment time t and Lawson product nt as a function of time, together with illustration of the rise in value of both nt and T_i with time. [After R. S. Pease, *Culham Laboratory Annual Report*, 1976]

The tokamak magnetic field configuration is currently the most favoured because it seems to offer most promise of achieving the plasma parameters required in a reactor. In the simplest version the confining magnetic field consists of two components and these are indicated in fig. 66. The stronger toroidal field is produced by current in a set of coils placed around the toroidal vacuum chamber, and the weaker poloidal field is generated by passing current along the axis and through the plasma itself. The poloidal field is created by inducing current to flow by transformer action and this current assists at the same time in the heating of the plasma. Experiments with a

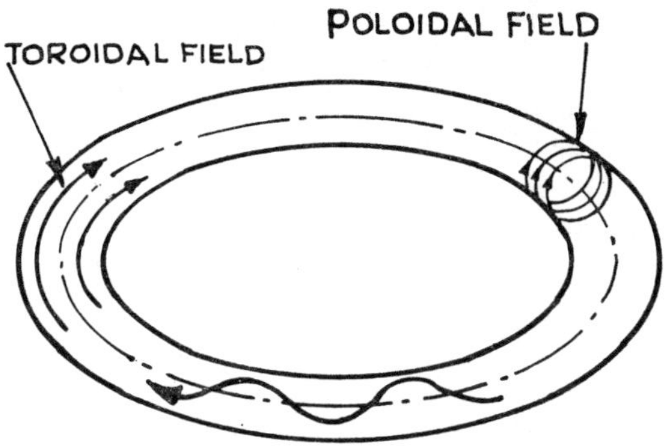

FIG. 66. Principal magnetic fields in toroidal geometry, as used in tokamak systems

number of tokamak arrangements are uniformly encouraging and hotter and denser plasmas have been confined for longer periods. As a result a large part of the European collaborative fusion programme will be concerned with the construction of a very large tokamak (the project is known as JET for Joint European Torus). With the construction of such a machine it is possible a positive yield of controlled fusion energy will be achieved for the first time in human history. Planning of the machine is now in progress and it should be in operation in the early 1980s.

If a magnetic bottle which successfully traps the hot material in plasma form is designed, the achievement itself will raise additional problems. Having trapped and consumed the fuel, we will then have to inject additional material and perhaps remove unwanted impurities. The powerful magnetic field which prevents the escape of ions also presents a barrier to injected hot fuel. Among the ingenious ideas advanced to deal with this problem is production of beams of energetic neutral atoms, first produced as fast ions, then neutralised and injected. Within the 'bottle' the energetic neutral atoms are easily ionized again to become part of the hot plasma.

Another trying problem is the heat lost from the plasma by radiation. In fig. 67 the effect of this loss is shown diagrammatically. We see that when dealing with a fifty-fifty mixture of deuterium and tritium the reaction power exceeds radiation loss only when the temperature is higher than 30×10^6 K. The radiation loss appears as heat at the walls of the surrounding vessel.

In the table below are set out the approximate data for specific fusion experimental machines, all involving the toroidal pinch. Five sets of important parameters, measured between 1955 and 1976, are listed, including t and T_i together with the value of nt shown in the fourth column; in the fifth column the appropriate sustainment time is given, that is, the time for which the machine provides the necessary current, voltage, and so on, to maintain T_i provided failure of magnetic confinement does not occur. The values of t show that energy can escape in less than the desired time. The last row of figures gives a full set of values that represent the target of a self-sustaining reacting system. As examination of the data shows, in the most recent work with tokamak, nt is approximately 10^{13}, a factor of 10 less than the required value of 10^{14} cm^{-3} sec, and the temperature T_i is less than a factor of 10 below the stated required value of 10^8 K.

Finally the behaviour of the structural material of the reactor is a matter for concern. We can readily appreciate that if a controlled thermonuclear device with worthwhile output was in operation, the neutrons reaching the walls of the container could be sufficiently intense to cause extreme changes in the structure itself. There are no ready means available at the present time for properly simulating the enormous fluxes of neutrons that would be produced in a large-scale fusion reactor so the problems that might arise from the bombardment of conducting and insulating substances cannot be explored realistically.

Further Possibilities: Power Extraction, and Laser Heating. The fusion process, if proved successful, might be able to supply electricity more efficiently than fission or thermal generating plant. Since the fusion reactor

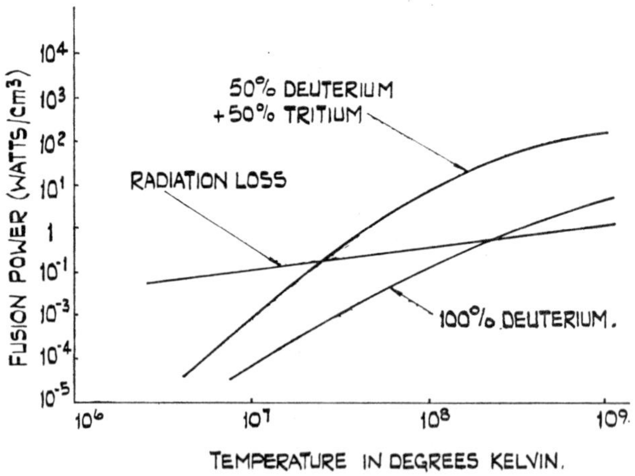

FIG. 67. Rate of energy production as a function of temperature, and for comparison rate of loss of energy by radiation

TABLE 17.1

PLASMA PARAMETERS ACHIEVED IN TOROIDAL PINCHES
(From R. S. Pease, *Physics Technol.* **8**, 1977)

Date	Energy containment time t sec	T_i Kelvin	nt cm^{-3} sec	Sustainment time sec
1955	10^{-5}	10^5	10^9	10^{-4}
1960	10^{-4}	10^6	10^{10}	3×10^{-3}
1965	2×10^{-3}	10^6	10^{11}	2×10^{-2}
1970	10^{-2}	5×10^6	5×10^{11}	10^{-1}
1976	10^{-1}	2×10^7	5×10^{12} (10^{13})	10^0
For Reactor	10^0	10^8	10^{14}	10

would probably consist of a kind of magnetic bottle in which sophisticated arrangements resulted in containing electric and magnetic fields, the plasma contained by means of these fields can be expected to react upon the fields themselves as the plasma expands and contracts, and from the machine the output energy might be obtained directly in the form of electrical power. Avoidance of the inefficient conversion from thermal to electrical energy would of course be a very attractive feature. No similar situation seems likely to emerge in dealing with fission reactors, for although fission energy appears mainly as particle kinetic energy which has its origin in electrical forces, it is extremely difficult to imagine a scheme in which the primary fission energy release is electrical.

The search persists for ingenious means of side-stepping known difficulties. The discovery of the laser opened up a possible additional avenue of research. Solid pellets of a suitable mixture of tritide and deuteride (compounds in which hydrogen has been replaced by tritium or deuterium) can be heated by intense pulses of laser light to very high temperature. The sudden heating of small amounts of solid which are fed in steady fashion into the laser beam avoids the containment problem. This kind of pulsed release of fusion energy is the subject of continuing research and is one of three or so possibly promising methods. It would, if successful, provide us with pulsed power which is a very different kind of power source from those to which we have become accustomed, but this in itself would not be regarded as a formidable problem. Practical means of using short-duration bursts of radiation are not easily conceived but an overall gain of energy in the machine would give great encouragement to those faced with the difficulties of harnessing the energy.

Conclusion. A few words should be said about the relative significance of fission and fusion in energy programmes. There are some who wish to delay the expansion of fission power programmes on the grounds that it is much more environmentally undesirable than fusion power. The fact remains that fusion power is still at a difficult, if exciting, research stage. It is conceivable that success in releasing it in a controlled fashion might elude the investigators and therefore the argument that we can delay the use of fission because of the possibility of introducing fusion is unrealistic. The stakes are very high, however, and continued investment in fusion research effort is wholly justified. We noted earlier (page 182) that if the process can be made self-sustaining and commercially viable with deuterium rather than tritium expansion could be practically unlimited, and, even if development turns out to be tied to the supply of lithium, a fusion programme could still provide massive amounts of energy. Work proceeds at an increasing rate and an up-to-date account is given in the *Status Report on Fusion of the International Fusion Research Council* (IFRC), supplied in the April 1977 findings. An account of the findings is also provided in *Atom* **260**, 1978, p. 150.

Part Four

THE PROVISION OF ENERGY

Chapter 18

Energy as Electricity: Turbines and Generators

The turbine is a variety of rotating machine which converts into mechanical energy part of the energy available as heat in a supply of steam or hot gas. The generator is also a rotating machine. It is coupled to the turbine in such a way that it receives via a shaft the mechanical energy of the turbine rotor, and it converts a very high proportion of it into electrical energy. In the turbine, a high speed jet of the steam or gas plays upon the blades of a rotor, causing it to turn at high speed against the drag of the generator. The rotor of the generator carries many coils of wire which are excited to provide magnetic fields. As the rotor turns, these strong magnetic fields produce electrical voltage and current in the stationary armature winding, and the output current is drawn from fixed terminals on the frame of the machine. In this way useful electrical energy is produced by the turbo-generator from some of the energy originally available in the steam.

A. Turbines

History of Turbines. The large modern turbine is one of the most important machines in the whole history of technology. Walter Hossli traces some of its history in *Scient. Am.*, **220**, no. 4, April 1969, p. 100. In the early 1880s Charles Parsons conceived the first steam turbine and soon began to demonstrate its superb performance in practical ways. His first major version appeared in 1884, and in 1891 he proved the value of condensing the exhaust steam to attain low pressure and increased efficiency. By 1897 he had built a turbine of 2,100 hp for the 44-ton vessel *Turbinia* which showed a speed of 34 knots, far exceeding that of any of the vessels of the British fleet assembled at Spithead. In 1904 the Admiralty launched the cruiser *Amethyst* with turbine engines, and in 1905 took the decision to order no new reciprocating steam engines. Indeed, the Dreadnought battleships were designed forthwith to use 23,000 hp turbines, and the Invincible class of cruisers were made to achieve 28 knots with 41,000 hp machines. By 1907 the Cunard ships *Lusitania* and *Mauretania* were equipped with Parsons turbines of 70,000 hp and these enabled the liners to run at speeds up to 25 knots. In a mere sixteen years, from the year 1891, when he added the condenser, to the year 1907, Parsons had increased the horsepower of his turbines from some hundreds to 70,000. The story of development of turbines for ship propulsion is one of extremely rapid growth followed by about a half century of relatively modest increase in output. The largest units, such as those for the liner *Queen Elizabeth II*, produced 110,000 hp while those for large modern oil tankers of between 250,000 and 500,000 tons deliver around 40,000 to 50,000 hp.

In the year that he introduced the condenser Parsons first coupled steam turbines to electric generators. Since then turbines have supplied the motive power for generators on an ever-increasing scale. In 1900 a turbine delivered 250 kW or 325 hp equivalent. By 1970 turbo-generators delivering 1,300 MW of electrical power were being built. But although the story of steam turbines and also gas turbines has been one of great progress, especially in the past two decades, A. Frankel, in the Parsons Memorial Lecture (*Trans. NE. Cst Instn Engrs Shipbldrs* **117**, no. 4, p. 799, 1970) concludes that there has been no really novel technological innovation since those of Sir Charles Parsons nearly one hundred years ago. In the same series of Memorial Lectures, N. C. Parsons (*Trans. NE. Cst Instn Engrs Shipbldrs*, **89**, 1972–73, p. 31) brings the story up to date with his analysis of the development of large wet-steam turbines suitable for use with water-moderated nuclear reactors such as those of the CANDU type. He concludes that massive wet-steam turbo-generator sets supplying 800 MW or more show the basic design philosophy still applies successfully.

Design is advancing so that it will be possible for a single machine to give an output of 2,000 MW. The incentive to produce machines of larger and larger output stems almost entirely from the desire to maximize the efficiency of generation of electrical power.

At the lower end of the power scale Diesel engines compare reasonably in performance with turbines but they are usually used to provide 30 MW or less. Gas turbines are limited to about 100 MW and they are generally less efficient than steam turbines. They are often used as stand-by units in large generating stations to enable the supplier to respond to sudden variations in load. In such gas turbines the combustion gas products take the place of the steam in steam turbines.

Practically every nuclear reactor supplies its thermal energy as electricity via turbo-generator equipment. This close connection between nuclear reactors and the turbines which convert their energy links the future growth of both closely together. As their names imply the Boiling Water Reactor and the Pressurised Water Reactor deliver the nuclear energy to the machine in the form of steam generated from the light water coolant. In the case of gas-cooled reactors, like Magnox and the Advanced Gas-cooled Reactor, the carbon dioxide coolant is used to generate steam in a heat exchanger arrangement.

The conversion of thermal to electrical energy is as yet an inescapable feature of the nuclear option. We know this conversion involves a substantial amount of energy wastage but the modern methods of conversion are very highly developed and the turbines and generators of modern electricity generating stations are extremely high performance machines. Intensive research and development work aimed at further improvement continues in order to minimize all sources of loss. The capital investment is high, so reliability of operation is crucial and any improvement is most rewarding. The future of coal as a fuel depends to an increasing degree on its use in generating stations and this provides an additional incentive to enhance the standard of operation.

Design of Modern Steam Turbines. It is not easy to visualize the contents of a steam turbine although the principle is simple to grasp. It consists of a shaft bearing a rotor and the shaft rests on bearings mounted within a cylindrical vessel. Round the inner wall of the vessel are apertures which are fashioned to direct incoming steam at rows of closely spaced blades mounted on the rotor. The steam, at elevated pressure, is supplied axially and flows against the blades of the rotor. Between each row of moving blades is a row of fixed blades which project inwards from the walls of the surrounding vessel angled in such a way that they direct the flow of steam at the best angle and velocity on to the blades of the rotor. The designer has to shape the blades as the aircraft designer would airfoils. Since the pressure of the steam drops as it passes down the turbine to the outlet ports, the length of the blades and the twist of the blades, from root to tip, are varied to suit it. At the inlet they are stubby with little twist. At the outlet they are much longer and twisted to a pronounced degree. The way the rotor blades change in length and twist is clear in the photograph due to Parsons (plate 4). The energy of the steam is converted, to a large extent, to the rotational energy of the turbine rotor. The turbine rotor is coupled to an electrical generator to provide as much as 1,200 MW on a single shaft.

A cross-section of a typical turbine is shown in fig. 68.

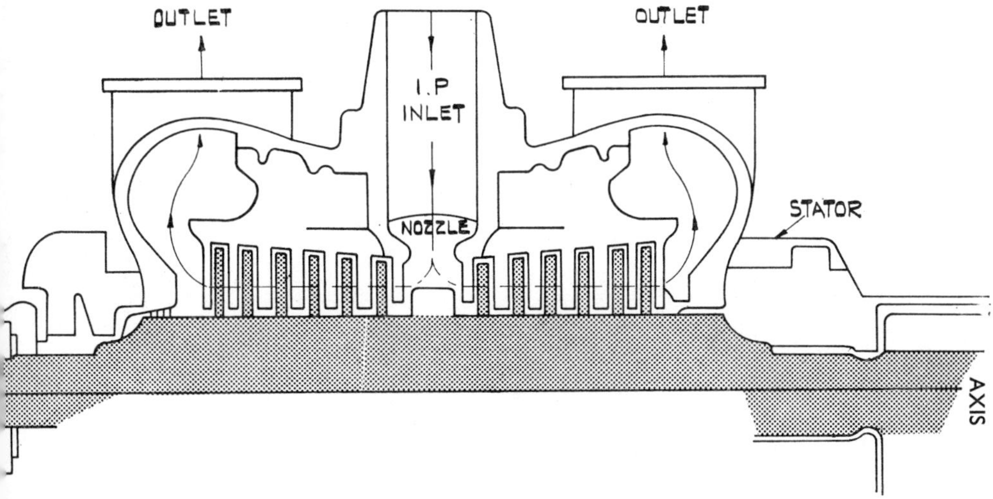

FIG. 68. Schematic cross-section of an intermediate pressure stage of a turbine. One half only is shown, as the blade arrangement is symmetric about the axis. The rotor blades move between the stator blades. The path of the steam is axial through the stage. The rotor shaft is usually continuous through low pressure and high pressure stages.

In a representative modern turbine set three turbines, designed to operate at high, intermediate, and low pressures respectively, are commonly used in sequence. The steam from the boiler is superheated and supplied to the high-pressure turbine at a pressure around 2,500 lb per square inch (almost 180 atmospheres or bars) and a temperature of approximately 1,000 °F (about 540 °C). In these conditions each pound of steam has a volume of 0·3 cu.ft. By the time it leaves the low-pressure turbine the same pound of steam has a volume exceeding 300 cu.ft, so the expansion is around a thousandfold. In velocity terms the input steam is supplied at about 150 ft/sec and it accelerates in the first stage. After successive accelerations, arising from the expansion, it can leave the low-pressure turbine at supersonic velocities as high as 1,000 ft/sec. The variation of steam temperature, pressure, and volume in the different stages of a typical steam plant are shown in fig. 69 due to W. Hossli.

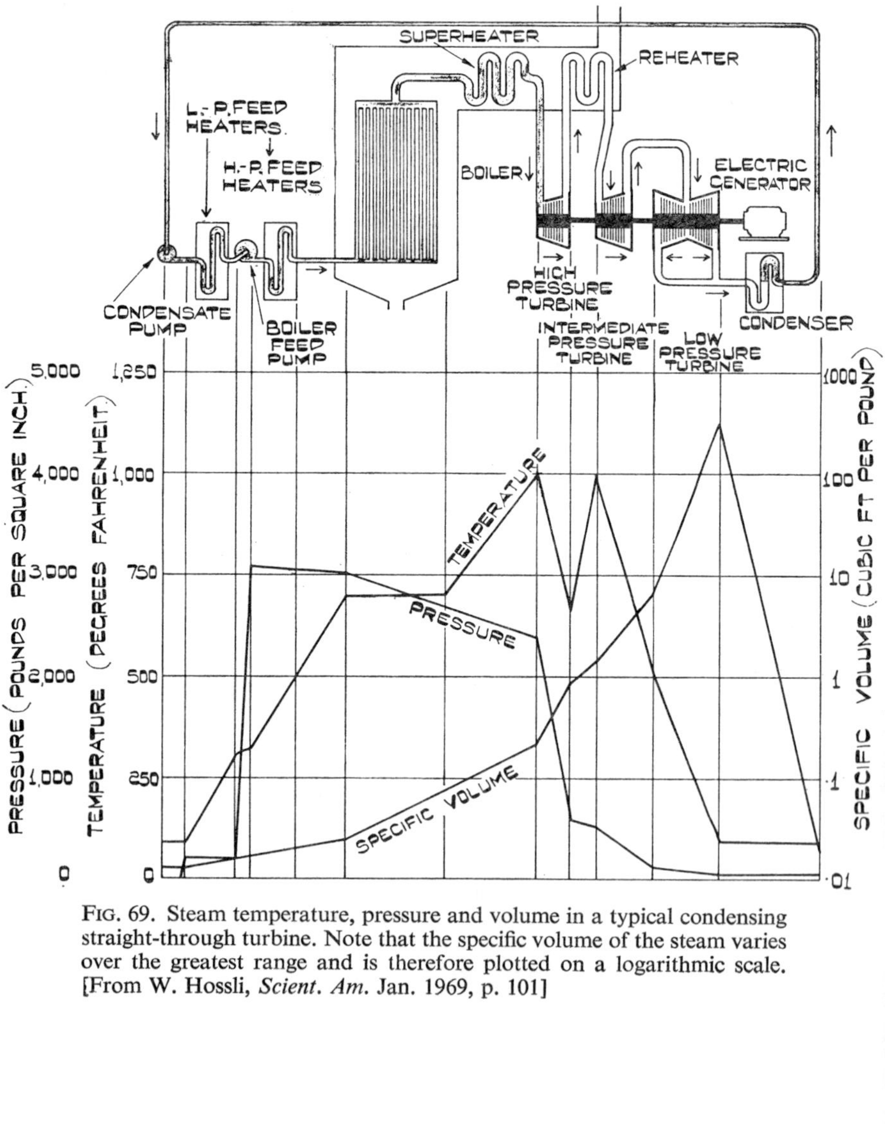

FIG. 69. Steam temperature, pressure and volume in a typical condensing straight-through turbine. Note that the specific volume of the steam varies over the greatest range and is therefore plotted on a logarithmic scale. [From W. Hossli, *Scient. Am.* Jan. 1969, p. 101]

Different classes of turbine are manufactured. When the entire machine, turbine and generator, is mounted on a single shaft it is known as a tandem compound turbine. When the turbine has two shafts, each with its own generator, it is said to be cross-compound.

Sometimes the turbine exhausts steam, after it has expanded within the machine, to the atmosphere or to a heating system or to some subsidiary equipment. The turbine is non-condensing in this instance and such machines are often used in industry where the exhausted steam can be usefully employed. Other turbines pass the exhaust steam to a condenser where the water in the tubes of the condenser absorbs the heat of the steam; the condensation of the steam that ensues results in the maintenance of a low pressure in the condenser and hence a high value of the expansion ratio of the steam. The higher the value of the expansion ratio of the turbine the higher the working efficiency of the turbine as a prime mover and the greater the yield of power. The expansion ratio is defined as the ratio of the volume of the fully expanded steam to its input volume. In electricity generation the turbines are almost universally of condensing type as the electrical power output is the prime requirement.

Some turbines are made so that a portion of the steam can be extracted on the way through the turbine itself. If no extraction is possible the machine is said to be a straight-through turbine. In other types of turbine steam may be extracted at certain stages of its passage and used to heat the water that provides input steam. Such turbines are described as bleeder types, and they are more efficient in principle than the straight-through. The steam extracted and used for heating the feed water retains the latent heat of steam and less exhaust steam is delivered to the condenser with a consequently reduced loss of latent heat to the water of the condenser. The machine can be more compact at the exhaust end. Generally bleeder turbines are used in electricity generation as the conversion efficiency can be raised by between 10 and 15 per cent as compared with straight-through. The fraction of the steam of the main stream which is taken to heat the feed-water is usually about 30 per cent. In bleeder turbines the pressure of the withdrawn steam varies with the load on the machine, but in some industrial applications of turbo-generators the steam removed by bleeding is required at constant pressure whatever the load. The automatic extraction turbines are designed to provide such steam in variable amounts from a number of extraction points. Extraction in this way involves complex governing equipment.

Within the turbine itself there are two main varieties of blading. In the first, or impulse type, the moving blades extract kinetic energy from the fast-moving steam and the amount of energy transferred is proportional to the change in absolute steam velocity. In the second, the reaction type, the pressure drop of the steam is equally divided between the fixed and moving blades, and in the fixed blades steam is accelerated to a velocity just greater than that of the moving blades. Further expansion in the moving blades provides thrust and the steam is given a relative velocity equal and opposite to its former absolute velocity. In this type the fixed blades and the moving blades are practically identical in design and function, and each accounts for roughly half the total pressure drop of one stage. In the impulse design the pressure drop occurs mainly in the fixed blades and little drop occurs in the moving blades. It appears as if impulse blading would offer a considerable advantage in needing fewer stages but the problem of sealing, to prevent steam leaking through the narrow spaces between the rotor and stator, complicates design. The fixed blades of the impulse type have to be larger than those of the reaction type to accommodate the larger pressure drop in the fixed blades. This means larger seals are required.

At the end of more than fifty years of experience of building and using large turbines neither impulse nor reaction machines can be claimed to be definitely superior in performance to the other. In practice compromise systems are often used.

For a substantial part of the twentieth century turbine development was empirical in character and extremely successful. In the last two decades advanced technological techniques have been increasingly applied to improvement of design. The sciences of aerodynamics, metallurgy, and thermodynamics have been the base on which research rested, and precision engineering has been essential to the advancement of the manufacturing skills. An example is the design of the long blades used in the low-pressure parts of a modern machine. These blades have the highest tip speed and the maximum vibration. The manner in which the steam flows through the low-pressure stages is complicated, with both radial and axial variations of pressure, and the velocity of the steam changes from subsonic to supersonic along the length of the blade. The blade has to be held in the rotor against centrifugal forces of 250 tons or more and knowledge of stress distribution is essential. Resonance can cause destructive vibration of the blades so the resonance frequencies must be known. Laser holography has been successfully applied in vibration tests. Obviously, however, the vibration cannot be completely eliminated by design, so the properties of the steel of the blades are most important. It should have high damping characteristics and at the same time it should lend itself to machining and forging. Moreover, the steam at this stage is near to condensation and the water droplets with their erosion possibilities involve the use of erosion-resistant steels or coatings of erosion-resistant alloys like Stellite. Since the blades should be able to work efficiently and reliably for at least twenty years the problems are formidable.

Turbines, and also generators, require careful cooling and an extensive literature is available on this specialised subject. Rotor cooling is essential and this is generally done by means of steam. For example, creep deformation (that is, permanent deformation caused by continuous stress) at the bore of an intermediate-pressure rotor has to be less than 0·2 per cent in 100,000 hours. The cooling steam is arranged to flow along the intermediate-pressure shaft and up the inlet face of each first stage disc, removing heat from both shaft and disc. Isothermal diagrams are studied to ensure that temperature changes are well controlled. Authors like G. R. Astin of AEI have provided detail on these aspects of design. It is interesting that the 'cooling' steam may be typically at 497 °C and flowing at 10,000 lb/hr when the main steam is supplied at 565 °C. In such a case the maximum bore temperature is held at 503 °C.

A range of other critically important features of the design is explored by Astin. They include the relationship of thermal stress to creep strain (and the provision of steam from the high-pressure stage to the intermediate-pressure stage to obtain the correct temperatures), the matter of critical speeds and the dependence of these on flexibility of bearings, the stiffness of the lubricating oil film, the rigidity of the foundations, and so on. The transient vibrations are considered and likewise the high torques arising from failures in the external load or in the electrical generator.

Additional important elements in turbines include the lubrication system and its associated emergency oil-pump arrangement. A governing system is required, as is also an automatic steam-sealing system which is designed to operate in accordance with load on the turbine. The design of the boiler feed pump may itself require a steam turbine of 13,000 hp to drive it; the incorporation of condensers, especially where they have combined functions, has repercussions on a number of other parts of the assembly. It is obvious that a marriage of scientific knowledge and experience is essential in designing such sophisticated machinery.

In large steam turbines, of output 500 MW or more, the room for mistakes is negligible. Stoppages are very expensive and the complexity of the machines can mean they are lengthy. For the same reason the design of the nuclear reactors producing the steam is of moment. A fast reactor using sodium as a coolant for the core must be free of any possibility of sodium coming into contact with water, for their chemical interaction can be violent and such as to cause extensive damage. P. R. Bolt has outlined some of the main considerations in this field (*J. Br. Nucl. Energy Soc.*, **15**, 1976, p. 147).

Efficiency, Cost, and Size. For large turbines with reheat systems better than 88 per cent overall efficiency can be achieved. As we have seen earlier, however, the Carnot theoretical study of heat engines showed the ultimate thermal efficiency is given by

$$\text{Efficiency} = \frac{T_1 - T_0}{T_1}$$

where T_1 and T_0 are the initial and final temperatures of the working fluid. For $T_1 = 1{,}000$ °F or 811 K and $T_0 = 100$ °F or 311 K the efficiency is

$$\frac{(811 - 311)}{811},$$

that is just over 60 per cent. The best conversion that can be achieved is 88 per cent of this thermal efficiency, approximately 53 per cent. Steam temperature problems seem to involve certain inevitable reductions in this value and just over 46 per cent would appear to be the more probable limiting value of the overall efficiency of the steam generating plant, with reactor or fossil-fuelled boiler, plus the turbine.

Turbines are usually designed to work at 3,600 revolutions per minute or at half speed, that is 1,800 rpm. These speeds are suitable for generating electricity at 60 Hz (the US frequency). In the case of generation at 50 Hz the corresponding speeds are 3,000 rpm and 1,500 rpm. The half-speed turbines are larger in weight and cost per megawatt than the full-speed machines. The aim for the future is to design full-speed turbines with still larger low-pressure last stages. Success in this would help to provide more electricity at less cost per unit.

Package sets are offered by a number of manufacturers and are proving acceptable, particularly those with output in the range 50 to 100 MW. These machines offer advantages such as ease of installation, ease of control, and adaptability in meeting the customer demand for power. In this last respect the gas turbine, as distinct from the huge 1,000 MW or larger nuclear turbo-generator, has an important role. Such gas turbines, driven by the combustion gases rather than steam, can deliver to the grid in ten to fifteen minutes from start-up.

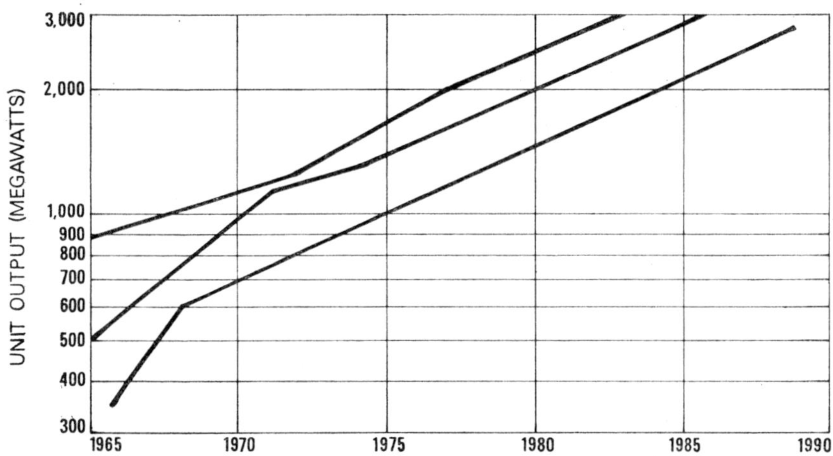

FIG. 70. Projection of turbine sizes made by Brown, Boveri, reflects the fact that US nuclear power plants (top curve) run larger than US fossil-fuelled plants (middle curve), and that both are larger than fossil-fuel plants contemplated in Britain (bottom curve). [From W. Hossli, *Scient. Am.* Jan. 1969]

It is clear that with the large capital costs involved in providing electricity in the 1,000 MW to 2,000 MW range, and in doing this with full regard to variations in demand, the cost of a delivered unit is affected by many factors. These include the slow start-up to full delivery of large turbo-generators and the associated thermal plant, nuclear or fossil-fuelled. Times of many hours are required and these base-load installations should not be expected to respond to rapid changes of load. The efficiency is obviously important and the use of waste heat involves questions of siting as does the supply of fuel. In the case of coal it is desirable to site the plant near the supply. In the case of nuclear plant, the site is generally away from large centres of population and transmission costs have to be considered as an important element in the final analysis. The performance of the electrical generator coupled to the turbine is clearly important to cost calculation. In view of this wide range of factors the large manufacturers are obliged to set up power systems departments whose main task is to optimize performance and obtain the minimum cost per kilowatt-hour supplied. On the whole, however, as size is increased the capital cost of the machine does not rise in proportion to the output in megawatts. Thus true economies of scale are obtained, and the figure for megawatts per ton gives a good measure of the improvement in performance. There has been a rapid increase in the size of machines and in some fifteen years the megawatts per ton have doubled so that the figure is close to 1 megawatt per ton for a machine of 1,300 MW output. A projection of turbine sizes has been made by A. G. Brown, Boveri & Cie of Baden, Switzerland (see fig. 70). The company anticipates that turbines built in the US will be bigger than those in the UK.

All in all, the great value of the various forms of the steam turbine, our prime mover and energy convertor, is such that the annual production in the 1960s and early 1970s reached a value corresponding to a combined output of more than 65 million hp or 50,000 MW. The large dependence on turbines indicated by these figures shows no signs of abating and increased utilization of nuclear energy to provide electricity will lead to a rapid increase.

Waste Heat, and Combined Heat and Power Schemes. As we saw above, though steam turbines are excellent energy conversion machines the maximum overall efficiency of conversion to electricity is around 40 per cent. It is sometimes contended that the remaining 60 per cent could be profitably utilized, but such an argument is open to doubt. Much of this thermal energy is of low grade: in the illustrative case cited on p. 199 the final temperature T was only 311 K, only about 15° to 20° above the probable surrounding temperature. It is indeed difficult to see how a large saving can be effected through utilization of the waste heat of low grade held in cooling water of condensers. The scheme most often advocated is Combined Heat and Power (usually known as CHP) where the heat and power are supplied within a single factory. Conversely, in situations where much high-pressure steam is required, for example in some chemical industries, electricity may be generated economically, almost as a by-product. Supplying the heat to a district heating system, such as that discussed on p. 177 in connection with the economics of nuclear power, can be useful, but it is costly in capital investment and it is restrictive to architects and planners concerned with good siting of commercial and housing schemes. We have noted that steam is bled to improve the thermal efficiencies of turbines, and that if heat is one of the desired products of the turbine the need to provide it can have adverse effect on the level of performance that the designer sets as his goal. The argument against supplying heat is more valid in the case of nuclear plants where safety is of paramount importance; on that account reactors are not placed very close to large centres of population.

For more detailed studies of this subject readers should refer to *Efficient Electricity Use*, ed. Craig B. Smith (Pergamon Press, 1976).

B. Generators

Fig. 71 (a) demonstrates the principle of electromagnetic induction on which the generator depends. A coil of wire (C) is rotated in a magnetic field (H) and a current drawn off through copper rings at A and B via brushes. The magnitude of the voltage varies with H, with the area of C, and the number of turns of wire in C. The frequency of alternation is directly proportional to the rate of rotation of C. Fig. 71 (b) shows schematically the principle on which modern alternators are constructed. The field-magnet frame carries an even number of coils and forms the rotating part of the machine (the rotor). The current is passed to the coils via slip-rings (not shown) on the axle. Positive and negative poles alternate round its circumference. The armature coils of the stator consist of one or more stout copper conductors wound alternately in opposite directions and finally connected to fixed terminals TT on the frame. Each coil is shown in the maximum magnetic field. Rotation puts a north pole to a south pole position so that fluxes change sign and this corresponds to a half cycle.

The rapid growth in size of steam turbines has had its counterpart in the equivalent growth of generators. The Parsons lecture of A. Frankel, to which reference has been made above, traces the similarity in rate of development. In the earlier 60 MW sets indirect cooling with hydrogen through ducts in the insulation wall was adopted; in later machines the coolant was taken through ducts, or the conductors themselves, to the inside of the main insulation. At

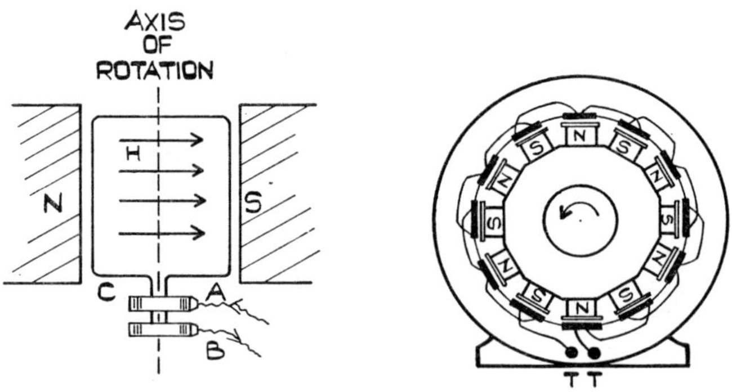

(a) (b)

Fig. 71. (a) Schematic diagram showing generation of alternating voltage. (b) Large alternators designed so that the current is collected from stationary windings.

200 MW the machine had hydrogen cooling of both stator and rotor. For machines of 660 MW, hydrogen cooling of the rotor was combined with water cooling of the stator. The pressure of the hydrogen increased from around 0·5 lbf/sq.in in 60 MW, to 60 lbf/sq.in in the 660 MW units. In the majority of sets for 50 MW or less direct air-cooling is usually employed.

In water-cooled generators the individual tubes in stator conductors are supplied with demineralised water flowing through them from one end of the stator to the other. The windings form part of a closed water circuit, which is arranged to operate at a water pressure less than the hydrogen pressure and thus prevent water leaking into the generator. The rotor windings, formed from hard-drawn silver-bearing copper, are frequently directly cooled with hydrogen which may be passed radially in the lower-speed machines (1,500 and 1,800 rpm) but which is axially supplied in the larger 3,000 and 3,600 rpm generators.

Because of the large size of the generating sets made for modern power stations, transport difficulties may arise between factory and station. The stator in these cases is usually made in two parts, the outer casing and an inner frame supporting the core and windings. This is evident in the photograph taken during assembling operations (plate 5). The stator casing is a rigid explosion-proof gas-tight housing. Balancing of completed rotors is essential.

The pattern of changes in the design of generators during the past twenty years is illustrated by the figures in table 18.1 (from A. Frankel, *Proc. I.E.E.*, **117**, 1970, p. 801). The rotor is the major factor limiting increases in both

TABLE 18.1

CHARACTERISTICS OF VARIOUS SIZES OF GENERATORS

Turbine-generator rating, MW	60	200	660
Generator rating, MVA	75	222	776
Power factor	0·8	0·9	0·85
Voltage, kV	11·8	16·5	23
Hydrogen pressure, lbf/sq.in	0·5	30	60
Flux density (gap), Wb/sq.m	0·8	0·9	0·95
Weight of heaviest component, ton	115	172	232
Total weight of generator, ton	141	245	475
Output relative to weight, MW/ton	0·425	0·82	1·39
Specific weight of copper, lb/kW	0·29	0·13	0·073

the magnetic field and electrical loading of the generator; and the latter two factors together determine the specific output coefficient, that is the output relative to weight (MW/ton). The design of the rotor becomes critical through limitation of space for accommodating windings, flux paths, and ventilation ducts within a cross-section curtailed by high mechanical stresses. Here the physical properties of large forgings are particularly important.

Future requirements include 2-pole generators in the range 1,000–1,300 MW and still larger 4-pole units for use with wet-steam nuclear sources. Outputs of 2,000 MW 2-pole and 2,500 MW 4-pole will be called for. Frankel has given an excellent survey of design features to be explored for such large outputs. He notes that the maximum ratings have roughly trebled in ten years during the two decades or so prior to 1970, and that the same rate might well continue to 1980 or even 1990.

The Royal Society discussion on recent advances in heavy electrical plant, organized by J. S. Forrest (*Phil. Trans. R. Soc.*, **A.275**, pp. 33–253, 1973), provided much information on the work done up to 1973, particularly on machines, transformers, switchgear, and system control. The papers on power transmission in the same volume complement those on machines. C. Concordia of General Electric Company, Schenectady, N.Y., predicts that ratings of 2-pole and 4-pole generators will continue to increase for many years. The 2-pole type, manufactured by his company, doubled in size every 7·6 years for more than thirty years, and the 4-pole doubled every 6·2 years for fifteen years. He observes that a machine of output 6,000 MW would provide only 2 per cent of the total generating capacity of 300,000 MW in the US. His remarks on alternatives to extensions of present technology show there are both benefits and disadvantages, for example in the use of superconductive field windings—that is, windings held at temperatures so low that the metals lose practically all their electrical resistance, a phenomenon known as superconductivity. The reduction of size and weight (possibly to about one third), the higher efficiency (about 0·4 per cent increase) and somewhat improved stability are among the benefits. Forrest stresses, however, that the loss of one extra day per year through faults cancels savings effected by an increase of 0·4 per cent in efficiency.

The accompanying paper (p. 51) by K. Abegg of A. G. Brown, Boveri and Cie, provides descriptions of the 772 MW–3,600 rpm Cumberland generators for the Tennessee Valley Authority and the 1,333 MW–1,800 rpm Cook generator for American Electric Power. Their consideration of the possibility of employing a superconducting rotor winding is particularly relevant. In the design proposed the rotor is a rotating Dewar vessel containing a field winding of Nb Ti (niobium titanium alloy) kept superconducting by liquid helium at 4 to 5 K. The stator winding is directly water cooled.

A separate treatment of superconducting machines (p. 77) is due to A. D. Appleton of International Research and Development Co. Ltd., Fossway, Newcastle-upon-Tyne. He gives an outline of the development of d.c. (direct current) machines and discusses their applications in, for example, marine propulsion. The author indicates the substantial advantages to be gained from successful development of superconducting a.c. (alternating current) generators and he considers the principal problem areas. The d.c. generators are feasible up to at least 200 MW. As regards a.c. generators, Appleton gives a curve of growth of ratings (see fig. 72) and considers the limit of conventional designs is not clear but might be at 2,000 MW. He stresses reliability near the

PLATE 5

One end of the core and windings of a Parsons 660-MW hydrogen/water-cooled generator stator. Each conductor in the stator winding is built up of hollow rectangular strips through which cooling water (demineralised) is circulated. The coolant flows through all conductors in parallel, from one end of the stator to the other, being fed from, and discharged to, manifolds which are connected to the conductors by the flexible tubes shown. The core and windings weigh some 300 tons, constituting the heaviest part of the whole turbine-generator set. The generator, which is driven by its turbine-line at 3.000 rev/min, supplies power at 23,500 volts. [Photograph from Parsons]

PLATE 6

Air-source heat pump. The unit on the right is designed for US conditions. The appropriate condenser size for Britain is on the left of the photograph, and is approximately twice the size of the condenser in the US unit. [Photograph from *Electron and Power*, Oct. 1976].

limit is the important question. He contends that for superconducting a.c. generation the limit is much higher than 2,000 MW, and is probably at least 5,000 MW with reliability an open question. He contends that there is a reasonable probability that because the design limit is higher they may be more reliable than conventional machines near their limit. He also thinks manufacturing costs can be less for superconducting machines and that they can be more efficient. The efficiency of large 1,000 MW machines is near 99 per cent so no appreciable percentage increase is possible. However, for a 2,000 MW machine a gain of 1 per cent in efficiency is an increase of 20 MW which is a great deal of power.

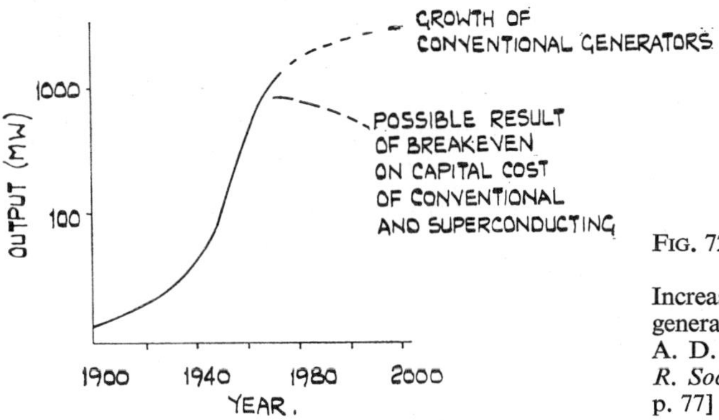

FIG. 72

Increase in growth of A.C. generator ratings [based on A. D. Appleton, *Phil. Trans. R. Soc.* London. A275, 1973, p. 77]

Chapter 19

Transmission, Distribution and Transportation

We have referred in preceding chapters to the difficulties and cost of transporting different forms of energy and emphasised that in assessing the value of any energy source it is advisable to take into account the means of providing the energy to the consumer.

Organic Fuel. C. Marchetti has stressed the important role that transport plays. Approaching the problem historically he writes convincingly of the severity of the problem of supplying wood to industries as they burgeoned in the twentieth century. Wood is extremely bulky and, as the adjacent forests were used rapidly, the wood had to be transported with difficulty and for ever longer distances at considerable extra cost. Coal is also heavy and bulky to transport. Oil and gas, however, are very much easier to deal with, and thus have a great advantage over other organic fuels. Oil can be piped into a tanker and afterwards piped to storage tanks on the shore. Thereafter it lends itself readily to easy transportation by pipe if the consumer is close to the shore-side depot, or by road or rail tanker. With natural gas the same techniques apply, with refrigeration being used when it is more economical to convey it as a liquid. To date liquefaction is not used in piping natural gas over long distances. It could be done if it proved essential but it is unlikely to be necessary since there are well-established techniques of piping the natural gas and synthetic gas over distances of hundreds of miles. In West Germany and in the US, particularly in California, extensive networks of pipes for transporting gas have been used for several decades. The network in Britain is well-developed and being expanded rapidly.

Apart from the cost in money terms of transporting energy, the cost in energy terms for transportation and distribution must not be allowed to rise to unacceptable levels. It has been estimated that in the US with its indigenous oil resources the energy cost of transporting oil is in excess of 25 per cent of the energy transported and that the fraction will soon exceed 30 per cent. Indeed it is with such figures in mind that experts tend to advocate gas as a most attractive form of energy. Once the investment in a network of large diameter pipes has been made the system proves flexible, requires little upkeep, and generally offers reliability and long life.

Electricity. In many instances the distribution of energy, even that derived from fossil fuels, is effected in the form of electricity. In spite of their transportability, gas and oil are used in large amounts in generating stations and the energy loss in conversion to electricity has been, until now, accepted in many instances. This is particularly true of coal, especially in countries such as the UK where coal of very good quality is no longer easily mined and much is of medium to poor quality. Such coal is nevertheless well suited for use in large generating stations, and in 1977 about 70 per cent of all coal mined in the UK was used in the electricity industry.

In addition, electricity is in itself an extremely convenient form of energy. In offices and homes, and in smaller industrial companies, the power can be switched on and off as required, and this in itself can be set against the loss of energy at the generating plant. The construction of grid systems in the advanced countries, and coupling between countries as, for example, in the United Kingdom–France undersea link, means that energy savings can be made. The high-voltage distribution network can give close to 100 per cent coverage of consumers. To illustrate the point fig. 73 shows the location of the power stations and the supergrid development for England and Wales in March 1977. With such a distribution network in operation there are few places where consumer demands cannot be met.

The output of the generators at a large power station is usually at about 20 kV. The generator supplies the energy to a large 3-phase transformer or a bank of three single-phase transformers. These large transformers step up the generator voltage from the 20 kV level to the desired level, which usually falls in the range 275 to 700 kV, for as voltage is increased energy losses in transmission become much less. In Britain the blocks of power are usually transmitted at 400 kV and this is reduced to 275, 132, 66, and finally to 11 kV. Not all these voltages are necessary in practice and as load densities in cities and certain urban districts increase, the need for intermediate voltage between 132 and 11 kV decreases: where loads of about 60 MW exist within areas of one or two miles radius it is more economical—in capital, land, and transformation losses—to provide bulk supply points at which the voltage is stepped down direct from 132 kV to 11 kV (see P. J. Freeman, *Electrical Power Transmission and Distribution*, Harrap, 1974). The final voltage of power supplied to the consumer is 240 volts in the UK, and 110 volts in the US.

The distribution systems normally radiate in several directions from each power station; the primary supplies from a group of stations are generally interconnected at grid stations to constitute the National Grid Network. From each grid station the feeders can be arranged to spread radially, or they may form a ring. The ring appears to offer some advantages—failure of one feeder need not interrupt the supply to any of the substations and the grid substation need not be at the hub of the feeders.

Transformers. A transformer is an apparatus for converting alternating voltages to different voltages of the same frequency. Basically it consists of an input coil and an output coil insulated from each other and both wound on a core of magnetic iron. The current in the input coil induces an alternating magnetic field in the iron, which in turn induces a current in the output coil. The voltage difference between the two currents depends on the number of turns in each coil. In all the primary transmission and distribution of electricity transformers play a major role. The design of large modern power

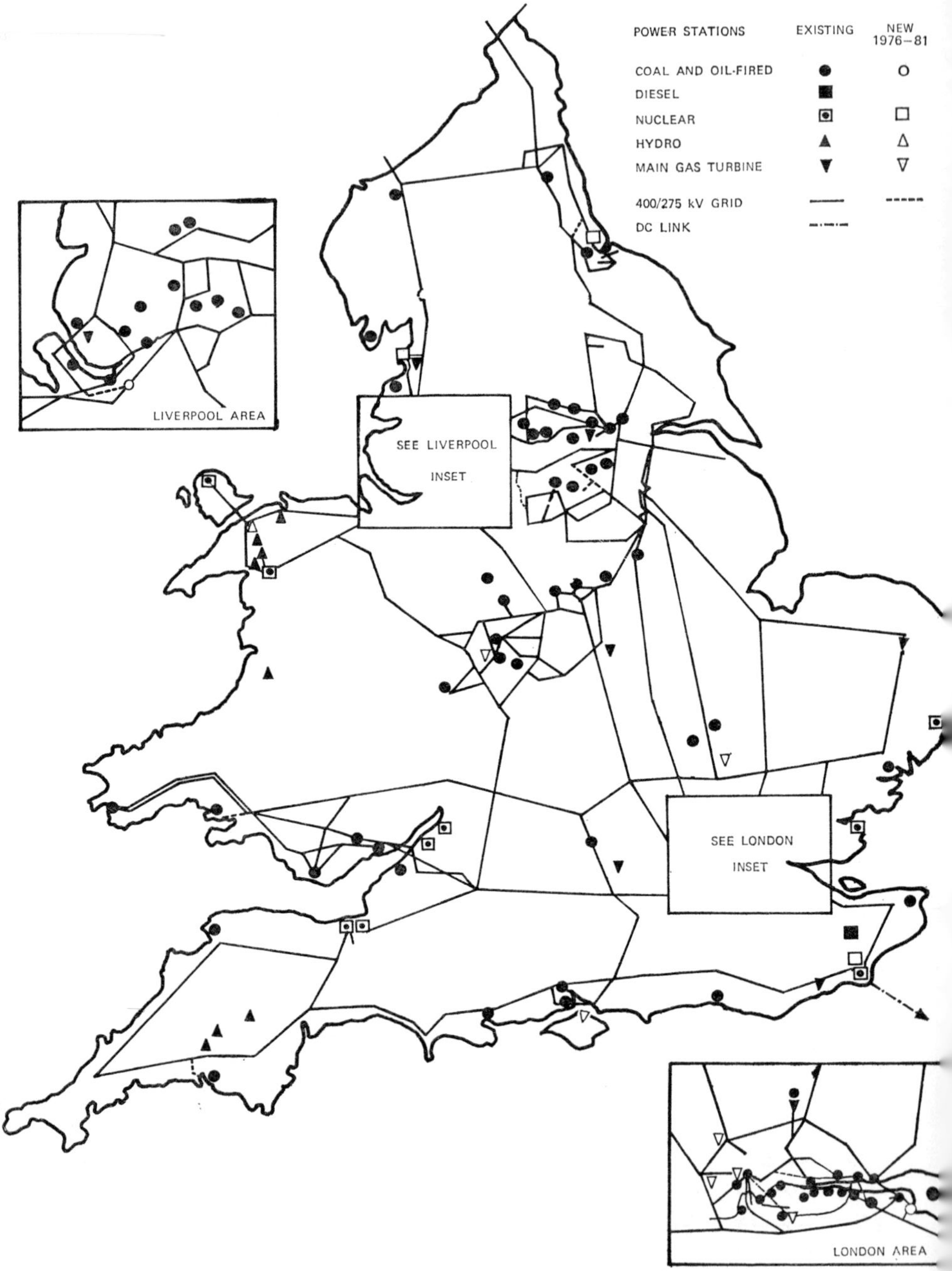

FIG. 73. Location of power stations and supergrid development. [From *CEGB Annual Report and Accounts* 1976/77]

transformers becomes increasingly complicated, and recent papers have outlined the way future developments are likely to go (see P. Bailey, A. B. Madin, and L. L. Preston, *Phil. Trans. R. Soc.* **A275**, 1973, p. 95). There is a substantial incentive to design so that materials operate successfully at higher levels of stress. The case for increasing the operating voltage suggests that it might reach the megavolt range. In turn, the problems of leakage flux losses, and of providing sufficient insulation strength to prevent short-circuiting, can be very severe at the high voltages and loads now in use. Another difficulty is weight: transformers at the station output points and at grid stations may weigh hundreds of tons. Large units of 260 tons have been transported in the UK, and a new transporter for 420 tons is available. A 3-phase unit of 420 tons weight would handle about 1,000 MVA (megavolt-amps or MW) at 400 kV.

The energy losses in power transformers are mainly resistance losses in the output winding calculated by the formula I^2r, where I is the output current and r is the winding resistance measured in ohms. Eddy current losses which arise in the iron are restricted by careful design of the core to less than 20 per cent of the ohmic losses. The power transformer gives an extremely high level of performance, as can be realized from the fact that a 67 MVA single-phase at 75 °C loses 0·22 per cent of its power and a 400 MVA single-phase loses 0·19 per cent at the same temperature.

Clearly with turbo-generators converting thermal energy into electricity with about 90 per cent efficiency (about 1 per cent of the loss being in the generator) and transformers performing at only 0·2 per cent loss, the thermal energy is fed into and out of the grid very successfully. The question of I^2r loss in the overhead and underground cables used in distributing the energy has also to be taken into account in examining the performance of a complete electrical power system, and efforts are made to keep total loss to a minimum.

Network Switching. As we mentioned, in the UK bulk transfers of power from generating stations to load-centre substations are done at 400 kV and 275 kV. The corresponding figures in US practice are 500 kV and 345 kV. The term 'supergrid' is sometimes applied to the network of such high-voltage lines and the lines are fed through transformers. The improvement in performance of switchgear and circuit-breakers for use in handling gigawatt (GW) amounts of power has more than kept pace with development in electrical power generating capacity. In recent publications (see for example *Phil. Trans. R. Soc.*, **A275**: C. Kessler, p.109 on switchgear and M. P. Reece, p. 121, and T. Kelsey and H. C. Petty, p. 131, on circuit-breakers) it is made clear that the reliable operation of these important elements of networks is assured for the likely period of growth of the electricity supply industry. Circuit-breakers have to carry the required current when closed and provide effective insulation against the operating voltage and overvoltages that can occur with the contacts closed or open. Additionally a reliable breaker has to be able to switch under conditions of both normal operation and fault. Currents of values as high as 60,000 amps can occur in fault situations and the temperature inside can reach 30,000 K. With proper design such a current can be broken in a time between 5 and 20 msec.

Different types of circuit-breakers are employed successfully; among them are oil breakers, air-blast breakers, vacuum breakers, and sulphur hexafluoride (SF_6) breakers. The last is especially well suited to high-voltage circuits and for use in metal-clad switchgear which is becoming more readily acceptable for environmental reasons. It would seem certain that voltages as high as 1,300 kV could be handled if required and research is proceeding vigorously in the range 1,000 to 1,500 kV.

Scale of Networks. Some interesting facts about distribution in Britain and North America are given by B. M. Weedy (*Electric Power Systems*, John Wiley & Sons, 1972). He expects that a simultaneous maximum demand of 70,000 MW will obtain in Britain by 1980. In 1969 the main transmission circuits consisted of the following lengths of overhead line in route-miles: 1,769 at 400 kV; 1,475 at 275 kV; 5,945 at 132 kV; and 58 at 66 kV. The corresponding lengths of underground cable were 6, 134, 765, and 183 at the respective voltages. The number of towers was 52,213 and substations numbered 884. For distribution systems the route-miles in 1969 were 12,346 at 33 kV; 73,634 at 11 kV; and 2,150 at 11 kV to 6·6 kV.

For North America the position is not so easily summarised as both private companies and government agencies are in operation. There are interconnections that enable a single massive synchronous network to be formed. The total system is easily the largest in the world and a single utility can equal most European countries. The long distances involved have brought higher voltages into use. Thus some companies use 765 kV lines, and before long voltages exceeding 1,000 kV may be adopted.

The circuit miles in 1968 were: 4,000 at 500 kV; 9,000 at 345 kV; 37,000 at 230 kV; 47,000 at 138 kV; 77,000 at 115 kV; 82,000 at 69 kV; and 136,000 below 69 kV. The consumer supply is 120 V single phase. In 1973 the peak demand was 330,000 MW, with a load factor of 63·1 per cent.

Clearly, in countries like the US and the UK, there is an enormous investment in the electrical industry, and until 1977 the rate of growth has continued to be steep. The same is true, though to a slightly less extent, in the Federal German Republic and France.

The importance of the various major elements in the electrical industry can be appreciated from the fact that, in the US, in terms of total plant investment, generation accounts for 40 per cent, distribution 40 per cent, transmission 17 per cent, and offices and the like 3 per cent. Underground transmission accounts for 10 per cent of the total investment in transmission (about 20 billion dollars). Although it is more costly than overhead transmission, its environmental advantages continue to be stressed in the US and most other countries. Corona loss—that is, the energy lost by ionization of the air surrounding overhead cables—is appreciable with high voltages and it may be that it adversely affects people living under and near power lines. This is the subject of research.

Overall Efficiency of Transmission and Distribution. It has been made clear that on the whole delivery to the customers is a reasonably efficient operation with losses depending on many factors, including the siting of the major generating stations relative to population centres. High voltage is helpful in reducing resistive losses in cables but, as we noted, it may in turn introduce greater corona loss. The average distance over which major blocks of power are conveyed is about one hundred miles and this results in resistive loss of between 2 and 3 per cent.

From the South of Scotland Electricity Board Annual Report for 1976–1977 it is possible to deduce a figure for the total transmission plus distribution efficiency. The units used up in delivery in 1977 amounted to 1,892 and a total of 19,416 was delivered to the system. The loss was therefore 9·74 per cent. The corresponding figure for 1968 was 9·52 per cent. The overall efficiency of delivery to the customer appeared therefore to be nearly constant at about 90·5 per cent. For more densely populated countries than Scotland the figure could be better but it is probably a fair average for most industrialized regions.

Effort has been spent in research on the use of superconducting cables but it appears unlikely on cost grounds that this technique will come into use for normal transmission purposes in the near future.

Though energy losses for electricity networks are low, initial costs are high and maintenance can prove expensive. Transmission and distribution in 1972 in the US accounted for around 45 per cent of the delivered cost of electricity or about \$2.22 US out of a total cost of \$4.89 per million btu. This is an unexpectedly large part of the cost and serves to underline the importance of the distribution element. By way of contrast transmission and distribution costs for natural gas averaged only \$0.47 US per million btu.

Transmission of Heat. As far as transportation and distribution are concerned, heat is one of the less tractable forms of energy. Losses waste energy, increase costs, and can damage the environment. Two schemes—one small-scale but practical, and the other large-scale but speculative—are described below.

Heat Pipes. This term is ambiguous since it is used to describe parts of a number of different arrangements, but here we use it to designate a device which utilizes the latent heat of vaporisation of a liquid. Every substance requires a certain amount of heat to raise a given mass of it through a given temperature difference. The amount required per unit mass, which depends on the substance in question and is known as its specific heat, is commonly expressed in joules/kg/K. When, however, a substance reaches its melting point or boiling point it requires heat to transform it to the liquid or gaseous state at the same temperature. The amount of heat required for the solid/liquid phase changes is known as the latent heat of fusion, and for the liquid/gas phase changes as the latent heat of vaporisation. The latent heats of substances are much greater than their specific heats; for example in the case of water the specific heat is $4{\cdot}2 \times 10^3$ J/kg/K while the latent heat of vaporisation of water is 540 times this value. One way of storing, or producing, heat is therefore to allow a substance to undergo a phase change. The principle is widely applied, for example the liquefication of gases in domestic refrigerators and the proposed liquefactions in the ocean thermal heat engine described below.

We know that metals, particularly copper and aluminium, are used extensively as conductors of electrical current, and usually the metals that are good conductors of electricity are good heat conductors. Thus copper or aluminium rods or tubes can transport heat to a site remote from a source of heat tolerably well. However, when they are used as pipes and designed to make use of a liquid-to-vapour phase change, the transport of the thermal energy is greatly enhanced. The efficiency of the transportation permits a considerable increase in the distance of separation of heat source and the heat sink (the place wherein the heat is to be disposed) and delivery at the colder end is readily achieved.

In fig. 74 we see that the arrangement consists of a double cylindrical structure with a wick or liquid-conducting material contained between the walls of the two cylinders. For a small heat pipe the outer diameter may be 1 cm or so. When one end of the pipe is heated the liquid absorbs from the source the latent heat of vaporisation. The vapour so produced passes freely along the length of the inner tube and when it reaches the sink (where the temperature is lower than that of the hot source) it delivers up its heat, and particularly its latent heat, becoming liquid in the process. The wick serves to transfer the liquid rapidly back to the hotter end of the pipe and the process proceeds continuously. With the liquid chosen to match the temperature range involved, the transport of the thermal energy can be done efficiently over a considerable distance.

Up to the present little large-scale use of the heat pipe seems to have been made, but it has proved valuable as a means of cooling parts of equipment subject to over-heating. It could probably be applied with advantage to improving the performance of turbines or heat-exchangers as it is well fitted to return any waste heat to its source and thereby improve the overall performance of a system. Any devices that assist, however marginally, in increasing the efficiency of large power plants are most worthwhile. Among

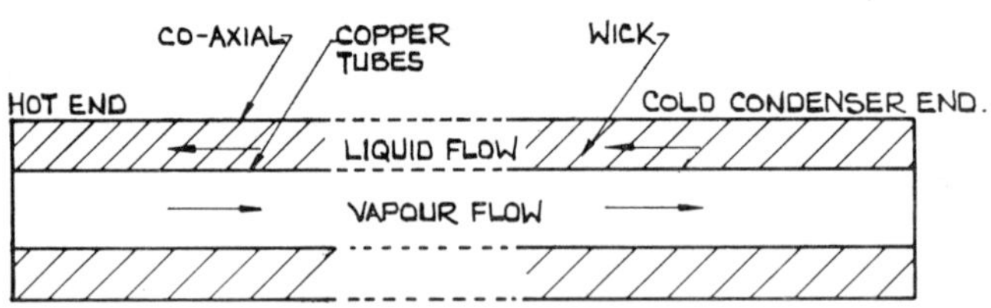

FIG. 74. Cross-section of heat pipe. Conduction of heat can be 400 times that of an equivalent solid copper rod. Pipes usually operate at a temperature within the range −40 °C to 350 °C, but this may be extended with careful choice of components.

the more speculative fields in which heat pipes might be beneficially involved are the direct conversion of solar radiation—the radiant energy could be focussed mainly at one end of a heat pipe which in turn would transfer the energy to the reservoir—and the conduction of geothermal heat. In all new applications it is clearly important to choose the liquid so that it offers suitable temperatures for vaporisation and condensation.

Elimination of Thermal Pollution. In a scheme designed to avoid thermal pollution Marchetti considers a very large energy source based on the successful development of ocean thermal gradient plants (illustrated in fig. 25). These plants use the gradient to run an ammonia engine which drives a turbine and so compresses a gas or air, thus transferring into an ordered form of energy ('negative entropy' or simply 'negentropy') in the gas the entropy of mixing waters at different temperatures. The thermal balance is zero. The compressed gas or liquid air is carried to the shore by tankers or pipes and distributed to cities by pipes. The heat absorbed in the release of the gas when it is used in apparatus and equipment of all kinds is replaced by the thermal energy, that is, the final form of the energy that has been deployed. At the consuming, as well as the manufacturing, end the thermal balance is zero so that in such a system there is no thermal pollution problem whatever. It is an imaginative concept and to illustrate it the schematic diagram of fig. 75 is provided.

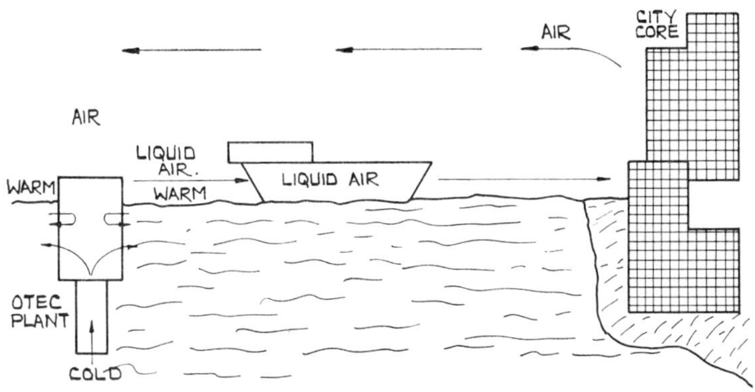

(a) LIQUID AIR CYCLE TO CARRY NEGENTROPY TO THE URBAN CORE.

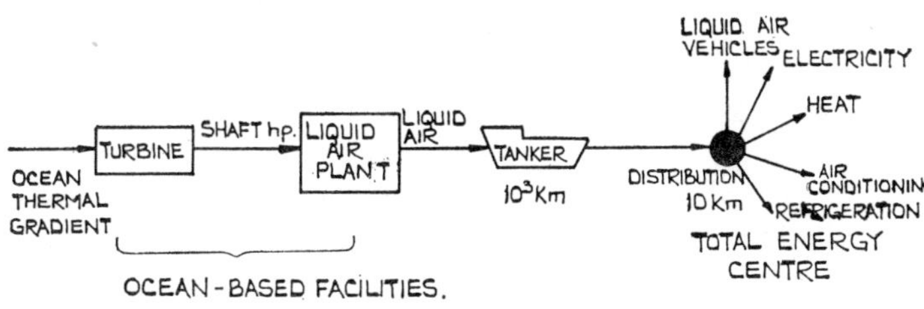

(b) NEGENTROPY SYSTEM FOR THE CORE CITY.

FIG. 75. Energy system without thermal pollution

Chapter 20

The Hydrogen Economy and Other Possible Future Trends

The Hydrogen Economy. Throughout this book we have continually stressed the losses occasioned when one form of energy is converted into another, and also problems of energy storage. We have also mentioned hydrogen manufacture as a possible solution to some of these difficulties. Hydrogen, like natural gas, whose distribution and storage properties we mentioned in the previous chapter, is easily and cheaply distributed, so in the sphere of distribution, too, it could be advantageous.

Water is, of course, the source from which hydrogen is easily and abundantly obtained. It can be directly dissociated into hydrogen and oxygen by electrolysis, or by very high temperatures. The electrolytic method makes good sense where electricity is the first product of the energy-releasing process, for example in the conversion of sunlight to electricity in photovoltaic cells. Up to 60 per cent of the thermal energy is lost, however, if the electricity is the product of a prior conversion—for example, in nuclear plants where nuclear energy is converted to heat, and heat to electricity by turbines and generators—so thermal dissociation is to be preferred if it can be achieved on a commercial scale.

Conversion of Thermal Energy to Hydrogen. Since the temperature for direct thermal dissociation is about 2,500 °C, well above that generated by nuclear reactors, many groups of scientists are investigating chains of chemical reactions which would bring about dissociation at much lower temperatures. Gianfrancode Beni and Cesare Marchetti have worked on this problem at the Euratom Research Centre at Ispra, paying particular attention to reaction chains which could be used at temperatures below 730 °C, well within the scope of such reactors as the High Temperature Reactor which can provide energy at temperatures of around 800 °C. Theoretical studies of a large number of cycles seem to support the view that an efficiency of 40 to 70 per cent might be achieved—and if an efficiency as high as 70 per cent could be reached the rewards in energy saving would be notable indeed. The difficulties, though, are formidable. The individual stages in a single cycle require detailed study and the work is time-consuming. Furthermore, although the higher the temperature of the source the more feasible the approach on chemical grounds, the less is its value from the practical and commercial point of view, for there are substantial difficulties in operating large reactors above 800 °C. By 1977 no one cycle had shown real promise. As an example of the kind of cycles that have been examined we give the vanadium one. Vanadium (A.N. 23) is a rare metal with chemical properties not unlike manganese or iron. During the cycle water and chlorine are first used to make hydrochloric acid and free oxygen (equation 1) and then the affinity of vanadium dichloride for chlorine is exploited to free the hydrogen from the acid (equation 2). During the last two processes (equations 3 and 4) the vanadium trichloride remaining after the hydrogen is released is converted to dichloride and chlorine, which are thus made available for re-use in the chain of chemical reactions.

(1)	$Cl_2+H_2O \rightarrow 2HCl+\frac{1}{2}O_2$	700 or 800 °C
(2)	$2HCl+2VCl_2 \rightarrow 2VCl_3+H_2$	100 °C
(3)	$4VCl_3 \rightarrow 2VCl_2+2VCl_4$	700 °C
(4)	$2VCl_4 \rightarrow 2VCl_2+Cl_2$	100 °C

$$\underbrace{H_2O+Cl_2+2VCl_2}_{\text{(INPUT)}} \rightarrow \underbrace{H_2+\tfrac{1}{2}O_2+Cl_2+2VCl_2}_{\text{(OUTPUT)}}.$$

This cycle has also been investigated by the Allison Division of General Motors Corporation (see C. Marchetti in *Chem. Econ. Eng. Rev.* **5**, 1973, p. 7).

In principle, then, the dissociation of water by thermal energy is extremely attractive but, as said, it is by no means solved. The concept of an hydrogen economy is well illustrated by a diagram due to Marchetti and shown in fig. 76.

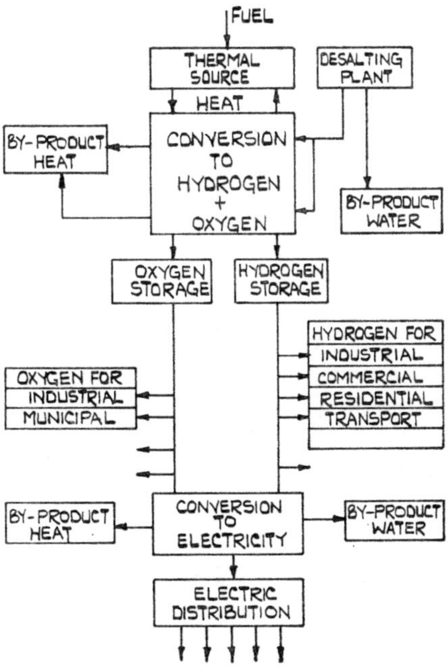

FIG. 76. An hydrogen economy as seen by Marchetti

Conversion of Electrical Energy to Hydrogen. Even though electrolysis can be wasteful of energy, it does appear to be used less often than might be justified. In 1977 commercial electrolysers operated at efficiencies of between 60 per cent and 70 per cent but special high-pressure types were expected to reach about 85 per cent efficiency since in this type the operating voltage is higher. In theory a maximum electrical efficiency of about 120 per cent can be attained, the extra yield being due to the absorption of heat from the surroundings and the effective conversion of this heat in the electrolytic process. It would thus be entirely reasonable to assume that 100 per cent efficiency in energy terms could be achieved in a large-scale commercial plant designed to convert electrical energy to hydrogen fuel. If this were to happen, and if the efficiency of generation of electrical power were raised from its present average of some 35 per cent to 40 per cent, the overall system could be attractive and competitive in many applications. Nevertheless an extremely large increase in electrical power generation would be required if enough hydrogen was to be produced by electrolysis to replace natural gas. With the present 35 per cent efficiency the amount of electrical generating plant necessary to produce hydrogen equivalent to US natural gas annual consumption of 1968 would be 10^6 MW, that is 1 TW. This total of 1 TW has to be compared with the total US generating capacity in 1968 of about 0·3 TW. Further discussion of electrolysis is given in chapter 21 where possible trends are explored.

Properties of Hydrogen as a Fuel. The advantages of hydrogen as a fuel form are considerable. Its primary source, sea water, is virtually inexhaustible, and its gradual substitution for natural gas would be an obvious way out of the great difficulties that depletion of natural gas will cause.

When hydrogen is burned in air the only possible pollutants are oxides of nitrogen derived from the nitrogen in the atmosphere but their concentrations are generally lower than those produced by burning other fuels. If hydrogen is burned in oxygen the product is water.

The ignition energy of hydrogen is only 0·02 mJ, less than 7 per cent that of natural gas, and it can therefore be used in low-temperature catalytic burners that also produce no pollutants.

The energy content of hydrogen gas is 325 btu per standard cubic foot (1 btu/scf $= 2{\cdot}98 \times 10^5$ J/cu.m) and this is less than one third that of natural gas ($\sim$1,000 btu/scf). Against that, the low viscosity of hydrogen gas allows a threefold increase in the flow capacity of a pipeline and a given pipe therefore carries nearly equal amounts of energy as natural gas or hydrogen. The hydrogen does require greater pumping capacity.

In terms of energy per liquid unit mass hydrogen is impressive being 2·75 times greater than hydrocarbon fuels. This means that it is very suitable for providing the power for aircraft or rockets. Equal masses of liquid hydrogen and hydrocarbon have very different volumes since the specific gravity of liquid hydrogen is only 0·07.

Thomas Maugh (*Science*, **178**, 1972, p. 849) has explored some of the interesting features of hydrogen as a fuel. He assesses the risks of hydrogen in comparison with natural gas and suggests that the less favourable properties like the higher diffusivity of hydrogen (in practical terms a propensity to leak faster) are counterbalanced by other features of its behaviour as a gas. Marchetti has reached similar conclusions.

While hydrogen can be used in hydrogen fuel cells, discussed more fully in chapter 21, and used in this manner to power automobiles, it can also be simply burned in an internal combustion engine. In such engines it compares reasonably well with petrol as regards torque power and efficiency. Some work has been done on the conversion of automobile engines to run on hydrogen and it has been found that the efficiency may be 50 per cent higher than a petrol engine. The pollution is much reduced.

Storage. The storage of hydrogen in large amounts is one of the more serious difficulties. At present hydrogen is usually stored in cryogenic tanks, that is tanks refrigerated to such low temperatures that the hydrogen is maintained as a liquid. This is a questionable practice on environmental and safety grounds. Nevertheless the present practicability of storage of liquid hydrogen at large power plants is a major advantage even if the problem of small-scale storage should not be satisfactorily solved in the near future. With the possibility of storing hydrogen against peak demands much less surplus generating capacity at electrical generating stations would need to be provided, with economic advantages following. The principle here is the same as that which makes pumped storage attractive. A minor advantage is that such hydrogen can be used as a coolant, for instance in the case of the coils of large generators.

More research work on storage could be done with advantage. One of the most attractive methods under investigation is the storage of hydrogen as one of the metal hydrides (chemical compounds containing only hydrogen and a metal). Gaseous hydrogen penetrates readily the lattice structure of many solid metals and alloys and, for certain metals like titanium, can reach a concentration per unit volume exceeding that of liquid hydrogen. The hydrides are formed by exposing the metal to pressurised hydrogen gas and, as the process is exothermic, it is readily reversed by applying heat. Since waste heat is usually available in any process in which the hydrogen is used as a fuel, the waste heat can be used for the release process. The presence of an oxide layer on the surface of the titanium can complicate the release of the gas from the metal. In the absence of such a complication the metal can, in theory, be used time and time again.

Intermetallic compounds of hydrogen, in which the hydrogen is loosely held within the structure, can be produced with the lanthanides or rare earth elements (elements with atomic numbers from 57 to 71) and nickel or cobalt. The gas can also be used to combine in hydride form with vanadium, niobium, and their alloys. The metals are expensive, and more extensive investigations must be undertaken before an improvement on the usual form of storage in liquid form can be claimed.

Chemical Uses of Hydrogen in Industry. Hydrogen already has many industrial uses as a chemical, rather than as a fuel, and if it were in abundant supply these uses could be extended. At present it is used principally in the synthesis of ammonia (NH_3) for fertilizer manufacture, accounting for 40 per cent of the present consumption, and in the direct reduction of iron ore.

Hydrogen might also be employed in coal gasification (see p. 119). About 1,500 scf of hydrogen is needed to produce 1,000 scf of methane, CH_4. The coal itself can produce the required hydrogen but clearly if hydrogen can be made available from another source the saving in coal could be considerable.

In addition hydrogen could be used with advantage in the petroleum industry, for example in removing sulphur which occurs in petroleum as an impurity. Its role in the shale oil industry could prove to be still more important, particularly in extending the range of hydrocarbons that could be manufactured from the crude oil obtained from shale deposits.

Hydrogen as a Source for Energy Carriers: Ammonia and Methanol. Already widely used in the manufacture of fertilizers, ammonia can also be used as a fuel. Marchetti in particular has written on this subject and considered both it and methanol as possible secondary energy carriers. Ammonia is easier to liquefy than hydrogen (and therefore easier to store and transport). It also burns easily but its true properties and smell seem to rule it out as a material for use by small consumers. On the other hand methanol can be transported and distributed and used in much the same way as oil and thus has some advantages over hydrogen. In cars it serves well when minor modifications are made to the carburettor and engine. It can be easily transformed to high quality gasoline by heating in the presence of a catalyst (a substance which speeds up a chemical action yet itself emerges unchanged). For example, a hydrated silicate of calcium and aluminium called Zerlite is used in a process developed by Mobil.

Marchetti takes the limitations on our coal supply to argue for the manufacture of a liquid fuel which contains a maximum of 'added energy' per carbon atom. This phrase refers to the amount by which the heat of combustion of the new molecule exceeds that of an atom of carbon. On this basis, methanol is the favourite with 82 per cent more energy per carbon atom than the coal used in the synthesis. Another simple hydrocarbon, ethanol (C_2H_5OH) 'adds' 74 per cent, and petrol (which is a mixture of more complicated hydrocarbons) 77 per cent.

Other factors in favour of methanol include its value as liquid fuel for fuel cells (which could open up a promising route to the fuel-electric car); the ease with which it can be 'cracked', the technical term for decomposition by heat, to provide gas with a low thermal value suitable for domestic use; and its suitability as feedstock in manufacturing plastics.

Assuming hydrogen as the starting-point, carbon and oxygen are required in making methanol by either of two reactions:

$$CO + 2H_2 \rightarrow CH_3OH - 21{\cdot}7 \text{ kcal} \quad (1)$$

$$CO_2 + 3H_2 \rightarrow CH_3OH + H_2O - 11{\cdot}8 \text{ kcal} \quad (2)$$

The first of these reactions can be carried out with coal or hydrocarbons providing the required carbon. The use of carbon dioxide in (2) above is conceptually more attractive as it could be obtained from the air or from sea water which contains small amounts in solution.

The carbon monoxide required in making methanol, by the method of reaction (1) above, must be pure and it is difficult to produce pure CO from coal. For this reason and others the method proposed is the Molten Iron process under development by Humboldt Wedag Industries.

The key operation consists in dissolving coal in molten iron covered by a thick layer of slag. The impurities in the coal move into the slag and the carbon stays with the iron. By blowing oxygen through the melt, carbon monoxide can be created. The reaction is exothermic and the heat of reaction of 30 kcal/mole maintains the melt at a temperature of about 1,400 °C. The heat can also be applied in various other processes including power generation. The carbon monoxide produced has only about 50 parts per million of impurities and it can be further purified rather easily.

It is thought a plant could achieve a throughput of about 100,000 tons of coal per year per square metre, so one handling a total of 100 million tons of coal per year could conceivably prove practicable. The operation would have similarities to modern steel-making procedures. Even the slag could prove valuable, particularly for making cement.

Further modifications to the scheme are mooted by C. Marchetti, including the ideas of, firstly, using nuclear reactors to provide hot water heated to around 600 °C, and, secondly, of utilizing the 'spare' heat of the Molten Iron process (carried in the carbon monoxide) to heat the water to still higher temperatures at which water dissociation yielding hydrogen would be possible.

Marchetti's scheme is proposed in the context of international centres at which huge amounts of power would be produced by nuclear reactors. It is at the same time not essential to consider it in such a context. It is an imaginative way of making methanol as a substitute for oil products by employing power from reactors. The constraints on the contribution that fission can make to the total energy requirement would be then greatly reduced for, as we keep stressing, although electricity is a clean and versatile energy form it is not able to satisfy some important existing needs (e.g., fuel for motor vehicles), and storage of electrical energy in massive amounts is an intractable problem. Electricity and methanol would together form a flexible combination.

The International Scheme (Plutonium Islands). Some members of the IIASA team use the methods of systems analysis to evaluate energy schemes, and, on a global view, attach great importance to the choice of locations, and to managerial schemes for utilizing depositories of irradiated fuel from nuclear stations. (See W. Häfele, *Energy Strategies*, Third General Conference of the European Physical Society, Bucharest, 1975; C. Marchetti, 'Primary Energy Substitution Models', internal paper of the IIASA; also

IIASA Research Report 75–38, Nov. 1975.) They consider an island controlled by an international body to be a suitable depository. The island would be the site of a reprocessing centre to procure plutonium from irradiated fuel and to convert waste to durable solid form. The plutonium would, in due time, be used on the island to operate reactors with plutonium-enriched cores. Some of the electricity could be used on the island as seemed best (e.g., aluminium production) and, to enable the energy to be exported, part of the total electrical energy would be transformed into a fuel form such as methanol. The total yearly oil from the Persian Gulf area amounts to the equivalent of 1,700 GW (thermal). The 1977 world total energy consumption was 7,500 GW (thermal). Of this total, oil accounts for 3,500 GW and gas for 2,000 GW. At the present rate of progress it is expected that nuclear energy will provide about 160 GW (electrical) by 1980 which corresponds to about three times as much thermal energy, say 500 GW (thermal). These figures give a clearer picture of the practicability of having a nuclear island with a methanol output comparable with the oil production of the Persian Gulf, say 2,000 GW (thermal) per annum. Wolf Häfele and Cesare Marchetti regard such an island as representing a man-made oil-field based on intelligent use of plutonium.

The island could be chosen suitably to enable the waste heat to be dissipated without environmental hazards. Canton Island in the Pacific and Ascension Island in the Atlantic are examples.

The scheme can be extended to include the analysis of various nuclear fuel cycles capable of satisfying and sustaining projected world electricity and synthetic fuel demands in the light of the limited world uranium endowment which we discussed in chapter 5. We can visualize fast breeder reactors on island sites producing excess fissile material as plutonium or as U^{233}. The regional reactors would be supplied with denatured nuclear fuel in the form of U^{233}/U^{238} and no weapon grade material would pass from the islands to the regional sites. The islands would be recognised centres for storage, reprocessing, and deployment of plutonium.

The scheme is an imaginative and ingenious answer to the problems posed by the required large world-wide expansion of nuclear power generation. It hinges on success in industrial and political collaboration on an international scale. Most of the organisations involved in energy supply have to consider day-to-day operation, the market response to their prices, and other mundane facts. Clearly the gap between their situation and a long-range scheme is very substantial.

Electrolysis. C. Marchetti explores the evolution, the present state, and the possible trends of the technology in view of the critical importance of electrolysis to the nuclear island concept (private communication). Independently of the island idea it seems certain that electricity conversion will be a crucial element in the future as electricity supply becomes a much more substantial part of the total energy supply.

Pure water is practically non-conducting, therefore ways have to be found

Plate 7

High stack flue gases emerging from the 200-metre high Fawley power station stack on the south bank of Southampton Water. [Photograph from CEGB]

PLATE 8

Model of vitrified radio-active waste. If all the electric power that one man used in his lifetime were generated by nuclear power alone, the long-lived radio-active wastes so produced could all be incorporated in a piece of glass this size. [Photograph from *UKAEA Annual Report & Accounts 1976/77*.]

of increasing its conductivity. The most practical ways of making water into a useful electrolyte, or conductor of electricity, is to add some suitable acid or soluble salt to the water.

Monopolar Cells. Water electrolysis was the subject of a patent granted to Dimitri Latchinoff in 1888. The monopolar cell, using the same material for positive and negative electrodes, was discovered by Pompeo Garuti and was the first industrial device; it was successfully developed with iron electrodes in caustic soda (NaOH) though it first took the form of lead electrodes in sulphuric acid. Very recently cells based on solid state electrolytes of acidic type have emerged. Such cells have been developed in Switzerland, and filter press electrolysers introduced in the 1950s and now commercialised by the Lurgi Company are possibly the best. The largest unit of Lurgi has a rating of 3 MW though General Electric (sponsored by Brookhaven National Laboratory) are involved in design for a 75 MW unit. The G.E. design indicates a definite trend towards increased power density expressed as kW/litre. The present achievement is 2 kW/litre; Marchetti suggests a target of 10–20 kW/litre.

The research effort on high temperature solid electrolyte cells operating between 1,000 °C and 1,500 °C could give valuable results. Electrolytes of doped zirconium oxide (ZrO_2), providing oxygen ions of high mobility, are among the most promising. The high temperature in itself is helpful in reducing the energy required for direct dissociation which occurs at 2,500 °C and above. The amount of energy which has to be supplied for dissociation is effectively reduced by operating the cell at an elevated temperature.

The Bi-polar Solid-Electrolyte Electrolyser. Marchetti has emphasised the trouble that arises when the ions move over long paths. Large currents 'compress' ions at the electrodes and in this process there is energy loss which is evident in the need for 'overvoltage' to drive the unit. This overvoltage is proportional to the logarithm of the current with the proportionality coefficient dependent on the properties of the electrodes. The noble metals—silver, gold, and platinum—generally perform well as electrodes on account of their catalytic activity.

Marchetti finds, however, encouraging evidence for the successful development of large-scale electrolysis. In new developments which aim at high current density the electrolyte is made solid and the ion is attached to the structure of a resin in such a way that water molecules can move into it. Ions do not move but their charges provide high ion concentration and hence good conductivity. In the General Electric scheme the resin which plays a most important role is sulphonated teflon with the SO_3^- ions constituting the electrolyte. Thin films of this plastic, one mm thick, have the electrodes pasted on the surfaces. The water is replenished by keeping the O_2 electrode in pure water. The method is described as Solid Polymer Electrolyte (SPE). There would appear to be scope for substantial advances.

Marchetti sums up the position in a table reproduced here (Table 20.1). He concludes that present-day electrolysers are unsuitable for very large-scale work being bulky ($<0{\cdot}2$ MW/m^3), costly, and somewhat inefficient at 70 to 75 per cent for energy efficiency and 60 to 65 per cent for thermodynamic efficiency, but that the data of the table show they have evolutionary potential which could be secured by increasing research and development investment.

The table shows the SPE type of the General Electric Company to be particularly promising. Among the advantages it appears to offer are:

1. Operation with high differential pressures, above 70 atmospheres, between the O_2 and H_2 compartments so H_2 at high pressure is delivered.
2. Corrosion freedom of electrolytes and operation with pure water.
3. Low power consumption per unit gas produced with the possibility of 100 per cent energy efficiency.
4. Possibility of high power density corresponding to compactness and reduced capital and operating costs.

We have discussed electrolysis in some detail because it is so closely linked to the future of nuclear energy. It is very unlikely that any form of energy other than thermal or electric will be available from nuclear reactor stations. The conversion of electricity to hydrogen and the production of methanol from the hydrogen is one possible large-scale route by which to achieve the goal of substituting a liquid fuel for oil. The sketch in fig. 77 outlines the scheme which could remove the constraints on the electricity industry, and so make it more versatile in respect of storage and transport and more attractive to consumers.

It will be seen in the figure that the part of the electricity output that is fed to the methanol plant has to be rectified (converted from a.c. to d.c.). A supply of about 1,000 volts d.c. could be provided by rectifying the a.c. output of the usual turbo-generator, and such a power conditioner can be highly efficient. Alternatively there may be in the future new machines, such as a homopolar generator or even a superconducting d.c. generator, which would be turbine driven and their d.c. output would be fed directly into the electrolyser.

Marchetti's Analysis of Transportation and Storage. Returning to less speculative matters, Marchetti has analysed problems of transportation

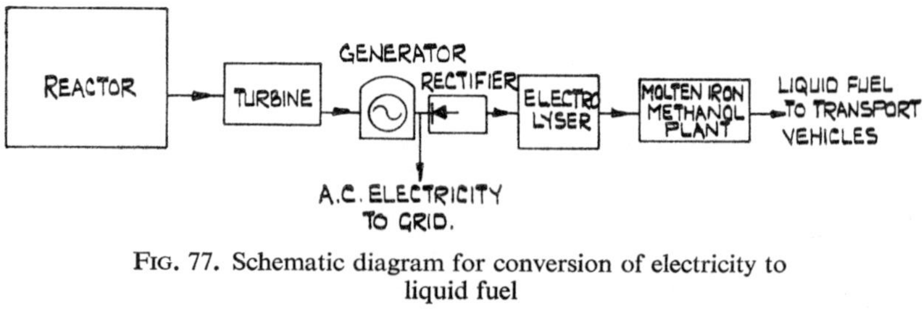

FIG. 77. Schematic diagram for conversion of electricity to liquid fuel

TABLE 20.1

COMPARISON OF ELECTRICITY-TO-HYDROGEN EQUIPMENTS (1985 PROJECTIONS)
(From Report of Meeting at IIASA, Laxenburg, 1st Oct. 1977; paper of C. Marchetti entitled 'On Electrolysis')

Electrolyser type	Cell efficiency	Unit efficiency	Total efficiency	H_2 production rate in Nm^3/day per GWe input	Electrolyser specific cost in \$/kWH_2 produced	Compressor cost in \$/kW H_2 produced (e)
Bi-polar alkaline electrolyte (a)	90	89·4	87·2	$5{\cdot}9\times10^6$	118	19
Uni-polar alkaline electrolyte (b)	82	81·9	78·6	$5{\cdot}3\times10^6$	100	33
Bi-polar alkaline electrolyte (c)	85	84	84	$5{\cdot}6\times10^6$	180	8
Bi-polar solid polymer electrolyte (d)	86·45	86·3	86·3	$5{\cdot}8\times10^6$	30	0

Based on information from (a) Teledyne Energy Systems, (b) The Electrolyzer Corp. Ltd., (c) Lurgi 1973 costs, (d) General Electric Company, (e) Ingersoll-Rand Company.

and storage and suggested possible future developments. He notes that the electrical industry, in the near-complete absence of storage (excepting pumped storage), has an apparent load factor (the average amount of the total generating capacity required) on expensive equipment of between 0·4 and 0·6 for most countries. He emphasises the disadvantages of storing coal in heaps, and oil in tanks, and the comparatively favourable position of natural gas, where large underground spaces can be used at low cost and where piping with pressure changes, and aquifers (underground water-bearing strata) can be turned to advantage for storage over shorter periods. These advantages of a gas system reinforce the case for a hydrogen economy.

Marchetti discusses the historical trends in both storage and transport of energy and concludes that over broad ranges a relationship exists between cost C and size S of the energy system in the form

$$C = aS^b$$

where a and b are constants for each type of system. Applying the idea to chemical plants, nuclear islands, supertankers, and pipelines he finds b lies in the range 0·5 to 0·7, showing it is not subject to much variation from system to system. For pipelines the transportation costs vary roughly as the inverse of the pipe diameter, and the amount transported with the cube, so $b = \frac{2}{3}$. In this case an order of magnitude increase in scale means a reduction in unit costs of 50 per cent providing a powerful incentive to deal in ever larger amounts (fig. 78). Such arguments also lead to the construction of huge

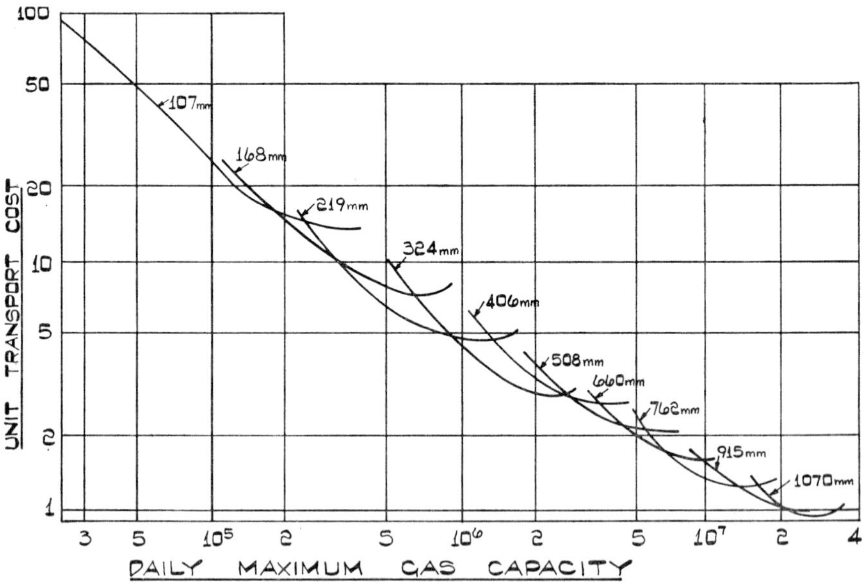

FIG. 78. Cost curves of pipelines of various diametres.
[C. Marchetti, IIASA Research Report 75-38, Nov. 1975]

tankers. Oil, however, is difficult to convey to a multitude of consumers. Gas has no such disadvantage so that both in respect of storage and easy delivery, gas is in a strong position. Furthermore, the fact that the cost of its transportation decreases with increasing scale is clearly shown in the curves of fig. 78. The reduction in unit transport cost with increasing pipe diameter (and consequently with maximum gas capacity) is impressive.

Marchetti pre-supposes that it is possible to construct a pipeline distribution system for the gas with a mean range of about 1,000 km. In contrast with this he notes hot water distribution might be done economically over a mean range of 10 km and he takes 100 km as about the economic mean range for electricity. The curves of fig. 79 show the relationship between the transportation costs of hot water and the temperature of the water at source. From this figure the effect of distance of transportation of hot water on the costs per gigacalorie of energy in the hot water is shown to be less than directly proportional to the distance involved. Thus, for water of temperature 140 °C at source (cost 6·7 Deutschmarks (DM) per Gcal) the costs at 10 km, 30 km, and 50 km become 11 DM, 17·5 DM, and 24 DM per Gcal respectively, for a rate of transfer of 200 Gcals per hour. The results are relevant to plans for district heating schemes, and combined heat and power schemes.

Marchetti advances his arguments to show that it could be advantageous to establish an energy distribution based on hydrogen; in such circumstances, whatever the primary source of the energy, the output would take the form

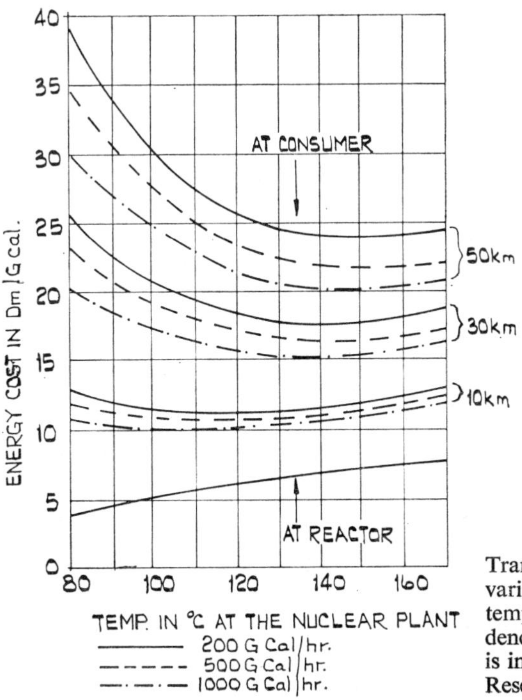

FIG. 79

Transportation costs of hot water over various distances as a function of the temperature of the source. The dependence on the amount of heat transported is indicated. [From C. Marchetti, IIASA Research Report 75-38, Nov. 1975] of the fuel hydrogen and he envisages a large-scale and complete system of power generation, hydrogen manufacture, storage and distribution.

It is true that the transportation and distribution of one form of energy rather than several would prove extremely advantageous especially if the transportation and distribution system is environmentally acceptable, such as underground pipes. However, it is early to assume such an outcome. The problems of conversion are considerable and also, as we have often remarked,

TABLE 20.2

ENERGY FORMS WITH TRANSPORTATION AND STORAGE MODES

(From C. Marchetti, IIASA Research Report 75–38, Nov. 1975)

	Form of energy	Physical nature	Mode of transportation	Mode of distribution	Range of distribution	Mode of storage	% Ener budge (world 1
Primary	Wood	Chemical free energy	Bundle Cart Truck	Truck	—	Log Heap	15% (wi farm wa
	Coal and lignite	Chemical free energy	Unit train Ship Slurry line	Truck	—	Heap	28%
	Oil	Chemical free energy	Pipeline Tanker	Truck	—	Metallic Tank	38%
	Gas	Chemical free energy	Pipeline Tanker	Pipeline	~1,000 km	Porous structures Cryotanks	17%
	Radiation	Electromagnetic	Space propagation	—	—	High Q resonance	—
	Mechanical	Mechanical	Wires Ropes, Bars	—	—	Flywheels	1%
	Heat Negentropy (e.g. compressed air)	Thermodynamic Thermodynamic	— Cryotanker Pipeline	— Pipeline	~10 km —	— Hole in ground	1% —
Secondary	Oil products	Chemical	—	Truck	—	Metallic Tank	35%
	Electricity	Electromagnetic	—	Wire	~100 km	—	10% a consum
	Misc. chemicals	Chemical	—	—	—	—	—

there are objections voiced against the idea of extremely large energy sources such as large nuclear plants even when they are offshore. The final form of the sources has still to be determined and until this question is clear it is difficult to resolve the matter of transportation and distribution in generalized terms. However, the strong case for hydrogen does remain and it is probably beginning to affect the course of research studies. In the meantime we show in the form of a table (20.2 above) a number of the more mundane facts of the present situation. It should be stressed that just as geography affects the question of recovery of energy resources, it has a considerable effect on transportation and distribution. Some important aspects of this topic are examined by Daniel B. Luben in his paper 'The Economic Geography of Energy' (*Scient. Am.*, Sept. 1971, p. 164).

In general terms the idea that large-scale operations allow energy to be distributed economically runs counter to the views of protagonists of the 'small is beautiful' view. Nevertheless they are often environmentally, as well as economically, desirable; for example, large pipes can be put underground and they result in major economies in transportation. Also small power plants can be as unsightly as large ones, and their greater number means that they offend the aesthetic sensibilities of many more country and urban dwellers.

Chapter 21

Energy Storage: Batteries, Fuel Cells, and Compressed Gas

We have already discussed storage in certain specific circumstances such as pumped storage near generating stations. We propose in this chapter to amplify the subject and also to relate it to the question of the hydrogen economy. When the initial supply of energy is electrical, storage presents special difficulties and this was in part the reason for the introduction of the idea of converting thermal energy of fission in reactors to hydrogen. We have also seen there is much yet to be done before large-scale plants to effect this change are entirely satisfactory. Research into ways of storing electrical energy therefore continues.

Storage Batteries. In recent years, a great deal of work has been done towards a substantial improvement of electric cells or batteries of various kinds. With the growth of the electricity supply industry continuing at a comparatively high rate in many countries, and the notable variation in demand over a period of, say, 24 hours, the need to maintain a reasonably constant base load on generating plant becomes more pressing. If some of the power could be used to charge batteries in periods of reduced demand and the batteries used to contribute to the supply at times of high demand, more constancy of generating level could be secured.

The heavy acid-filled batteries that are used in many applications today, and especially in powering the electric circuits of automobiles, are only capable of providing a limited amount of energy though involving a severe penalty due to the weight of the store. Typically such a cell is made with plates of lead and lead oxide in sulphuric acid and the total charge carried is usually between 40 and 60 ampere-hours at 12 volts. The charging efficiency of such stores is about 60 to 70 per cent, and this modest figure tends to offset to some extent their convenience. Nevertheless the fact that they can undergo charging and discharging and have a useful life of some years guarantees them a most important role. They can be extremely useful and satisfactory in applications requiring a supply of a few watts to several hundred watts.

Efforts have been made to improve the energy/weight ratio of the cells and there is some reason to believe research and development will bring a good degree of success. Some of the battery types in current use include those in table 21.1.

It is clear that good weight reduction per unit of energy gives rise to a considerable rise in cost and a loss of life. However, while the costs might be

TABLE 21.1

ENERGY CONTENT, LIFE, AND COST OF VARIOUS TYPES OF BATTERY
(From Intertechnology Report to the National Science Foundation, Washington, D.C.)

Type	Energy content in watt-hr/kg	Max. number of cycles	Cost in US $/kwh
Lead/acid	22	1,000	55
Nickel/cadmium	33–44	2,000	320
Silver/cadmium	55	200	415
Silver/zinc	110–145	100	470

reduced with much increased demand, continued research is required to increase the life of the most attractive types. The introduction of electric cars on a substantial scale is unlikely to occur until such difficulties as are evident are overcome. The trend in research at present is towards storage batteries that show promising performance at temperatures between 300 and 600 °C. These include the lithium/sulphur (Li/S) type and the sodium/sulphur (Na/S) type. In this the electrical energy is stored by producing molten sodium and sulphur from sodium sulphide. The charging reaction is

$$Na_2S \rightarrow 2Na + S.$$

On discharge the reaction occurs in reverse.

A combination of an alkali metal (such as sodium or potassium) electrode and a halogen (fluorine, chlorine, or bromine) electrode offers the highest cell electromotive force but the reactivity of alkalis and halogens gives rise to difficulties. The sodium-sulphur cell is an attractive compromise. It requires rather high operational temperatures to maintain the constituents in the liquid state and construction problems and gasket difficulties have to be overcome. All the same, in the light of recent research it appears that it is not unrealistic to hope that a sodium-sulphur type of battery of comparatively modest weight will achieve a range of from ten to a few tens of kilowatts—beyond the range of a lead acid battery—and one where there are many potential applications, especially if the output can be maintained for some hours.

Fuel Cells. It is difficult to classify the fuel cell entirely satisfactorily. It can be regarded as a conversion device which makes electrical energy directly from the fuel. On the other hand with a readily transportable fuel (such as liquid hydrogen, for example) as the feedstock of the cell, it can be regarded as a device for storing and transporting electricity.

In a typical fuel cell chemical energy released in the reaction between the 'fuel' and the 'oxidant' is converted, via an electrolytic medium, into a supply of electricity. Some aspects of the process have been reviewed by L. G. Austin and B. C. Almaula (*Chemical Engineering*, **76**, 1969, p. 85). They point out that if the cell is regarded merely as an energy converter it is not unlike a storage battery but that there are basic differences nevertheless. Unlike the storage battery, which contains the reactants and which can be regenerated by a recharging current as opposed to the discharging current, the reactants for the fuel cell are usually stored outside the reaction zone. The reactants produce the electricity and are consumed themselves in the process. The reactants are supplied from the storage only when electrical power from the fuel cell is required. As they are used up a fresh supply is added and the energy output immediately starts again. In principle the working parts of the fuel cell remain unchanged as a result of the operation of the cell, and their real function is that of catalysts which assist the oxidation-reduction process. In contrast, in a storage cell, the electrodes are an integral part of the total device and chemical transformations of their material occur in both charging and discharging.

The work on fuel cells goes back a long way. In the decade 1839–49 W. R. Grove established many of the basic facts about the behaviour of such cells. His work on the hydrogen fuel cell is one of the romances of amateur science but unfortunately his early researches were followed by a long period of little interest in the subject.

Grove established clearly the need for exposed electrodes and for a large area of contact between the gases used and the electrolyte and the absorbent. L. Mond and C. Langer extended the work of Grove and did a great deal to establish the fuel cell as an all-purpose source of electrical energy. Later, around 1905, F. Haber elucidated the thermodynamics of the cell and his studies of many different electrolytes and gases extended knowledge of the device. It was also shown that solid electrolyte cells could be made to operate successfully. In the present century some of the research and development has proved notably successful but on the whole the story of the fuel cell is complicated and in several respects disappointing. The work of F. T. Bacon has done much to maintain faith in more recent times and his 6 kW high-pressure hydrogen-oxygen cells, later developed for the Apollo spacecraft, showed there are good grounds for optimism.

It is difficult to choose a single form of representative fuel cell. A good many have been designed for particular applications such as the provision of electrical power for experimental apparatus in spacecraft or for the propulsion of electrically-powered automobiles. Cells can also be differentiated according to design features; they may be classified as hydrogen-air or hydrocarbon-air or zinc-air type, for example. In itself this tends to underline the fact that at present there is no completely successful type of fuel cell capable of being used in a wide range of applications.

The possibility of a hydrogen-air fuel cell is a particularly interesting one. Clearly if hydrogen can be made in a nuclear fission reactor with an efficiency higher than that with which the reactor makes electricity, and hydrogen can in turn make electricity efficiently, then an overall system capable of providing, firstly, hydrogen as a fuel, and, secondly, electricity from the hydrogen only when required, is a most attractive one.

In fig. 80 a fuel cell, using hydrogen as the primary fuel and oxygen as the oxidant, is shown in schematic form. Potassium hydroxide (KOH) is the electrolyte. Oxygen supplied at the cathode takes electrons from the cathode to form hydroxyl ions, which migrate towards the anode and react with water present in the electrolyte.

$$\tfrac{1}{2}O_2 + H_2O + 2e_- \rightarrow 2OH^-.$$

Incoming hydrogen is absorbed at the anode catalyst, where it dissociates. It then combines with a hydroxyl ion which has migrated to the anode through the electrolyte. Electrons are released onto the anode.

$$H_2 \rightarrow 2(H)$$
$$H + OH \rightarrow H_2O + e_-.$$

With the electrodes of the fuel cell connected to an external circuit, the electrons are able to flow from anode to cathode making good the loss of electrons occurring in the formation of the hydroxyl ions. Thus the overall reaction amounts to:

$$H_2 + \tfrac{1}{2}O_2 \rightarrow H_2O.$$

In theory, as in the reverse process of electrolysis of water to hydrogen and oxygen, the reaction of the hydrogen and oxygen to produce water can proceed with high efficiency in a good fuel cell.

In practice, however, it is found that there are severe complications, among which is the loss of electrical potential due to polarisation when hydrogen coats the anode. This can limit severely the operational efficiency and power rating of the machine. Mass-transfer, the removal of material from electrode to electrode, can prove an impediment to reliable continuous operation and reduce the level of performance of the catalysts incorporated in both of the electrodes, which may consist of metals such as silver and copper though many types have been used. A good many catalysts have been tried,

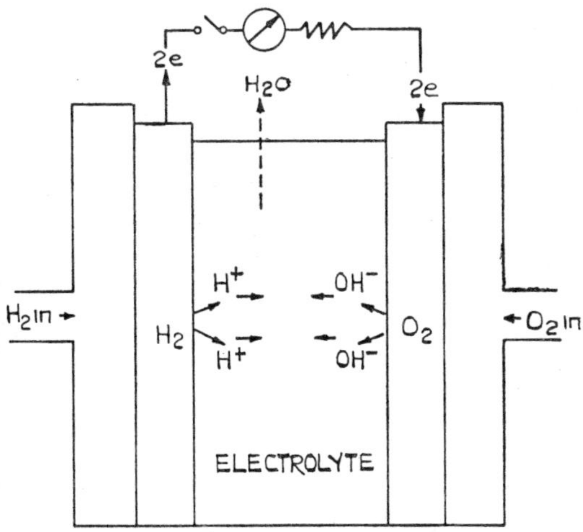

FIG. 80

Diagrammatic form of hydrogen-oxygen fuel cell

including platinum, palladium, and silver, but these are expensive materials that can be regarded as practicable only when efficient and reliable performance of the fuel cell is secured in a device of long useful life. For the large machines they represent an undesirably high initial investment, but possibly even more serious at high output levels is the relatively low value of power density that can be achieved.

During the last decade the fuel cell has been increasingly used to provide electrical requirements. No large-scale plants in the megawatt range of output have been manufactured but devices providing several hundred watts are on the market, though these are generally made to meet rather highly specialized needs. Studies of the advantages and disadvantages of using fuels other than hydrogen continue; among fuels tested are methanol, hydrazine ($H_2N\ NH_2$) which is sometimes used as a rocket propellant, and decane ($C_{10}H_{22}$).

The success of the fuel cell has been long delayed. A machine to replace the more conventional diesel-powered generator, which has only a moderate conversion efficiency, could find wide acceptance. As we saw in chapter 18, the efficiency of electric generators, normally driven by turbines or diesel engines, is intrinsically very high—being about 99 per cent for large generators—which means that the efficiency of the mechanical means of driving the generator is what essentially determines the overall efficiency of the system. To illustrate the point, and also to clarify ideas about the efficiency of many of the devices discussed in this book, we show in fig. 81 an instructive diagram by Claude M. Summers. We observe that a diesel engine can convert thermal to mechanical energy with an efficiency of about 37 per cent. If such an engine drives an electrical generator, the overall efficiency for conversion of the chemical energy of the fuel to electrical energy is between 36 and 37 per cent. The same figure shows that a fuel cell converts chemical to electrical energy with an efficiency of 60 per cent—and theoretically this figure could be much higher. If hydrogen is used as a fuel to make steam to drive a turbine, or simply used as in an aircraft gas turbine, the conversion process is reduced to an overall efficiency of around 35 to 40 per cent.

Clearly, therefore, there is good reason to strive for more complete success in the design of a large fuel cell of high conversion efficiency. An excellent account of the present situation is given by A. McDougall (*Fuel Cells*, Macmillan, 1976). He shows that theoretical analysis has been verified in practice with the production of fuel cells with high efficiencies and long working lives. As well as high efficiency their good points include the absence of noisy moving parts and of pollution. McDougall considers, however, that the fuel cell is usually not good enough on economic grounds to supersede conventional systems where commercial performance is the main criterion of acceptability, and he considers significant technological advances in the future as unlikely, or at least no more likely, than in competitive methods of energy conversion. Nevertheless a substantial programme of development of a modular form of power plant to deliver 5 MW is proceeding in the US and is expected to be in operation in 1981.

Compressed Gas. In introducing the concept of negentropy and the possible elimination of thermal pollution (p. 213), the transport of energy in the form of compressed gas, notably in the form of compressed air or even liquid air, was discussed. Such a method of storage of energy is, in principle, widely applicable. For example, electrical pumps similar to those used in pumped hydro storage systems could be used to compress the air and, when the power demand rises again, the compressed air could be used to drive generators to meet peak demand. Again, generating stations usually deliver part of the generated power to gas turbines which compress air prior to delivery to the combustion chambers for the fossil fuels. The compressed air that is produced at off-peak times is used to replace or complement that provided in the normal mode of operation of the station. It can also be heated

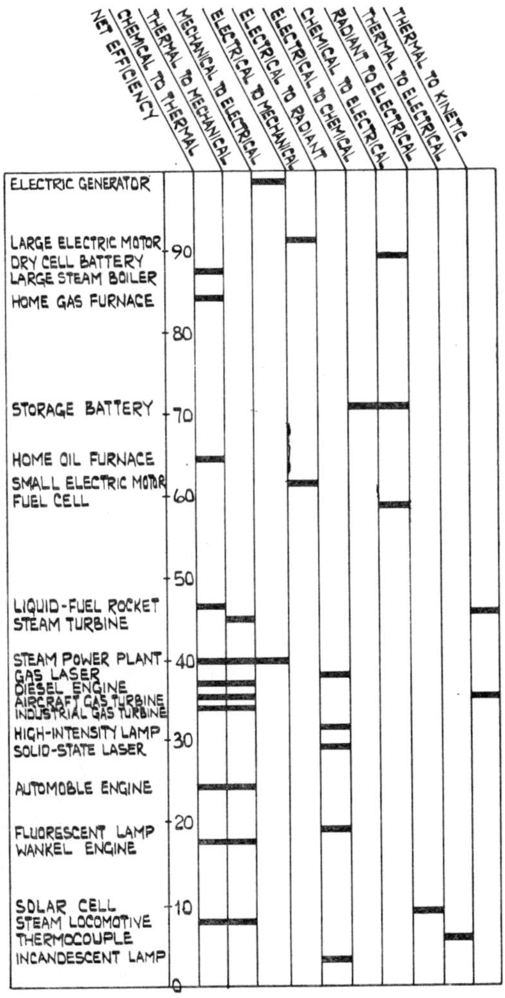

FIG. 81

Maximum efficiencies of energy converters presently attainable, with value for liquid-fuel rocket computed. [From Claude M. Summers, *Scient. Am.* Sept., 1971, p. 151] to some extent with the waste heat at the station itself. All of these moves could amount to a worthwhile economy of the fuel supply.

In such a compressed air storage scheme the reservoirs for retaining the gas at high pressure are vitally important. In the IIASA Research Report 75–38 Marchetti has studied this question and concluded that many suitable sites, which are leak-proof or which can be rendered leak-proof, are available. He proposes the use of exhausted natural gas fields or underground porous structures such as aquifers. A schematic diagram of an aquifer suitable for storage is shown in fig. 82 (a) and the practicability of the method is established by the curve (b) of the figure. This curve shows the rise in storage capacity in the US during the twenty years 1950 to 1970.

Although some pilot schemes for compressed gas storage are undergoing tests, and in the German Federal Republic one has been brought into operation in 1978, it is not regarded as a real competitor to pumped storage

(a)

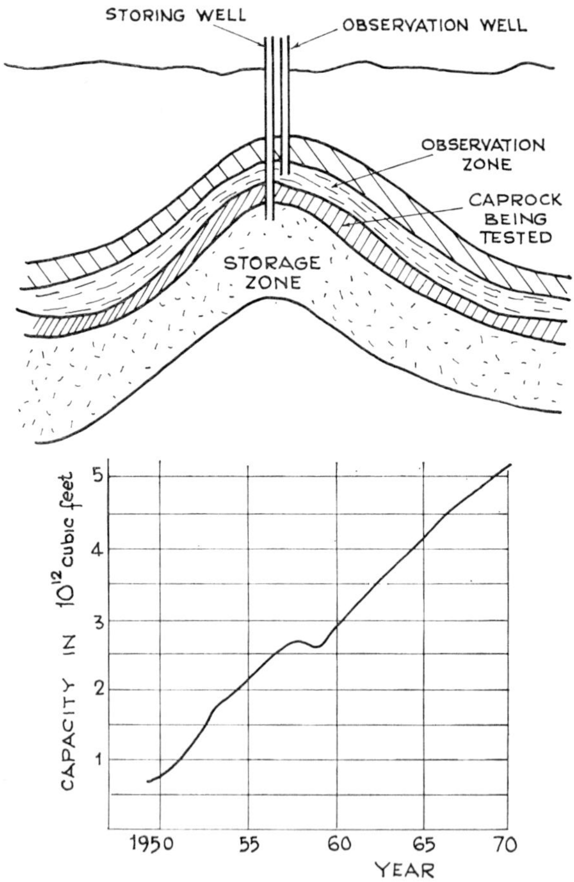

(b)

FIG. 82. (a) Principle of storage in an aquifer, (b) Growth of available underground reservoir capacity in US in the UK, for instance, where pumped storage sites are considered to be adequate for some time in the future. Nor are all experts convinced that compressed gas can be stored, for example in such places as abandoned salt mines, with 100 per cent safety.

Chapter 22

Heat Pumps (Air Conditioning and Refrigeration) and Thermoelectricity

Heat Pumps. A number of failures in the summer supply of electricity have surprised the public. Sometimes these failures have driven home the important lesson that in modern cities a very great deal of energy may be used in warm weather to cool the air in offices, public buildings of all kinds, in homes and in factories. It is during the same kind of weather that refrigeration plant is fully deployed and this adds to the load. While space heating consumes, in general, a much larger fraction of the electricity supplied to a city in the course of a year, the amount supplied to refrigeration and air-conditioning machines continues to grow rapidly. If the total electrical energy generated in warm weather is considerably reduced below the winter maximum for economy reasons the reduced total can be surpassed and failure ensues.

We start our discussion of heat pumps by reminding readers that a heat engine does useful work in the process of transferring heat from a source to a colder sink. Readers will remember that the efficiency of a heat engine is theoretically given, for an engine working in a Carnot cycle, by

$$\begin{aligned}\text{Efficiency} &= \text{useful work done/heat absorbed from source}\\ &= (Q_1 - Q_2)/Q_1\\ &= (T_1 - T_2)/T_1\end{aligned}$$

where T_1 = source temperature, T_2 = sink temperature.

For water such as is used for domestic purposes, say 132 °F (100 °F above the freezing point of water) we have a value of T_1 of about 56 °C or 329 K. If we operated a heat engine with such a source temperature and with a sink temperature of say 72 °F, that is approximately 22 °C or 295 K, we find the efficiency is

$$\begin{aligned}\text{Efficiency} &= (T_1 - T_2)/T_1\\ &= (329 - 295)/329\\ &= 34/329 = 0{\cdot}103.\end{aligned}$$

As expected this is a low efficiency, in percentage terms merely 10·3 per cent. However, if we use the heat engine in reverse by supplying it with energy—that is, putting work in rather than taking it out—we find that in taking an amount of heat Q_2 from the colder body (which is called the refrigerant effect) and adding heat Q_1 to the hotter body, the work performed, by the first law of thermodynamics, necessarily being $Q_1 - Q_2$ we can define

$$\begin{aligned}\text{Coefficient of performance of the machine} &= \text{refrigerating effect/work spent}\\ &= Q_2/(Q_1 - Q_2)\\ &= T_2/(T_1 - T_2).\end{aligned}$$

We now find with $T_1 = 329$ K and $T_2 = 295$ K, that

$$\begin{aligned}\text{Coefficient of performance} &= 295/(329 - 295)\\ &= 8{\cdot}68.\end{aligned}$$

A heat engine used in reverse in the way we have described is called a heat pump. The machine is driven so that it can move thermal energy from one place at a low temperature to another at a higher temperature by the following device. A volatile liquid whose boiling point is below that of the colder area from which heat is to be drawn, is introduced into that area. Naturally it evaporates, taking in from its surroundings its latent heat of vaporisation (see p. 211) and thus cooling them still further. The gas is then compressed, usually with an electrically driven pump, and passed to the heat exchanger outside the unit, where it loses its heat and condenses, before being returned as a liquid through a finc nozzle to the colder area. Heat pumps are probably more familiar to us in refrigerators and air-conditioning than in other devices, but they can also be used for heating if they are employed so that they draw in heat from the colder air outside, as if they were refrigerating the air outside. Heat can also be drawn from the ground or water in a lake or river or sea.

Given a supply of energy, usually electrical, to power an electric motor, we find that in theory the transfer of heat is done efficiently. In the example given, the motor requires to be provided with less than the equivalent of one-eighth part of the thermal energy that it pumps from the colder reservoir to the hotter source. Thus, while a heat engine designed to provide work is inefficient unless the source is a good deal higher in temperature than the sink or reservoir, a heat pump, because some energy is supplied to drive it, can be used as an efficient way of moving thermal energy from a relatively cool reservoir to a hotter one.

Let us turn to a typical air-conditioning situation. With an external temperature of 90 °F and a required internal temperature of a building of 70 °F, we are involved in pumping heat from a source at 70 °F (294 K) to a sink at 90 °F (303 K) so that

$$\begin{aligned}\text{Coefficient of performance} &= \text{heat removed/work done}\\ &= 294/(303 - 294) = 33.\end{aligned}$$

This suggests air-conditioning is an extremely efficient process, consuming in this instance only the equivalent of about 3 per cent of the thermal energy removed from inside to outside the building. In actual fact the air-conditioning equipments in use fall far short of this idealized level of performance, usually indeed by a factor of about 10, so that the energy required in pumping for air-conditioning is about one third or more of the heat equivalent removed.

When the heat pump is used to provide heat to a building and the heat has to be drawn from a colder environment, the attainable performance may also be much less than the theoretical value. In part this reduction in efficiency can be made good by the heat provided in driving the motor being usefully added to the heat transferred by the pump itself. Nevertheless the low temperatures outside during winter make it desirable, if possible, to have a large column or mass at a moderate temperature on which to draw, higher than that of the atmosphere which may be at the freezing point of water or lower; for example, if the heat of river water or the sea can be used the performance is improved. However, even without such a store of heat the thermal energy can still be pumped from the cold atmosphere to the inside of the building, with reasonably good efficiency. Such an air-to-air machine can perform well, but in many climates changes in the design of the evaporator and condenser are usually desirable if the machine is used for heating rather than air-conditioning.

While air-conditioning can be an efficient process there are circumstances in which the system is almost self-defeating. In a densely populated city with all the business premises and dwellings maintaining themselves at reduced temperature by ejecting heat into the extremely warm atmosphere, the temperature in the vicinity of the buildings is raised substantially. Warmer and warmer air enters the building, air-conditioning proceeds, and the load on the air-conditioning plant is continuously increased. We encounter here a particular facet of the heat pollution problem.

Similar remarks apply to refrigeration. As we have said, the refrigeration system is, in theory, like the air-conditioning plant in so far as both depend on heat pumping.

The variety of systems is very considerable, but in most cases the system consists of a compressor, condenser, throttling device, and evaporator. Such a system constitutes a cycle which is reversible; in other words it is capable of raising or lowering the temperature of a defined space. If the temperature has to be raised—i.e., the equipment is to be used for heating—the condenser is placed in the space itself and the evaporator is placed outside. When cooling, the evaporator is within the space which it has to cool and the condenser is outside.

For a given range of temperature, a system has a higher level of efficiency operating as a heat pump than as a refrigerator. In heating, the energy of compression provided as electrical power to the compressor motor is an additional useful component, but in refrigeration the same component is simply a nuisance. The coefficients of performance, as defined for the system used in the two different ways, show this advantage in heating as against refrigeration. Following Carnot we have in the case of refrigeration

$$\text{Coefficient of Performance (COP)} = \frac{\text{Heat absorbed in evaporator}}{\text{Heat equivalent of energy input}}$$

and in the case of a heat pump

$$\text{Coefficient of Performance (COP)} = \frac{\text{Heat ejected by condenser}}{\text{Heat equivalent of energy input}}.$$

Clearly the numerator in the second case contains the energy supplied to the motor in the form of heat and it is larger than the first.

A number of working systems are discussed by R. C. Gunther (*Refrigeration, Air-Conditioning and Cold Storage*, Chilton, 1969). There are many factors involved in efficient design, among these the choice of the refrigerant itself.

It would appear that much too little use has been made of heat pumps. They perform, in principle, an extremely desirable function in that they transport a part of the thermal energy, which is abundantly available in the atmosphere and in the waters of rivers, lakes, and seas, to the consumer; moreover, they do so at relatively little cost in energy terms. In addition to transporting it to the desired site they 'upgrade' the lower grade thermal energy. It seems clear that much waste heat in the modern world could be used with their help, and their improvement in performance and in reliability could mark the beginning of a most worthwhile energy conservation venture. The reversibility of the system, permitting the heat pump installation to be used in warm weather for air-conditioning, gives us a substantial degree of control of the comfort of our surroundings. It is not easy to explain the fact that there has been no widespread adoption of the heat pumping technique. In the case of domestic installations the installation can present problems, and if energy supplies are comparatively cheap it may be years before the capital investment is recovered in reduced charges. Till recently the relatively low cost of fuels has been the main reason that industrial companies have neglected heat pumping as a thoroughly practical means of usefully providing themselves with thermal energy from low-grade heat sources. Another reason for the slow adoption of the system could be that the desire to have a single system do both air-conditioning and heating has led to inferior results. Best results can be obtained by having two systems, with the compressor and evaporator in both designed for optimum performance.

The elements of a typical dual-purpose heat pump equipment, to operate both as an air-conditioning and a space-heating device, are shown in fig. 83. In this air-to-air system there are two heat exchangers with electrically-driven fans adjacent to them. On expansion the refrigerant is evaporated and heat is absorbed in the process. This heat is passed via the compressor to the indoor coil. Controls, usually within the dwelling, can be automatic or manual or a mixture of both. The commercial pumps generally incorporate a reversing valve as indicated and, together with the non-return valves, this effects a complete reversal of the mode of operation. Control of the expansion valves is automatic and may be by bi-metallic means, that is by the action of thermostats which are made from two metals with different coefficients of thermal expansion, enabling them to constitute temperature-sensitive switches.

The study of heat pumps and housing by R. D. Heap (*Building Services Engineer*, Conference on Housing and Energy, April 1975) notes that for many years heat pumps have been used in US homes to provide cooling in the summer and heating in the winter, and that this has been done with reversible equipment constructed as indicated. The factors exerting notable adverse influence on the level of this performance as a dual system can be examined, and changes made to equipments where in certain climates heating is much more required in winter than cooling in summer. In this mode of operation, where heating is the required function, the equipment is properly described as a heat pump (and in the other case it is a refrigerator). The breakdown by Heap of the factors reducing the coefficient of performance is interesting. He notes the theoretical Carnot cycle indicates a coefficient of performance of 21 for an indoor temperature of 21 °C and an outdoor of 7 °C, but this falls to 7·2 when temperature differences across the heat exchangers are taken into account. The heat exchangers operate properly only when assisted by fans; the energy provided to those and the losses elsewhere in the total system reduce the COP to 2·6 in Heap's example (see also E. R. Ambrose, *Heat. Pip. Air Condit.*, **46,** 1974, p. 77).

Heap gives the results of tests on some systems noting certain disadvantages from the point of view of a typical consumer on a domestic scale.

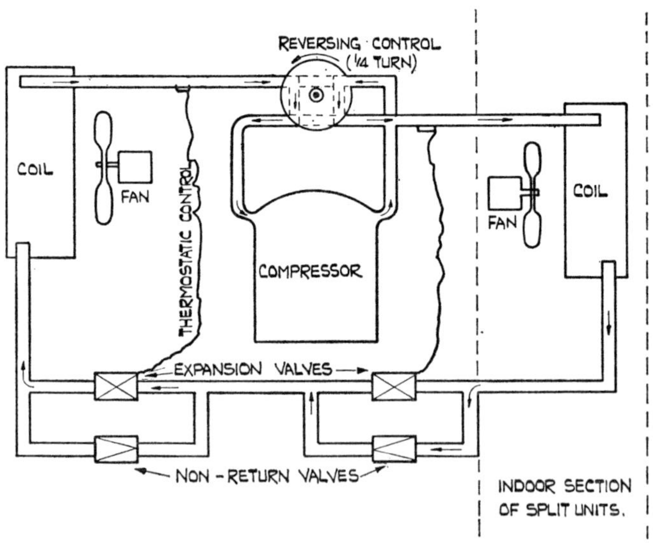

FIG. 83. Dual purpose air-to-air heat pumps system. [From R. D. Heap, *Building Services Engineer*, Conference on Housing and Energy, April 1975]

Many compressor motors are rather large, producing voltage surges on starting. They are sometimes wasteful as they respond to changes in temperature by demanding *in toto* more power than the equivalent supply, by simple radiant heaters, would temporarily involve. The tests with a system operating in a reasonably well insulated house of 235 sq.m area and 560 cu.m volume, with overall insulation coefficients of 3·6 W/sq.m °C and 1·5 W/cu.m °C and with a design loss of 850 W/°C are shown in fig. 84. The contribution of the heat pump during two winter months is a substantial one amounting to almost half of the total requirement (44 per cent).

The results were further improved in later work when the same house was cavity-insulated by injection. In similar conditions the heating energy required fell to between 72 and 78 per cent of the amount previously required for temperatures outside ranging between 7 °C and 10 °C. On the basis of his results, Heap considers a design temperature difference should be about 15 °C, implying a heat pump output of about 3 kW at a −1 °C outside temperature with a power input to the heat pump of 1·1 kW. He considers reversibility desirable for dehumidification and defrosting, and he regards infinitely variable speed of an interlocked fan and compressor system as well worthwhile. He stresses the need to insulate and seal ducts carefully so that the heat is properly distributed.

The design of domestic heat pumps for climates like those in Britain has been studied by C. J. Blundell (*Heat. Vent. Eng.*, June 1976 and the Electricity Council Research Centre (ECRC) Report N949, 1976) and G. W. Brundrett (*Electron. and Power*, October 1976, p. 686). Blundell considers the characteristics of the commonly used heat exchanger consisting of

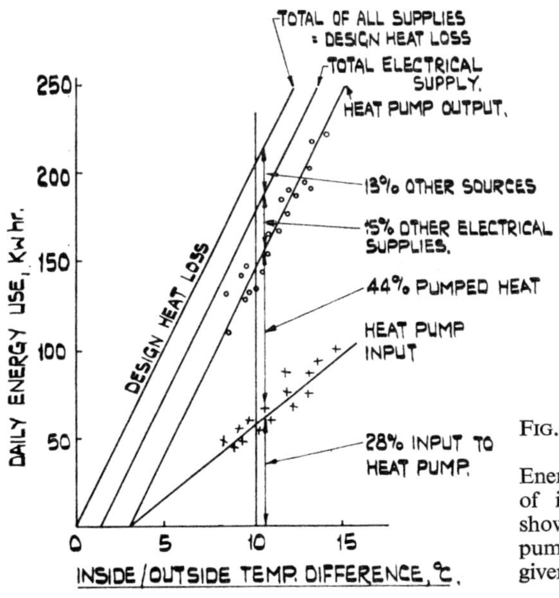

FIG. 84

Energy used in a house as a function of inside to outside temperatures showing 44% contribution from heat pump action. [From same source as given for Fig. 83] the cross-flow, plate-fin air to refrigerant type. He notes the amount of heat transferred is given by the equation

$$Q = VA\Delta T$$

with
Q = heat transferred
V = heat transfer coefficient (W/sq.m °C)
A = heat transfer area of exchanger (m^2)
ΔT = temperature drop across exchanger (°C).

The heat transfer can be increased by the air velocity provided by the fan but this is expensive in energy and in cash. Another factor, U, is a function of design and is generally optimised with

(*a*) Cross-flow, plate-fin air to refrigerant arrangement
(*b*) Circular refrigerant tubes on a staggered array
(*c*) Wavy, continuous fins with 2 mm pitch.

The value of A can be raised by increasing the exchanger depth or the face area but the fan power requirement means such increase of A must be optimised. In fig. 85 the results of calculations for a 5 kW capacity heat pump are shown. Fig. 85(a) shows a coefficient of performance (COP) increase with increasing evaporator face area is always obtained, but that it is at an optimum for a certain air flow rate. In fig. 85(b) the COP is seen to increase with the condenser face area (with the evaporator face area fixed, here at 0·2 sq.m) with the air flow adjusted for a definite value of the temperature of the heated air (35° C in this instance). As a result of such studies Blundell can arrive at curves like those of fig. 86(a) and (b) where the costs are shown as functions of the evaporator face area at a range of air flow values and of condenser face area. It is concluded that the improvement of the COP in increasing the size of the condenser (roughly doubling the size of certain manufactured units) is economically justified through savings in electricity consumption. The COP value of the units ranges between 2·8 and 3·0. The discussion is interesting in its own right and in showing that the climate must

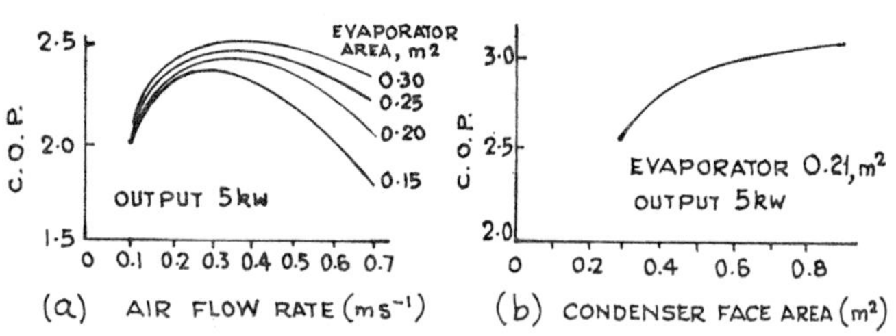

FIG. 85. (a) Coefficient of performance as a function of evaporator conditions.
(b) Coefficient of performance as a function of condenser face area, given constant air flow at a constant temperature (35 °C). [(a) and (b) both from C. J. Blundell, *Heating and Ventilating Engineer*, June 1976]

be borne in mind in designing the most acceptable and economically desirable heat pump for domestic use. This point is illustrated in plate 6 which shows on the right a complete equipment made to operate in average US conditions and on the left the condenser more suited for the conditions in the UK.

We noted some reasons why heat pumps have not been widely adopted in spite of their apparent advantages. Another may be that heat pumps, being not in themselves a source of heat, give a performance which varies with the temperature of the source. When the ambient temperature is low and heat is most needed both heat output and efficiency fall, facts made evident in fig. 87 which shows the characteristics of a heat pump designed around 1970 for use with outside air as the heat source. Clearly if river water were available at,

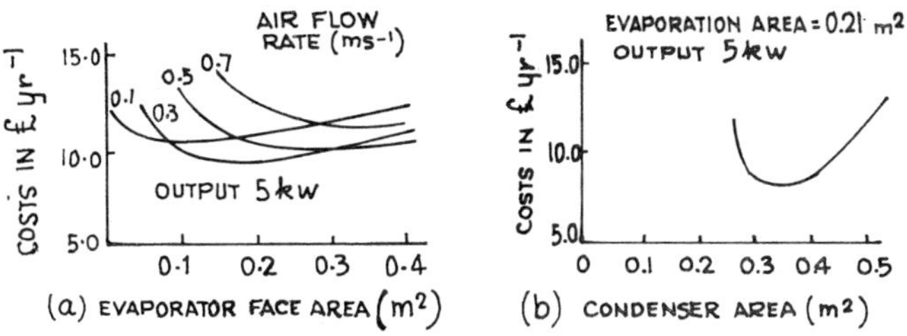

FIG. 86

(a) Annual costs of heat pumping as a function of evaporator conditions

(b) Annual costs as a function of condenser face area [Source of (a) and (b) same as given for Fig. 85]

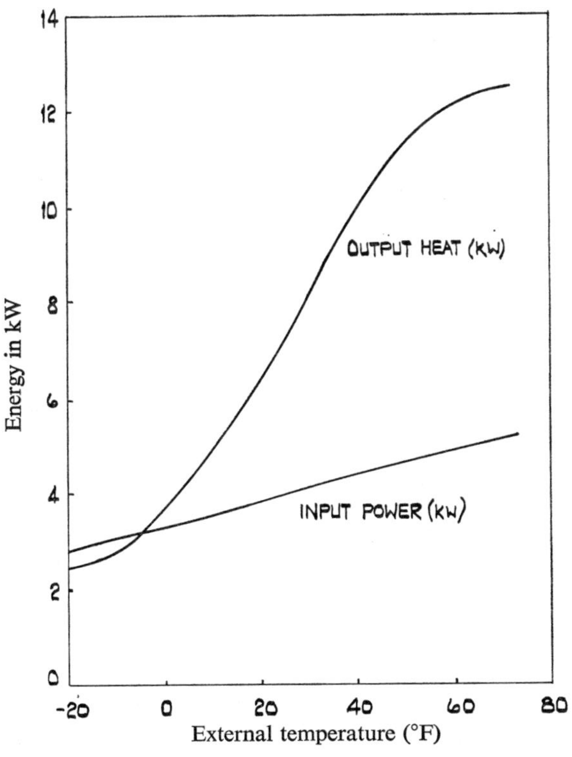

FIG. 87

Heat pump performance for 70 °F room temperature. [Source: G. W. Brundrett, *Electronics and Power*, Oct. 1976, p. 686] say, 40 °F for the heat source, much improved performance compared with that at low air temperature would be secured.

Heat pumping can increase the overall thermal efficiency of an electrical heating system to about that of oil or gas heating. Used in conjunction with gas or oil-fired boilers, heat pumps could play an important role in times of high fuel costs. Calculation shows that for a house of 100 sq.m area the energy used in constructing a suitable heat pump is about 20 GJ. This figure is roughly the same as that required in alternative schemes but because of their complexity the cost of heat pumps in money terms is greater and this militates against widespread adoption. Nevertheless as fuel costs continue to rise they seem certain to be used much more widely.

Thermoelectricity: Seebeck and Peltier Effects. The problems of producing electricity from thermal energy sources have been examined in considerable detail in earlier chapters yet a very direct connection between them was discovered as early as 1821 by T. J. Seebeck who first produced thermo-electricity. In his early experiments Seebeck showed a current flowed in a circuit composed of two different metals if one of the junctions was hotter than the other (see fig. 88), a phenomenon which came to be known as the Seebeck effect. The direction of the current depends on the pair of metals and on the difference in temperature of the junctions. The thermal electromotive force E varies with the temperature t of the hot junction (the cold junction being at 0 °C) according to the equation

$$E = \alpha_1 t + \tfrac{1}{2}\alpha_2 t^2 + \tfrac{1}{3}\alpha_3 t^3$$

In fact α_1, α_2, and α_3, which are constants for a given pair of metals, are all small and for a favourable pair of metals copper and nickel $\alpha_1 = 20{\cdot}39 \times 10^{-6}$ volt per degree and $\alpha_2 = 0{\cdot}0453 \times 10^{-6}$ volt per (degree)^2. The thermocouples, fig. 88(a), can be connected in series to give a thermopile, fig. 88(b). With about 100 such junctions in series and a temperature difference of some 500 K between the first and the last, the thermopile provides about 5 volts. In fact in any practical arrangements so far examined the thermopile appears to be a low efficiency thermal engine. The efficiency, at best, is around 3 per cent, so

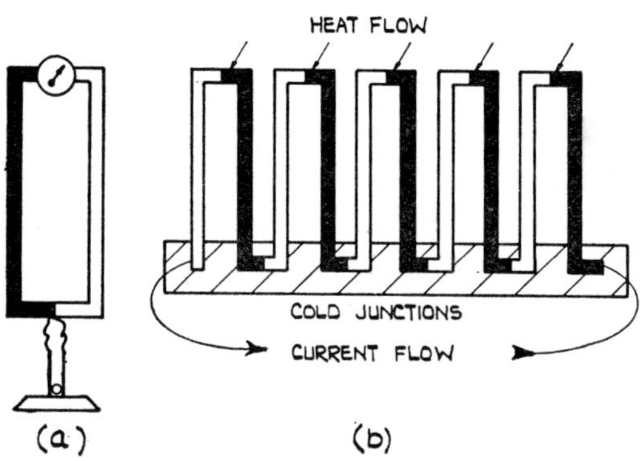

FIG. 88

The Seebeck effect

(a) Thermocouple arrangement

(b) Thermopile system of junctions in series

no large-scale application in energy schemes has been found for it. It has many other applications where the efficiency of conversion of heat to electricity is not an important consideration, for example in the construction of measuring instruments. It may occasionally be found useful for pumping thermal energy on a small scale. It has also been utilized for power supplies in space vehicles. A radioactive source, or even a small nuclear reactor, is used to provide thermal energy, and this is able, by means of thermocouples applied to the hot materials, to supply the necessary 100 watts, or few kilowatts in the reactor case, for energising instruments in the vehicles. Such a source of electrical energy is independent of the solar radiation which is often used to generate electrical power by means of arrays of photovoltaic cells.

The Peltier effect, named after its discoverer J. Peltier (1785–1845), is the converse of the Seebeck. If a current is passed through a circuit made with two different metals, or semiconductors, a difference in temperature is created between the junctions, and the cooling of one relative to the other can be applied in refrigeration. Again the efficiency of the process is low and there seems little possibility of application in energy conversion or transport. As with the Seebeck effect, the Peltier effect is valuable for instrumentation and small-scale processes.

If the Seebeck or Peltier effects could be increased in efficiency, for example by a major reduction in the internal resistance of the system (superconductors suggest themselves here), the simplicity of the thermal/electrical conversion could be extremely advantageous. It was reported (*Sunday Times*, 16th December 1973) that a method of increasing the efficiency of the Seebeck system to around 30 per cent had been discovered. It appeared that the effect might be used to power an electric car, since by heating one junction and keeping the other cool the electric current generated would produce the motive power at competitive efficiency. No further reports have emerged and the long-held view that the process of conversion is inefficient still unfortunately appears sound.

Chapter 23

The Affluent Society: Energy Conservation and Energy from Waste

In recent years, because of rising cost of fossil fuels and an appreciation of the finite nature of the sources of supply, much attention has been paid to the efficient utilization of energy in industry, commerce, and the home.

Industry, Transport, and Communication. The industrial sector consumes the largest share of most countries' energy supplies; fig. 35 demonstrates this point with regard to the US. Most companies now examine carefully every area of energy utilization. The greater part of their consumption consists of the direct burning of fossil fuels, with about 25 to 30 per cent provided in most countries as electricity. The electrical sector is primarily the responsibility of the generating boards or utilities. The substantial efforts to make generation and delivery more efficient have been described elsewhere in this book particularly on pp. 122 and 123 which describe the fluidised bed combustion technique and the magneto hydrodynamic generator, and in chapters 18 and 19 which deal with turbines, generators, and distribution.

Transport is an area where large savings could doubtless be achieved, particularly in the US where it consumes almost one third of the total annual energy. The downward trend in horsepower will bring savings. The average cubic capacity of 1977 models in the US was 4·6 litres as compared with a 1967 average of 5·24 litres, this being a reduction of 12 per cent in ten years. In due course an additional saving of over 30 per cent could be achieved: in many European countries the average engine capacity is less than 2 litres and the people are not notably less mobile. Furthermore, the average car engine is relatively inefficient (around 25–30 per cent) and diesel power units of low or moderate capacity whose efficiency is about 35 per cent are being introduced. All in all it is possible to make something between a 10 and 20 per cent reduction in overall oil consumption in the US by urgently attending to that wasted in cars. No doubt President Carter has this in mind in his wish to introduce new legislation. Even in Europe there is a renewed effort to reduce the energy spent on transport and, as in the US, there are signs that diesel power may become more customary.

Savings on both sides of the Atlantic might be achieved if public transport could be improved, and the public weaned away from their private cars and persuaded to use it. This difficult question, linked as it is with political decisions about subsidies to the public sector and taxes on the private one, is beyond the scope of this book. However, some of the need for people to move from place to place might be eliminated by still greater effort to improve methods of communication. Telephonic communication could conserve much of the energy that is frequently spent on journeys by road or rail or air. If it becomes more readily possible for people to be in visual as well as telephonic contact, meetings of many kinds could be reduced greatly in frequency. In industrial countries audio and visual communication has advanced rapidly in the last few decades and the amount of information that can be handled efficiently, and at low energy consumption, is very great indeed. Improved marketing of modern audio-visual means of conveying information could prove highly advantageous and should enable considerable energy economies to be made.

Domestic and Commercial Building. The domestic, commercial, and institutional consumption of energy in the advanced countries depends on the climate, the average standard of living, and several other factors, among them the social and environmental attitudes of the people. It is a sector which consumes about one third of the total annual energy supply and it is probably the one where the effort to economise has been most effectively concentrated.

Because of the small share of domestic energy they consume there is little to be gained in resisting the widespread adoption of TV, radio, record-players, washing machines, refrigerators, and other appliances of various kinds. It is in the use of space heating (and sometimes air-conditioning) and hot water that considerable economies can be made. Cooking, too, takes an important part of domestic energy and it could be that a close look at this element is justified. Roughly 20 per cent of the total energy consumed in a country such as the UK is used in homes. Of this 20 per cent, no less than 60 per cent is used in space heating, that is, 12 per cent of the overall national energy requirement. Ignoring the amount used by appliances as minimal, the remaining 40 per cent of the domestic energy total is equally divided between water heating and cooking, giving a figure of 4 per cent of the national energy requirement for each. These fractions are all large and savings can be most valuable.

A great deal of excellent work on the energy problems of some typical types of buildings has been done by a number of experts. The design of houses, offices, schools, and buildings of all kinds has been studied with both comfort and energy savings in mind. The task of introducing major energy savings is time-consuming and high in cost but one that can in due course pay substantial dividends.

Much of the research has been carried out in the UK and has been concerned with devices suitable for its particular climate. The energy savings in a cold climate are, on average, higher than those that can be achieved in warmer climates. Nevertheless the winter in many countries is as cold as, or colder than, Britain, even if not always so long, so doubtless the results of research in the UK can be applied advantageously elsewhere. The published work is so extensive that here we can only touch on a few of the many important works that have appeared in recent years. Our account is based largely on the results obtained from the substantial research programme of the Electricity Council Research Centre (ECRC) at Capenhurst.

Among the authors whose papers are particularly informative are J. B. Siviour (Aston University Conference, 7th July 1976; CIB-S17: Meeting in Munich, 28th Sept. 1977; ECRC Report 1070, Dubrovnik Seminar, 29th Aug. 1977; ECRC/M1042, May 1977); G. W. Brundrett (*Electron. & Power*, 4th Apr. 1974; Build International **7**, 1974, Applied Science Publishers Ltd; ECRC/M801, Mar. 1975; ECRC/M901, Feb. 1976; CIB Conference, Watford, 1976); P. Basnett (CIB-S17: Meeting in Munich, Sept. 1977; *Mathematical Models for Environmental Problems*, p. 353, Pentech Press, 1975); and D. McIntyre (*Appl. Ergonom.* **4**, 1973, p. 66).

The Building Energy Estimating Programme (BEEP) concerns itself with all uses of electricity and in particular with planning commercial, industrial, and domestic buildings to take advantage of the flexibility of electricity in cutting costs and conserving energy. In industrialized countries there are few buildings without electric lights and fewer without people. Each person, as we all realize in crowded rooms or in hot weather, produces a certain amount of heat even if involved in no strenuous activity. One electric lamp (of 100 watt power) and one inactive person supply an average of 200 watts energy input. If the person is working, e.g. typing, the total is nearer 250 watts. Thus around 1 kW is provided in an average room with four or more people present and good lights or equivalent light from the windows. Other sources of heat frequently in use include TV sets or radios. In fact, if heat loss from an occupied room can be made sufficiently low the amount of additional energy required for space heating can be drastically reduced.

In fig. 89 we show the variation with season of the heating requirement for a typical three-bedroom house. Three main curves show the heat gain arising from occupation and normal requirements like lighting; the heat

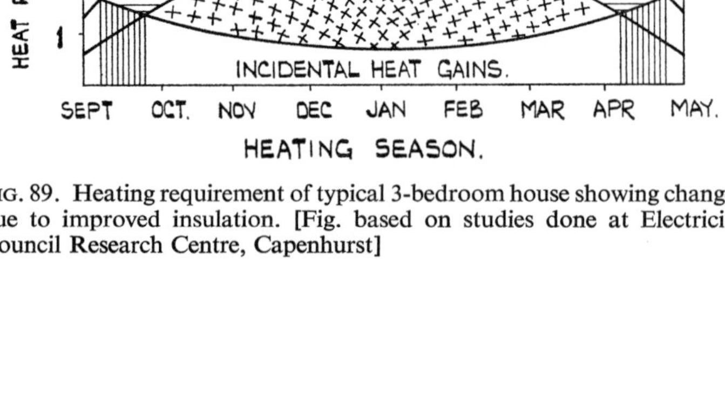

FIG. 89. Heating requirement of typical 3-bedroom house showing changes due to improved insulation. [Fig. based on studies done at Electricity Council Research Centre, Capenhurst]

required for a well-insulated house; and also the heat needed for a more typical house with the 1974 standards of construction. At the points where the incidental heating curve falls below the others, extra heating becomes necessary—i.e., from September to May. Since the incidental sources provide more than is needed in the summer (more than a six-month period) these months are not represented on our graph. By comparing the areas under the two insulation curves, the graph shows that over a twelve-month period the heat needed overall in a well-insulated house is less than half that needed in a house with normal insulation. If effective insulation could be achieved on a nation-wide basis, and similar savings contrived with water heating and cooking, no less than a 10 per cent reduction of the gross energy consumption of the UK would be achieved, representing a potential saving of about 30 million tons of coal equivalent. Obviously energy management in the home is of vital importance in the energy budget of the nation.

Great improvements cannot be brought about quickly. Home owners often cannot afford improvements; new regulations have to be introduced for new buildings; building materials and skilled labour are not in unlimited supply. There are also questions of ingrained habits to be overcome. People seem to like what is called fresh air, and opening windows contributes largely to heat loss. Ventilation in rooms is taken to be synonymous with cleaner air and is regarded as being healthier. There have been times when crowded homes had to be better ventilated, to reduce the scourge of T.B. infection for instance, but improved methods of accomplishing the same end can be adopted nowadays. Another deep-seated reason for opening windows appears to be the desire to correct humidity—kitchens become too steamy and other rooms, because of heating, become too dry.

The effects of thermal insulation on design have been examined in some detail by G. W. Brundrett who, in the papers previously mentioned, examined indices of insulation, fabric insulation effects, infiltration, energy consumption, and moisture control. His insulation study is particularly relevant to the design of windows, showing for example that for a wall of 40 per cent glazing an improvement in the insulation of the non-window part by a factor of three reduces the total loss of the whole wall by only 20 per cent. This shows the need to examine the whole structure of a building.

In order to investigate ways of reducing thermal fabric losses he considered a typical 90 sq.m floor area, three-bedroom, semi-detached house. The internal temperature was taken to be 19 °C and the mean external air temperature −1 °C. Five levels of insulation were involved, these being (1) minimum permitted; (2) external cavity walls with 25 mm roof insulation, single glazing and concrete floor; (3) cavity insulation, and roof insulation doubled to 50 mm mineral wool; (4) lightweight house construction and 100 mm of mineral wool in both roof and walls; (5) as in (4) with double-glazing. As insulation is increased the relative importance of unchanged components increased, and in the end the calculated loss of 7·2 kW is reduced to 4·4 kW with almost half of this final loss arising from infiltration of 1·5 air changes per hour. The results are well illustrated, following Brundrett, in fig. 90 (a) and (b).

A survey by J. B. Siviour (Aston University Conference, 7th July 1976) explores the possibilities of applying insulation by external cladding. This is a potentially important matter for many older properties, as the cladding can cover unsightly cracked and disintegrating old surfaces. Ideally the original external wall is provided with 100 mm or so of insulation material covered by a coating of breather paper; the outermost layer is of wood or concrete tiles or cement render on wire reinforcement. Alternatively, the surface layer may be glass or reinforced render. For an excellent account of external insulation reference may be made to the paper by A. E. Mould (ECRC/M1033).

Incidental heat gains arise from many sources. Over two-thirds of the houses in the UK have a refrigerator and many also a deep freezer. In winter television is watched on average nineteen hours per week and colour sets use more power than the less fashionable black and white ones. Water consumption rises and uninsulated hot-water pipes distribute between a third and a half of the heat lost by the water into the house. Lighting is used more in winter and finally, even in winter, sunshine and diffuse light add significantly to the total input. It has been estimated by N. S. Billington (B.S.E. **40**, 1972, p. 23) that the average value of the total incidental heat gains per house is a little over 2 kW and that this varies rather slightly with the seasons.

As we have seen from fig. 89 the gains constitute a considerable reduction in the actual heating energy required and field trials have verified this conclusion. With windows shut in the coldest weather the actual heating energy

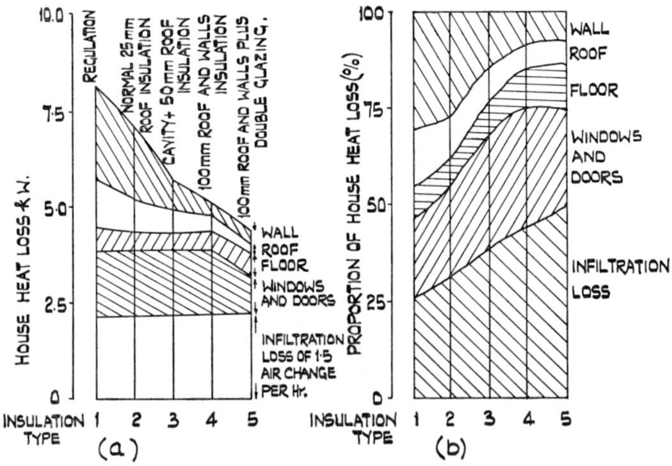

FIG. 90. (a) Reduction in heat loss of a 3-bedroom, semi-detached house with increasing insulation

(b) Proportion of loss through the various building components
[Both figs. from G. W. Brundrett, *Applied Energy*, 1, 1975, p. 7]

needed may be as little as half of the theoretical value, that is, the value without the incidental gains. In other words incidental gains and supplied energy are equal in these conditions.

Reduction of infiltration is a complicated matter. Windows are often so mounted that a good deal of cold air can pass into the heated room at their edges. On the other hand there may also be a gap because the window's edges are made better than the walls they fit in to. However, as we noted before, the reasons why people open windows is a special problem, related apparently to attempts to correct humidity, and more research is needed although a good deal of interesting work has already been done and an extensive literature on the subject is available (see D. McIntyre, *Appl. Ergonom.* **4**, 1973, p. 66 and references; also M. A. Humphreys and J. F. Nicol, B.R.S. CP14/17 of Building Research Establishment, Watford).

The possibilities of controlling moisture in buildings by using a heat pump dehumidifier have been explored by Brundrett (1977 International Seminar on Heat Transfer in Buildings, 29th Aug.). He regards the heat pump as a feasible means of dehumidifying and, at the same time, gaining a heat bonus for background heating from condensation.

The Thermal Wheel and Other Heat Recovery Devices. It is possible that useful dividends could be secured in a similar fashion by using the thermal wheel. The energy recovery, or thermal, wheel is diagrammatically shown in fig. 91. It is a rotary air-to-air or gas-to-gas form of heat exchanger, used to transfer energy from exhaust gases, drawn from hot or warm spaces, to the cooler supply gas. One side of the wheel is within the exhaust duct and the other is in the way of the inflowing gas. The wheel allows the gases to flow since it is a rotating filter in the form of a metal mesh, fibrous material or a glass ceramic, all with large surface areas exposed to the ingoing and outgoing gas streams. A special hygroscopic coating on the surface of the wheel absorbs and filters away moisture gleaned from the condensing vapours or gases in the exhaust stream. As they condense they yield up their latent heat of vaporisation and this surplus heat is taken up by the incoming gases. In many cases it is water vapour that is the source of the energy.

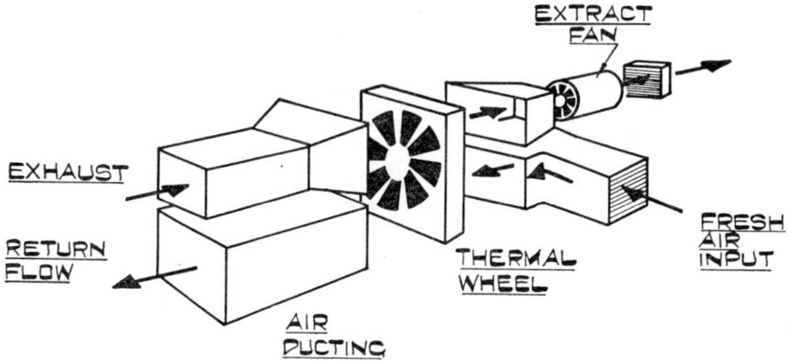

FIG. 91. Components of the thermal wheel

The speed of the wheel is adjusted normally in the range 10 to 20 rpm which permits optimum pick-up if heat from the exhaust gas and the flow rate produced by the fan are properly related to the diameter of the wheel itself. In practice, with the parameters suitably adjusted to the optimum design criteria, about 65 per cent of the total heat in the exhaust stream can be returned to the input.

The wheel performs well in a variety of industrial applications including the recovery of heat in metal furnaces, in printing works, paper drying, and spray drying. It could perform a useful role in recovering the heat of large or small kitchens since kitchens do require ventilation. The thermal wheel in fact may provide a way of making the best use of the incidental energy gains in houses or buildings generally. If ventilation could be secured with the return of two-thirds of the heat, the problem of opening of windows and their attendant heat losses would largely disappear.

Heat pipes (see p. 211) can sometimes be arranged to do a similar task to the thermal wheel. In addition, coil heat exchange systems, as well as parallel plate systems, can prove effective. In this last instance liquid-to-liquid recovery is accomplished well and 80 to 90 per cent of the available heat is conserved. The system is used in recovering heat from brew kettle exhausts, boiler flue gases, desalination processes, caustic soda manufacturing, evaporation plants, dry effluents, and in whisky distillation. The coil recovery method is used most in situations where the heat is to be returned to rather remote locations. A coil in the exhaust duct and one in the supply duct are able to exchange energy through water which is pumped through a circuit containing both coils. Something in the range 30 to 50 per cent of the heat is recoverable and the arrangement works well in large office buildings.

Costs. The amount of money saved on space-heating requirements compared to the cost of installations has been calculated by Siviour. He classifies such installations in terms of insulation, mechanical ventilation with heat recovery, and, thirdly, solar collectors. The data illustrated in fig. 92 show that, in the case of his chosen example, an 85 sq.m house, insulation is the most economic means of saving and that its financial advantages are considerable. All three installations are of course practicable in the same

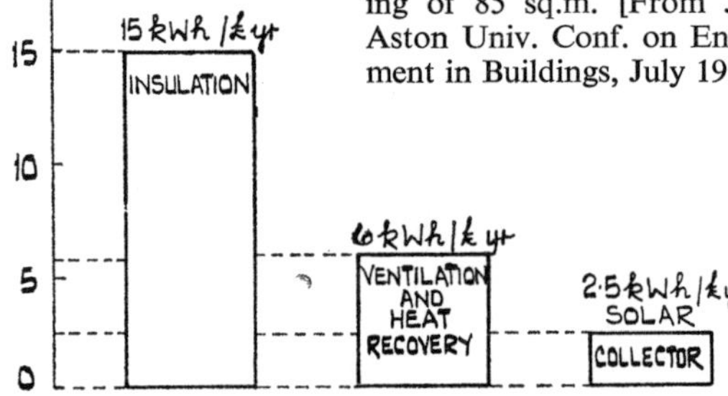

FIG. 92. Annual energy savings versus capital costs for three methods of reducing net space heating needs for a dwelling of 85 sq.m. [From J. B. Siviour, Aston Univ. Conf. on Energy Management in Buildings, July 1976]

dwelling but if used together they do not necessarily yield exactly the same economies as shown in the figure. Clearly, however, conservation can yield considerable gains both of energy and money if used with insight.

Energy Recovery from Waste. Another important aspect of energy conservation is the recovery of waste materials. In theory, though not so easily in practice, metals can be recovered as scrap thus both saving some of the raw material, say iron or copper, and also reducing the energy necessary to manufacture the new products—in a sense recovering part of the initial energy investment.

Not only metals but also the vast quantities of solid organic wastes in the more affluent societies represent a most useful source of recoverable energy. Thomas H. Maugh has provided an excellent review of progress in this field up to 1972 (*Science* **178**, 1972, p. 599). He emphasises that conventional methods of simply dumping the waste become more and more troublesome, and that organic wastes lend themselves well to processes of conversion to synthetic fuels. However, he is critical of some exaggerated claims about the amount of waste lending itself to economic methods of conversion to fuels and the ease of marketing it.

In the US the estimated annual total of inorganic mineral waste is $1{\cdot}1 \times 10^9$ tons and this is accompanied by 2×10^9 tons of organic wastes. On the basis of these figures it was argued, over-optimistically in the opinion of Maugh, that as much as $2{\cdot}5 \times 10^9$ barrels of oil could be produced annually. However, Larry Anderson (US Bureau of Mines Information Circular 8549, 1972) has pointed out that roughly half of the total mass is water, leaving a figure of $0{\cdot}88 \times 10^9$ tons per annum of dry organic waste. The chief difficulty of recovery arises from the wide dispersal of most of the waste, since this makes collection for processing extremely costly. In fact the estimate of dry organic waste that can be usefully collected for treatment falls to $0{\cdot}136 \times 10^9$ tons. After conversion to oil this amount would yield 170×10^6 barrels, which is approximately 3 per cent of the total annual consumption of crude oil in the US. The alternative route for conversion, the production of methane, would yield the equivalent of 6 per cent of the total amount of natural gas consumed in 1971. It would clearly be possible to raise the recovery levels noted here—in no sense negligible—to much higher values if the amount recovered could include the smaller yields in the less densely populated parts of a country. Work on the problem of successful recovery in small communities is required.

The major recovery, or conversion, techniques include hydrogenation, pyrolysis, and bio-conversion, but only the first two have been put into successful operation as yet. In a number of industrialized countries a good part of the bulk of waste material is simply burned in industrial plants as part of a fuel supply to give some heat or to yield some additional energy as electricity. This is not an efficient use of the material but the sophisticated methods of conversion are not readily introduced, even in the US, and the capital investment required is not inconsiderable.

In the hydrogenation system the organic waste and up to 5 per cent of its mass of an alkaline catalyst, such as sodium carbonate, are put in a reactor with carbon monoxide and steam and raised to a pressure between 100 and 250 atmospheres and a temperature in the range 240 to 380 °C. In the best circumstances almost 99 per cent of the carbon in the waste is converted to oil and the yield can reach 2 barrels per ton of dry waste. Part of the oil is needed to provide the carbon monoxide and the heat input to the reactor, so the net yield is 1·25 barrels per ton. The actual content of the oil varies somewhat with the nature of the waste and about 5 per cent may be nitrogen when manure is the initial starting material. Sulphur is usually low, about 0·4 per cent being about the upper limit, which is a point very much in favour of the product.

The range of wastes that can be handled in the recovery processes is given in table 23.1.

TABLE 23.1

AMOUNTS OF DRY, ASH-FREE, ORGANIC SOLID WASTES PRODUCED IN THE US IN 1971 (IN 10^6 TONS)
(From T. H. Maugh, *Science* **178**, 1972, p. 599, based on the work of L. L. Anderson)

Source	Wastes generated	Readily collected
Manure	200	26·0
Urban refuse	129	71·0
Logging and wood mill residues	55	5·0
Agricultural crops+food wastes	390	22·6
Industrial wastes	44	155·2
Municipal sewage solids	12	1·5
Miscellaneous	50	5·0
Total	880	136·3
Net oil potential (10^6 barrels)	1,098	170
Net methane potential (10^9 cubic ft)	8·8	1·36

Some of the rather serious problems of the hydrogenation technology are discussed by Maugh, including those of handling the solids and transferring them to the pressure vessel. Another problem is to minimize the pollution of the water used in the process and purify it afterwards. These difficulties have to be counterbalanced against the advantage of disposing of much waste safely.

The pyrolysis technique, in which the bulk material is heated to decompose it, is, in principle, somewhat simpler but the products are more diverse. Per ton of dry refuse they include about 1 barrel of oil, 140 lb of ferrous metals, 120 lb of glass, and 160 lb of charcoal, plus a variable amount of low energy gas, which usually has to be used as effectively as possible at the pyrolysis plant itself. The oil is somewhat low in energy content but an attractive product nevertheless. The introduction of the pyrolysis method is proceeding steadily and it may soon be fairly widely spread in the US.

The third method, bio-conversion, is at a much earlier stage of development but it could prove in due course the most successful. Since its success might be limited by the amounts of suitable waste available, it has been suggested that fast-growing algae could be used to supplement the supplies to the conversion plants.

In conclusion, the scientific conversion of the wastes of affluent societies can result in due course in at least a 1 to 3 per cent contribution to their total energy consumption. This is a well-worthwhile addition but not enough, of course, to change at all radically the prospects for the future.

Chapter 24

Energy Analysis and its Application

Energy analysis as a whole was discussed briefly in Chapter 11 and then related to agriculture. Here we consider it in more detail, particularly with regard to the manufacturing industries.

In energy analysis scientists, engineers, and statisticians attempt to evaluate the amount of energy required to produce or operate a piece of equipment or a whole manufacturing system. One of two techniques may be applied. In the most satisfactory method, known as 'process analysis', the overall process of manufacture—from obtaining raw materials to completion of the finished product—is assessed. The form or quality of the energy involved is taken into account because it may be high grade, as in the case of electricity from a power station, or virtually useless, as in the cooling water of the same station. In such circumstances it is difficult to generalize and analysis is used to maximum advantage in specific circumstances and for particular problems. In the second method the input-output tables of the economy of a particular product are analysed. The great variety of equipment used in the manufacturing industries makes this second approach less accurate, but it is simpler to use and it provides answers comparatively rapidly. Process analysis may be used to reinforce the results of the simpler technique.

We can learn a good deal about most energy schemes by applying these methods of energy analysis. Furthermore, the measurement and budgeting of energy yield much information about costs, and have only received little attention in the past because energy costs were low. Analysis techniques should prove increasingly useful in arriving at policy decisions. Their value is most readily demonstrated by considering some of the most important examples of their application.

Frequently the efficiency of an engine is regarded as much the most significant measure of its performance. The methods of energy analysis, however, enable us to look more comprehensively and more penetratingly at the whole range of factors that should be taken into account in comparing engines, or entire energy systems, involving machines of very different characteristics. We can thus reach rational conclusions based on proper comparisons. Heat pumps, for example, may raise complex questions (T. Byrley, 'Analysis Disproves Claims of Heat Pump Economics', *Fuel Oil News*, September 1974, p. 20).

An important area of energy analysis is its application to the nuclear industry. While nuclear weapon technology called for production of plutonium, the costs of nuclear energy as such did not need to satisfy stringent market tests, but this is no longer the case and studies of energy accounting in depth are most desirable and have a most important role in decision-making. The analysis techniques involved are illustrated in the work of the Energy Studies Unit of Strathclyde University in the report 'Dynamic Energy Analysis of the EEC Energy Transition Programme' (see *Energy World*, May 1977). The concept of a unit of gross energy requirement (GER) allows the assessment of the repercussions of an energy policy or changes of energy policy to be more exhaustively assessed (see *Electrical Review*, **198**, 1976, p. 16).

Primary Materials. The energy contents of the various materials used to manufacture articles and appliances enter into most energy analysis studies, but unfortunately there is often disagreement about the values to be assigned to them. The use of recycled material, as in the case of aluminium or magnesium for instance, can produce much reduced figures. Magnesium may be put between 200 GJ/tonne and 380 GJ/tonne depending on the quality of the process and the amount of recycled material involved. For steel, an outstandingly important material whose manufacture accounts for roughly 11 per cent of the total world energy consumption, H. M. Finniston gives a value of 25 GJ per tonne of liquid steel in the UK, excluding mining, crushing, transport, and subsequent forming (*New Scient.*, 11th July 1974). A study at the Illinois Institute for Environmental Quality where the aim was to consider the total manufacture in more detail yielded the figure of 42 GJ/tonne. For sheet steel a variety of estimates appears to justify a value of 45 ∓ 5 GJ/tonne.

In spite of uncertainties arising from differences in manufacturing methods, transport costs, and other variables, the approximate energy requirements of some materials are shown in table 24.1. Among the great amount of work done on this subject in recent years are papers by R. S. Barnes, *Proc. Inst. Mech. Eng.* **190**, 1976, p. 29; J. Gibson, *Chemistry and Industry*, Aug. 1975, p. 677; and J. S. Hunter, *Chemistry and Industry*, Nov. 1976, p. 967.

The Electric Car. For a wide range of articles the energy required to produce the necessary materials is the major component of the total energy required in their manufacture. According to the calculations of R. S. Berry and M. J. Fels ('The Energy Cost of Automobiles', *Science and Public Affairs*, December 1973), assuming 100 units of energy are spent on producing the raw materials for a car, a further 35 units are necessary to manufacture and assemble the components. The raw materials require a total of roughly 27,000kwh. Hence in the complete manufacture the amount used is $27{,}000 + (0{\cdot}35 \times 27{,}000) = 36{,}450$ kwh. The additional 35 per cent necessary for manufacture is a figure that applies approximately in many other industries.

Starting with this kind of figure for car manufacture, K. R. Williams

TABLE 24.1

ENERGY REQUIREMENTS

Material	Energy requirement in GJ per tonne
Steel	42
Sheet steel	45±5
Aluminium	260±20
Magnesium	375
Copper	60±6
Nickel	136
Chromium	144
Cobalt	86
High temperature nickel alloys	129
Silicon nitride	684
Low density polyethylene	81–86
High density polyethylene	86–90
Polypropylene	104–109
PVC	77–81
Polystyrene	127–132
Expanded polystyrene	132–136

('Energy Analysis and the Electric Car', *Automotive Engineering Congress and Exposition*, Detroit, Feb. 1976—760120) generalizes in discussing the energy content of products of a particular industry. He found that, in 1968, 38·5 kwh was used in the UK motor vehicle industry for each pound sterling value of the manufactured goods. In the electrical machinery industry the figure was almost the same, 38·1 kwh/£, and in the mechanical engineering field it was 42·2 kwh/£. For less energy intensive industries such as electronics the figure fell to around 23 kwh/£ (see O. J. Wright, *The Natural Resource Requirements of Commodities*, Department of Environment, London, 1974). Williams examines the energy advantages and disadvantages of electric-powered vehicles and concludes the electric car will be widely adopted only if a successful and economic fuel cell were to be produced. Such a development would at least halve the hydrocarbon demands for a given mileage. He stresses that up to 50 per cent of the total oil consumed in industrialized countries can be used for commercial and private transport, so a practical fuel cell, on his argument, is a most worthwhile if somewhat elusive research goal.

Williams finds that in the case of the motor vehicle industry the two principal techniques of energy analysis—namely those of process analysis and of input/output comparison—give figures that agree well. Clearly the agreement provides reassurance as to the validity of the analysis.

Energy Gain and the Provision of Power. Energy gain is a most useful concept in analysing many activities. We used this concept in our discussion of the agricultural society of the Tsembaga people but it applies equally in many more complex situations. For a Middle East oil well the energy gain may be as high as 1,500. The oil is rather easily obtained so the energy used in manufacturing the required equipment is a relatively negligible amount compared with the amount obtained from a typical well. By contrast, although in many countries solar energy is abundant, the solar heat panels needed to secure the diffuse energy are comparatively costly in energy terms to make and maintain in working order. Their energy gain is around 10. The acquisition of deep-mined coal has a gain figure, in the UK, of about 45, and strip-mined coal is roughly 160. This last figure applies also to North Sea oil because it is much more difficult to 'win' than oil from the Middle East. When UK coal is used as a fuel in electricity generating stations the overall gain is about 12. This figure applies to the consumption of a mixture of deep-mined and strip-mined coal.

The figures above are those provided by Shell International Petroleum Co. Ltd. in an unpublished report (PL/15, May 1975) and the authors are indebted to Dr. Keith R. Williams for drawing their attention to such an excellent example of the application of energy analysis.

This paper also considers the view that nuclear power stations use more energy than they produce. A thorough analysis establishes the fact that nuclear power stations can offer a higher energy gain than stations fired with UK deep-mined coal. This is of course consistent with the fact that, in 1977, nuclear stations supplied energy more cheaply to the UK grid than coal- or oil-fired stations. The estimates show that, in the case of a 1 GW nuclear station of AGR type, 6,300 GW hr are used in construction and fuelling and 4,500 GW hr are required in refuelling. Of the 6,300 GW hr used in building and operating, about 1,000 GW hr is the measure of the initial core requirement but the core can be used again at the end of the useful life of the station so the energy investment is returned. Complete calculation of the energy gain on a thorough and searching basis shows that whereas P. F. Chapman and N. D. Mortimer (*Open University Research Report*, ER 4005, revised December 1974) reached pessimistic results the true value of the energy gain of an AGR station, with an operational life of twenty-five years, varied between 18·5, when normal high grade uranium ore (0·3 per cent uranium) was the source of uranium for fuel fabrication, to 4·7 when sea water with pump extraction was the uranium source. With allowance for the Pu^{239} formed in the AGR core the value increased from 18·5 to 19·0. If the material of the core is not scrapped at the end of twenty-five years, the 19·0 could rise to about 22·7 and the sea-water value of 4·7 could be increased dramatically to 14·5.

This energy gain is particularly noteworthy because the rate of investment in new nuclear stations compared with the rate at which they come into useful operation obviously has a substantial effect on the value obtained. In the extreme case of a very rapid rate of reactor construction the energy investment must clearly outrun the energy return for a period and it can be some years before the true energy gain is secured. An example of this effect on the net energy balance for a construction programme of AGR reactors, with 2·25 as the average number of reactors on which construction is started each year, is shown in fig. 93. The peak energy deficit of 61,000 GWhr occurs in this case in the seventh year of construction but thereafter the situation improves and the energy credit begins to increase very steeply in the eleventh year and continues afterwards for a period of twenty-five or so years. Thus the construction of the machines is expensive in energy terms but the energy payback can greatly exceed the input after a decade or so, the delay in return depending on the scale of construction and its rate.

The nuclear industry, therefore, is intensive both in energy capital and in money capital but is clearly able to provide, in the longer term, an energy gain of 15 or more. Such a value might be obtained even if the uranium has to be obtained from sea water so the value is certainly not closely dependent on the grade of uranium ore that has to be used for the fuel fabrication. A large fraction of the initial construction energy is used in the fabrication of all building materials, including the housing and the reactor itself, the hall for the turbo-generator system, and in the multifarious operations at the site including transport and handling.

The problem of the time of 'profitless growth' is not unique to nuclear power. One can readily appreciate that it is common at the start of any capital-intensive form of industry, for example some of the chemical industries in a number of industrialized countries in the 1960s.

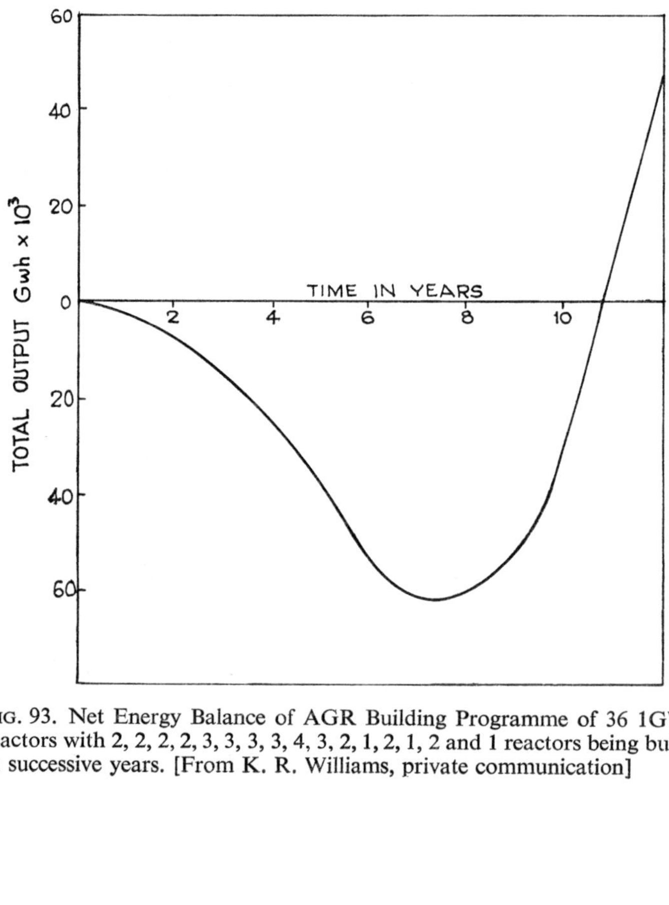

FIG. 93. Net Energy Balance of AGR Building Programme of 36 1GW reactors with 2, 2, 2, 2, 3, 3, 3, 3, 4, 3, 2, 1, 2, 1, 2 and 1 reactors being built in successive years. [From K. R. Williams, private communication]

Similar energy analysis techniques can be applied to a wide range of energy-producing machines and energy-consuming processes and systems. If photovoltaic cells are considered, for example those of the cadmium sulphide type (see p. 92), the potential energy gain when used in a suitably sunny country like the US lies between 20 and 40. For silicon cells the value is much less attractive.

Among more orthodox energy systems, the diesel electric generators of output 1 MW or more offer advantages in energy gain over central generating stations. The machines are as efficient as very large steam turbines in energy conversion and, because they can be placed closer to the consumer, the distribution loss is largely eliminated. Moreover, the 'waste' heat of such a 1 MW diesel machine can be obtained at 80 °C from the cylinder jacket and at 120 °C to 140 °C from the exhaust 'waste' heat boiler. In contrast to the steam turbine, this 'waste' heat removal does not reduce appreciably the efficiency of the diesel, making the overall combination eminently suitable for use in industrial and commercial premises where the spare heat can be usefully exploited in heating the buildings, providing hot water and so on.

The advantages which can be secured in using dispersed diesel units for energy production as against a comprehensive electrical grid system will become even more marked if photovoltaic electricity production becomes a substantial contributor to the total energy supply. Though photovoltaic cells produce no utilizable waste heat, the fact that stations are likely to be relatively small in output, and therefore numerous, means distribution costs will be minimal.

Where machines or appliances have a long life-time, the energy consumed as fuel generally far exceeds the energy deployed in their manufacture; where this happens, efficiency of consumption is much the most important criterion of performance. To achieve high efficiency designing for minimal thermal inertia (heat absorbed by the materials from which the machine is constructed) and for minimal heat loss is important, and in some cases may be found to be inadequate. For example, using synthetic natural gas derived from coal in gas-fired appliances in the home is less efficient than using coal in power stations supplying electricity for heating through the common electric fire. Where the efficiency of an appliance is the over-riding consideration, replacement of a low efficiency appliance with one of higher efficiency is well justified on energy-saving grounds.

Part Five

SOCIAL, ENVIRONMENTAL, AND INTERNATIONAL ASPECTS

Chapter 25

The Environment: Non-Radioactive Pollution

A large number of papers, reports, and books on safety, environmental protection and ecological stability have appeared during the past decade. It is impossible to do more than try to survey the more important aspects of pollution, and we have in fact touched upon a number of them in earlier chapters; for example the question of so-called heat pollution was considered and we concluded that it was not among the more serious.

Atmospheric Pollution. The emission of gaseous effluents of all kinds can have serious effects on the biosphere and there is no certain answer to some of the complicated problems that are posed. Not all such effluents have their origin in the production and conversion, as opposed to the consumption, of energy. Hugh Stephenson (*The Times*, 10th Feb. 1975) illustrates the nature of energy consumption troubles when he discusses Japanese regulations regarding car pollution which do not appear to be sufficiently strict. If exhaust emission standards are lax the motorist escapes the financial burden that possibly he ought to bear in the form of more expensive cars and less mileage per gallon of fuel. Lead tetraethyl as an anti-knock constituent of petrol certainly results in significant health hazards, and other exhaust gases give rise to an undue amount of harmful smog in the atmosphere. Blood sample studies carried out in Manchester showed that, in the city centre, lead concentrations of 970 parts per million were measured, as compared with 85 parts per million in the countryside nearby. This leads to people, particularly children, living in cities having a harmful concentration of lead in the blood and possibly being hyperactive in consequence. It has been suggested violence among adolescents may partly derive from this effect. Opinions on this question may differ but only in respect of the degree to which the lead is harmful.

Those with knowledge of cities like Manchester, Glasgow, and London in the UK, and Pittsburgh and Los Angeles in the US, will realize that the control of atmospheric pollution can bring enormous benefits. Many of the worst atmospheric effects of burning open coal fires have been eliminated through imposition of Clean Air Acts and there has been a fall in the incidence of illness and death from respiratory diseases. In Glasgow the fogs at the airport fell to less than half their previous frequency within a decade. Without doubt the control of effluent gases that is now practised in parallel with the centralisation of electricity production in large generating stations has resulted in purer air and its attendant benefits. Many of the cities formerly regarded as dirty and unhealthy now have far cleaner atmospheres throughout the year than in former times, and even in winter when domestic heating is at its highest this is still true. A typical oil-burning station at Fawley in Southampton (plate 7) is provided with a stack 670 feet high. If the effluent is not as perfectly clean as it might be, at least it is spread over a much wider area and is thus much more dilute, whereas in Manchester and many similar industrial cities, until some thirty years ago there were thousands of homes in closely-spaced rows pouring their smoke from their house chimneys and giving rise to frightening pollution. It cannot be claimed that 'small is beautiful' in these circumstances: each going his own way brought maximum harm to all.

Carbon Dioxide. The fact that fossil fuels burn to produce carbon dioxide and oxides of sulphur has given rise to considerable argument, some of an international nature. We mentioned briefly on p. 61 the problem of the increased rate of release of carbon dioxide into the atmosphere and this topic merits a good deal of detailed study. One of the main research lines of the IIASA energy programme is the analysis of the long-term consequences of release of very large amounts of CO_2. Such release is followed by sluggish processes including the conversion to carbonic acid, precipitation in rain, and passage to the ocean, until the gas is eventually trapped, in effect, in the deep ocean waters. The alteration of infra-red diffusion due to CO_2 in the atmosphere can lead to important climatic changes and has been carefully monitored in recent years. In fig. 94(a) the observed changes between 1957 and 1972, and in 94(b), the estimates back to the 1870s, are shown. It is clear the mixing in the deep oceans is so slow that a steady increase in the atmosphere occurs. It is not clear yet whether the average temperature is raised or lowered by the increase, but at least the equilibrium that prevailed for many thousands of years is disturbed. Expert opinion tends to the view that an increase in carbon dioxide means more blanket action of the atmosphere and an increase in average earth temperature. Even a small change will disrupt the growth pattern of the plants we depend on for food, and, if a rise persists, it will gradually melt the ice from the poles so that low-lying areas of land will disappear beneath the rising sea.

A study of the problem from a general point of view, including reference to the question of an increase in the radioactive isotope of carbon, C^{14}, is due to Friedrich Niehaus, also of IIASA (Research Memo RM-76-35, May 1976). This paper is related to a series of risk evaluations and it shows that the C^{14} production, if a very large world-wide nuclear power system were built up, would require attention, say, by 2100 A.D.

An interesting paper of Peter D. Moore (*Nature* **268**, 1977, p. 296) outlines some of the difficulties to be faced in finding an efficient sink for atmospheric carbon that would be effective more rapidly than ocean trapping. It is particularly informative on the question of forests as sinks, and on the whole the author concludes that the ecosystem disturbance is not easily controlled.

C. Marchetti of the IIASA has proposed positive geoengineering steps to avoid this global problem (see IIASA Research Memo RM-76-17, March 1976). The deep ocean is considered to possess a very large equilibrium capacity for CO_2 and in Marchetti's proposals the CO_2 is injected into thermohaline (warm and salty) currents that would carry it into the deep ocean; one such current entering the Atlantic at Gibraltar is estimated to have sufficient capacity to deal with all the CO_2 produced in Europe even in 2100. The paper gives an excellent account of this problem of ocean engineering and concludes on an optimistic note, arguing that the problem of the level of

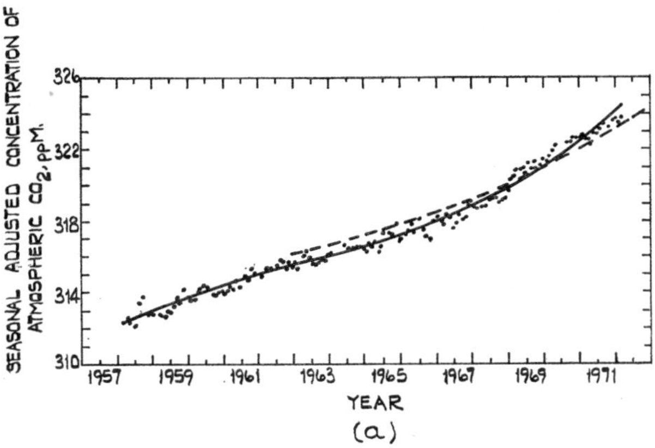

(a) The changes in the CO_2 content between 1957 and 1973

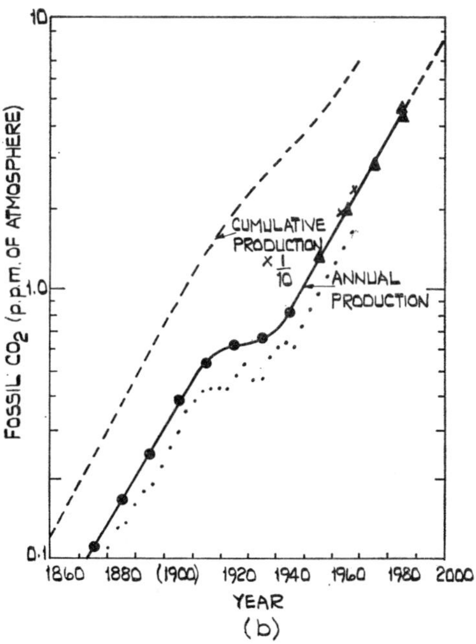

(b) The annual production and the cumulative production of fossil CO_2 since 1870
(The data of both (a) and (b) are discussed by C. Marchetti in IIASA Research Memo 76-17, 1976)

FIG. 94. Production of CO_2 and its accumulation in the atmosphere

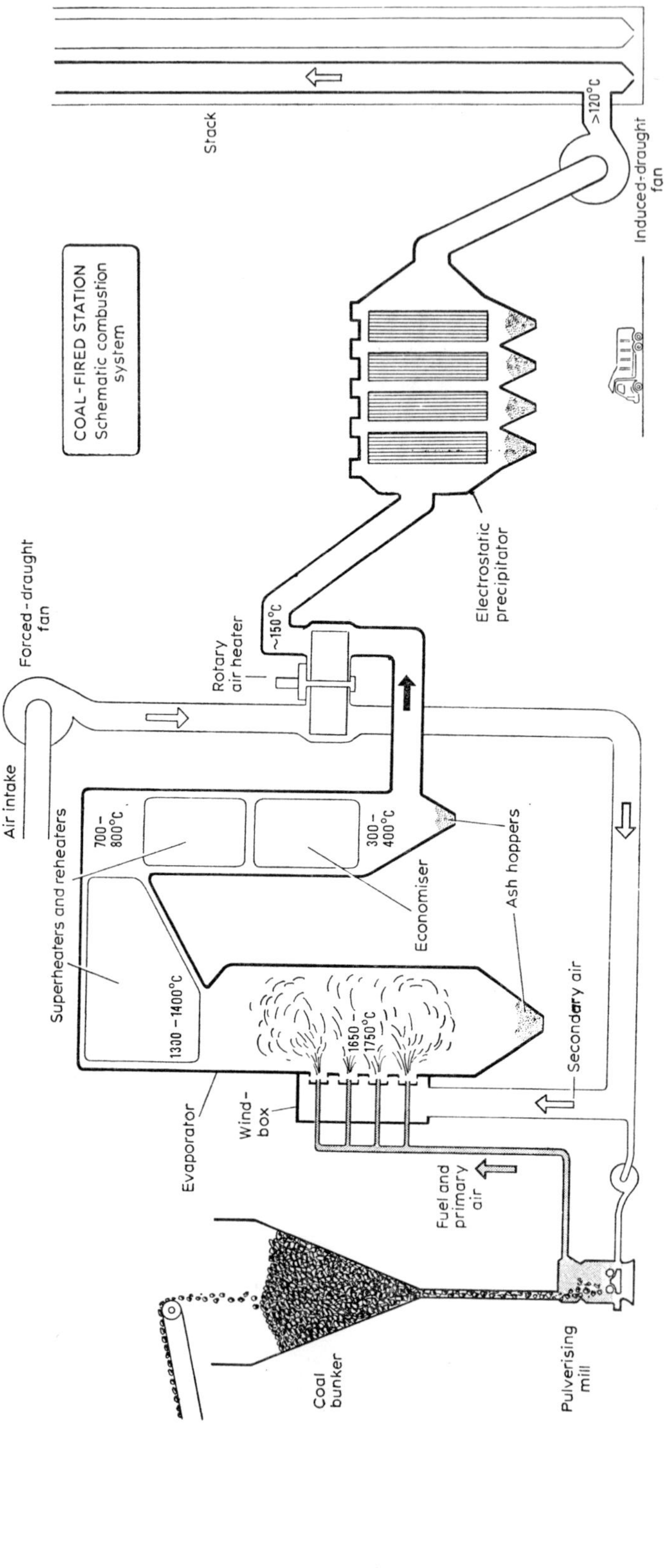

FIG. 95
Coal-fired combustion system. [From A. B. Hart and C. J. Lawn in *CEGB Research* 5, 1977]

CO_2 is much eased if equilibrium conditions with the total ocean mass are taken into consideration.

Sulphur, Hydrogen Chloride, and the Oxides of Nitrogen. Though the question of carbon dioxide is important it does not need to be treated as urgently as the pollution due to sulphur emissions.

An excellent account of the chemical nature of the combustion products of power-station boilers, of monitoring and control of flue gases, and the control of stack emissions, is given by A. B. Hart and C. J. Lawn. They consider the impurities in coal and in oil and their table is reproduced below.

TABLE 25.1

IMPURITIES IN COAL AND OIL
(Impurities in parts per million by weight)
(From A. B. Hart and C. J. Lawn in *CEGB Research*, no. 5, Aug. 1977, p. 4)

Element	Coal	Oil	Element	Coal	Oil
Sulphur (S)	15,000	20,000	Phosphorus (P)	1,000	4
Nitrogen (N)	12,000	1,500	Chromium (Cr)	60	3
Silicon (Si)	26,000	300	Zinc (Zn)	60	4
Vanadium (V)	145	150	Cobalt (Co)	47	3
Iron (Fe)	13,200	100	Manganese (Mn)	85	2·5
Nickel (Ni)	25	50	Copper (Cu)	130	2·5
Calcium (Ca)	9,500	100	Lead (Pb)	50	2·0
Potassium (K)	2,300	50	Selenium (Se)	7	1·0
Aluminium (Al)	25,500	75	Cadmium (Cd)	3	0·2
Sodium (Na)	1,470	50	Antimony (Sb)	2	0·02
Chlorine (Cl)	3,400	25	Arsenic (As)	10	0·01
Magnesium (Mg)	3,700	12	Mercury (Hg)	0·3	0·01

The main impurity of coal, and indeed oil, is sulphur, which typically occurs at around 1·5 per cent by weight but with considerable variations depending on the source. All of the sulphur burns to form sulphur dioxide (SO_2) and about 1 per cent of this may be oxidised further to yield sulphur trioxide (SO_3). The SO_3 combines rapidly with water vapour to give sulphuric acid.

A typical coal-fired combustion system as shown by Hart and Lawn is represented in fig. 95. An oil-fired is similar except for the coal-milling and ash-precipitation plant.

If it operates round the clock a 2,000 MW coal-fired plant consumes about 22,000 tonnes of coal and 200,000 tonnes of air per day. About 46,000 tonnes of CO_2 and 7,200 tonnes of water vapour are expelled into the atmosphere. There can be up to 6,000 tonnes of ash to be removed. The disposal of this ash in local tips can raise considerable difficulties. Some of the ash collects in furnace hoppers, most is recovered from the flue gases by precipitators and a small amount, 10 to 15 tonnes per day, leaves the stack as fine dust. Adhering to the particles are minute amounts of inorganic salts and oxides with sodium sulphate predominating. Operating with a typical grade of coal, the emitted flue gases carry to the atmosphere some 630 tonnes of sulphur dioxide per day (1,200 parts per million of total flue-gas volume). The fossil-fuel consumption of the CEGB since 1972 is shown in fig. 96. The move towards coal from oil is clearly evident.

The monitoring and control of flue gases according to UK practice is reviewed by D. J. W. Richards, W. S. Jones and J. W. Laxton (*CEGB Research*, no. 5, 1977, p. 18). The special problem of sulphuric acid is studied, and the dilution by high stack techniques to prevent high local concentration of flue gases and particles is examined. The gases contributing to atmospheric and ground-level pollution include hydrogen chloride and oxides of nitrogen. It is found that with 0·3 per cent of chlorine in coal there are about 250 parts per million of hydrogen chloride in the flue gases, which is less than the statutory limit. In the case of the oxides of nitrogen, the concentration in the flue gases varies between 200 and 600 parts per million and before these oxides reach the ground there is a dilution by a factor of between 20,000 and 50,000 for high stacks. This results in an expected ground-level concentration between 1 and 3 parts per hundred million. The CEGB estimates the total mass of nitrogen oxides to be between 25 and 30 per cent of the mass of sulphur dioxide.

One of the main differences between coal- and oil-fired plants is found in the proportion of ash and solids that the flue gases contain. Oil-fired plants are almost free of ash, but modern coal-fired stations must always incorporate

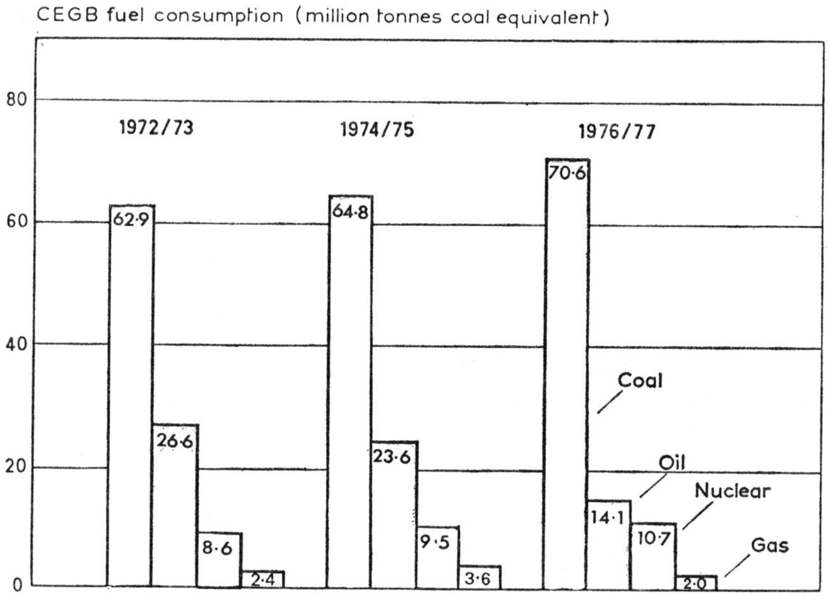

FIG. 96. CEGB fuel consumption, the years 1972–75 based on actual figures and 1976–77 on estimated ones. [From *CEGB Research* 5, 1977]

special collection equipment to deal with the substantial quantities of incombustible solid matter not retained in the furnace hoppers. Generally electrostatic precipitators of the type diagrammatically shown in fig. 97 are used, and these retain at least 99·3 per cent of the total weight of solids in the gas flow. In a large plant, a number of banks of parallel cells are placed in the gas flow with separate electrical supplies to each. Solid particles in the turbulent gas flow become electrically charged in the high-field regions near the wires, and are then attracted to the collecting plates. The amassed layer is removed by mechanical rapping. With coal of high ash content the precipitators sometimes require the injection of small quantities of acid materials, such as ammonium sulphate.

For further information on the factors influencing the dispersal of flue gases and the nature of the plume transport in the atmosphere, readers are referred to the paper by R. A. Scriven and G. Howells in *CEGB Research*, no. 5, 1977, p. 28. Their discussion of the possible ecological effects of industrial sulphur emissions in areas remote from their sources is particularly helpful. This has been a matter of concern for some time. An analysis carried out in 1968 showed that an area of highly acid precipitation, of *p*H value 3 to

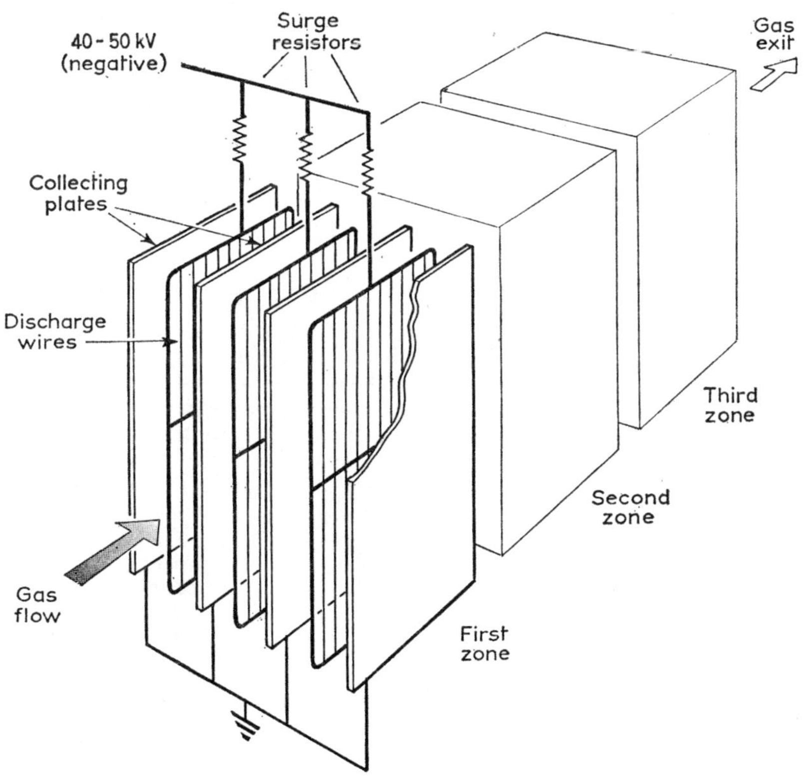

FIG. 97. Electrostatic precipitator. [From *CEGB Research* 5, 1977]

4, was expanding in the industrial heart of Europe. Moreover, in Scandinavia, the increasing acidity of inland waters, which was regarded as a cause of reduction in fish populations, and a reduction in forest production, were both attributed to sulphur dioxide emission from the industrialized countries of northwest Europe, especially from the UK. These factors gave rise to a study by the OECD. Their final report on the problem in northwest Europe was published in July 1977. The work was done by the European Air Chemistry Network and aimed at determining local and distant sources of sulphur compounds by monitoring the chemical composition of precipitation. A total of 11 countries agreed to participate (10 in 1972, and 11 from 1974) and the Norwegian Institute for Air Research analysed and interpreted the results.

From a total of 76 ground level sites, including 8 in the UK and 23 in Norway, mean sulphur dioxide and sulphate aerosol concentrations were observed to range from between 20 and 10 μg/cu.m respectively, close to the sources, to between 2·0 and 0·5 μg/cu.m in the remote areas ($\mu g = 1 \times 10^{-6}$ g). The sulphate in precipitation showed a similar pattern to the air concentrations, the value ranging from 50 to less than 1 mg in every litre. In regions of high precipitation both sulphuric and nitric acids were found to be present.

Airborne instruments were used by several countries to give above-surface measurements of air pollutants. These showed sulphur compounds were normally transported in the lowest 2 km of the atmosphere, generally speaking a few hundred metres above ground level. Distinct plumes of pollutant could occasionally be observed up to several hundred kilometres down-wind of a source. Modelling on a mathematical basis was used to

TABLE 25.2

ESTIMATED BUDGET FOR DRY PLUS WET DEPOSITION OF SULPHUR FOR 1974
(From R. A. Barnes, *Nature* **268**, 1977, p. 92)

Countries	Sulphur received (10^3 tonnes)	Sulphur emitted (10^3 tonnes)
Austria	300	221
Belgium	200	499
Denmark	100	312
Federal Germany	1,250	1,964
Finland	400	274
France	1,000	1,616
Netherlands	150	391
Norway	250	91
Sweden	500	415
Switzerland	100	76
United Kingdom	1,000	2,883
Others (Italy, Poland, GDR, Czechoslovakia)	11,000	—

obtain a more complete pattern of concentration and deposition in Europe but estimates were made for 1974 only. For the 11 countries man-made emission was found to amount to 9 Mt (million tonnes) in 1973. For the other European countries, and with less certainty, a total of 16 Mt was calculated. In fact the overall total of 25 Mt is roughly double that of 1950 (the UK rise from 1950 to 1973 was 30 per cent). Table 25.2 shows R. A. Barnes' estimated budget for dry + wet deposition. Barnes discusses some of the limitations of the two mathematical models employed and the weaknesses arising from the coarse grid. In particular the Trajectory Model is found to be incapable of estimating the dry or wet deposition in any one country due to emission in any other country to an accuracy greater than ± 40 per cent. The corresponding totals, dry and wet depositions, are not accurate for this Trajectory Model to more than ± 50 per cent. The survey nevertheless gives a useful indication of the extent and magnitude of transboundary transport of sulphur compounds. Within the area as a whole about 50 per cent seems to be dry deposited and 30 per cent wet deposited, with 20 per cent leaving the region. Clearly 25 Mt of sulphur is an enormous amount and it affects all countries in a complex fashion which requires more careful examination.

The costs of complete emission control at source are daunting. Many experts take the view that the costs of electricity would rise substantially if control at source was made obligatory (if indeed it were practicable). Instead the acidity of waters in rivers and lakes might be corrected by adding lime. This method is both practicable and cheap; the calcium sulphate formed by interaction of acid and alkali would be relatively harmless.

Pollution of Land and Water. Not only the atmosphere but land and water, too, are polluted during the manufacture of materials used in power production. Steel has to be used in constructing generating stations, whether it be for incorporation in the buildings themselves or in the turbines and generators. Steel likewise features prominently in the oil and gas industries, being used in large amounts in constructing derricks, platforms, containers, and pipes. Steel-making can have deleterious effects of many kinds, for example to water supplies since water is used in large quantities for cooling. After such use it has to be treated before it is returned to rivers or lakes and such treatment is not always as successful as is desired. The production of copper for electric conductors of all kinds, and in recent times the expanding use of aluminium in the electrical industry, means additional metallic extraction and refinement in which noxious byproducts are sometimes produced.

Oil spillages, whether these be on land, at sea, or during transportation by tankers, are another obvious problem. Much effort is now expended, and must continue to be expended, in devising improved methods of controlling the effects of such major disturbances of the biosphere. The gas industry presents its own special problems and, in particular, much care has to be exercised in transporting liquid natural gas. Large masses of LNG, or SNG, are very inflammable and they are particularly hazardous when vessels carrying cargoes of liquid gas are in port. In addition, gas leakages from pipelines not only is a serious loss of energy—it appears that amounts of up to 25 per cent of the total drawn from the source can be lost in transmission and distribution—but they can result in dangerous concentrations and possible explosions.

Undoubtedly the energy industry, as it operates today, is one requiring careful scrutiny. We shall have much to say on nuclear safety and the environmental problems of nuclear plants, but we have also wished to stress the scale of the hazards to which the public have become accustomed in the provision of coal, oil, and gas energy.

Chapter 26

Nuclear Power and Safety, Reprocessing and Waste Disposal

The political hazards allied to any nuclear programme—theft, sabotage, and the proliferation of nuclear weapons—are discussed in chapter 28. Here we describe the scientific aspects of safety problems. These fall into three categories—those concerned with the safety of the reactors themselves; those concerned with the hazards of processing and reprocessing; and, thirdly, waste disposal. The connection between these operations is illustrated in fig. 98.

Before beginning our discussion, however, we should remember that uranium mining has notable health hazards. The principal risk is of lung cancer caused by the gaseous decay products of radon, itself a daughter product of uranium. Good ventilation in mines greatly reduces the danger (see M. C. O'Riordan in *Geological Aspects of Uranium in the Environment*,

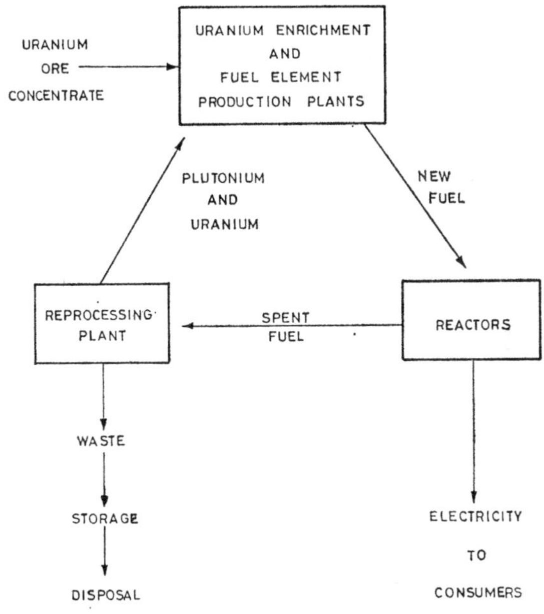

FIG. 98. The nuclear fuel cycle. [From D. W. Clelland in *The Breeder Reactor*, ed. J. S. Forrest, p. 65, Scottish Academic Press, 1977]

ed. P. McL. D. Duff, Geological Society Miscellaneous Papers **7**, 1977, p. 37). The numbers of affected miners have to be balanced against the numbers who would have been injured had they been mining enough coal to give the same amount of energy. It has been calculated (Newcastle University Study) that more than 300,000 miners would be needed in the UK if our estimated energy needs by the year 2000 were to be met with extra coal. The risk to the lives of such a work force are much higher than to the lives of the work force required in a corresponding expansion of nuclear power. We also note that winning the oil from the North Sea has already cost many lives—almost 300 by 1978. There is, alas, no known energy resource that can be tapped without hazard to human and animal life, and their use also has introduced substantial dangers to life and health. The alternative—survival with very inadequate supplies of energy—appears to be both unacceptable and hazardous.

A. Hazards of Radioactive Releases

In responding to the plea for sufficient supplies of energy, the possibility of a deliberate or accidental release of radioactive material from a nuclear plant has to be considered. In addition to its direct effect on human beings, the possible effects of the passage of radioactive waste into drinking water or the structure of plants or animals, and from thence into humans, have been studied closely and a good deal is known about possible repercussions.

In the UK the National Radiological Protection Board is responsible for many of the safety aspects of handling radioactive materials. In 1977 they issued a report, *Human Exposure to Radiation following the Release of Radioactivity from a Reactor Accident* by H. Smith and J. W. Stather (NRPB–R52, HMSO). The authors examine the effects of radiation on the population. They consider acute external exposure of the whole or parts of the body to a passing radioactive cloud, and also more continuous exposure from ground contaminated by radioelements deposited from such a cloud. Inhalation and ingestion of contaminated food and water are also studied as these lead directly to irradiation of lungs and gut, and some of the radioelements can pass to body fluids and thus reach other organs of the body.

The unit used in measuring absorbed radiation is the rad (1 rad = an absorption dose of 0·01 J/kg of affected material). Above a threshold of some 100 rad two pathological changes occur which lead to early and continuing effects. There is damage to cell membranes, causing leakage of fluids and electrolytes through the vascular system. Also, some of the stem cells lose their reproductive capacity, and this loss leads to a reduction of the number of mature cells and in turn to impairment of normal functions of the tissue. This second effect shows most in tissues with a rapid turnover of cells, such as bone marrow, gut and the basal layers of the skin. Such impairment of tissue function may become permanent at high radiation dosage and give rise to chronic illness. Cancer may appear after a variable period, which can be as long as some tens of years, and the risk of occurrence is assumed to be proportional to the dosage with no threshold for an effect. Among the non-malignant effects of radiation are cataract, sterility, and anaemia.

The concentrated alpha radiation to which lung tissue is exposed in the event of inhalation of a particle of plutonium or plutonium oxide, assuming the particle is trapped in the lung and insoluble, has been the subject of special investigation. There is no clear agreement on its effects and the truth may lie between two extremes: those who consider cancer highly probable may underestimate the natural capability of the tissue for self-protection. Nevertheless it is certain that inhalation of plutonium is one of the most serious hazards of a debris cloud.

The fission product strontium, Sr^{90}, is one of the most hazardous. It is readily assimilated in bone structure and the effects of the radiation on bone marrow are very dangerous. Another of the products, caesium, Cs^{137}, has to be given special consideration as it is very soluble. The radioactive isotope of iodine, I^{131}, is concentrated in the thyroid gland.

On the food side it has been discovered that some plants or parts of plants or animals concentrate fission products, increasing the hazard. Some fish and crustaceans, for example, concentrate elements like cobalt very strongly. Average values must not be taken to determine hazard levels in these particular cases.

B. Safety Problems of Reactors

By now the reader will have realized, if he did not do so before, that all nuclear power plants have to be extremely carefully designed because the uranium fuel elements, which may take a variety of forms as we saw in chapter 15, become highly radioactive within a very short time of 'start-up'. If the core material and its containers are well designed and carefully manufactured they need little attention for many years. While this is an attractive feature it means that the fabrication of every single fuel element must be done with elaborate safeguards and checks to prevent costly failures in operation. An even more serious possibility is that a reactor core will melt or be damaged if the cooling system fails as happened to a PWR at Harrisburg, Penn., in 1979; this hazard is made extremely unlikely by virtue of the many safety devices built into the reactor, but these in turn add considerably to the initial capital cost of the total plant. Also since used fuel elements contain so much intensively radioactive material as byproducts many of the operations are necessarily laborious and protracted.

The safety of the emergency core cooling system of light water reactors has been raised and argued strenuously. The Rasmussen Report of 1974 helped to allay some of the fears of major accidents but the debate continues and is likely to do so for some years until more thorough tests have been conceived and completed.

Apart from the core cooling system as such there is the question of the integrity of secondary cooling circuits associated with heat exchangers, for example. The various parts of such additional circuits may be exposed to radiation, and if neutron irradiation occurs they become radioactive. The handling of such associated parts of the reactor can then present substantial difficulties and long delays in repair work can ensue. This may be costly and it means the possibility of human error, such as that which caused the leak in the Hunterston AGR in 1977, has to be reduced to a minimum even when hazard to life is not in question.

Breeder Reactors. The safety of breeder reactors is a matter of much informed debate. The main criticism of the breeder, which as the reader will remember from pp. 167 to 172 is designed to utilize both isotopes of uranium as fully as possible, is based on two main points:

(1) The fast breeder reactor may be less safe than the thermal.
(2) Plutonium, which is both weapon material and extremely toxic, has to be used in substantial quantities.

It is indeed true that plutonium is a more dangerously radioactive substance than either U^{235} or natural uranium. It has a half life of 24,360 years and emits alpha particles which are intensely ionizing and therefore readily cause cancer. The alpha rays are stopped by a thin sheet of paper or a few centimetres of air but within the body they can damage the tissues in which they expend their energy. Ingestion or inhalation even of very small amounts can be lethal, so strict precautions in handling are essential.

We have described how there have already been years of safe operation of fast reactors in Britain, France, and the USSR, and how no unexpected hazards have arisen. From the start the safety features of the design have been regarded as of paramount importance. The use of plutonium, in quantities of the order of two or so tons in the core, is a new feature. However, for a large part of the life of a core of a thermal reactor comparable amounts are present (between one tenth and one half of the plutonium required in a breeder is created in the thermal core uranium), so there is only one order of magnitude change in the amount of plutonium within the breeder core.

Another hazard of the fast breeder is that we have such potentially high energy output per unit volume from the cores that we are required to employ relatively novel cooling materials, such as molten sodium. We have to regard such a coolant as comparatively untested, at least when our limited experience with it is compared with the longer time in which we have gained experience of water, steam, and certain gases. On the other hand the prompt power coefficient of liquid metal cooled breeders is negative: that is, if the core suddenly generates more than the required amount of fission energy, it then undergoes changes (which may include alterations to size and shape) such that the power output tends to drop. Thus upward excursions of power tend to be self-correcting. Against this desirable feature of core behaviour must be balanced the fact that if the sodium is allowed to boil virtual voids arise in the coolant mass and the power level tends to increase. Clearly the desirable effect must be made greater than the undesirable, and, indeed, the boiling of the sodium eliminated as far as possible.

Some far-reaching tests on the safety of breeder reactors have been entirely satisfactory. For example, a searching test was completed in 1977 at the time of official closure of the experimental Dounreay reactor. It provided realistic and reassuring evidence of the reactor's safety. After eighteen years of operation a few stainless steel canisters containing fuel were left in the core. The circulation of liquid sodium coolant was stopped. As the core temperature rose the sodium there began to boil, and later the containers and fuel melted and fell as a mass into the safety vessel. The experiment appeared to offer strong support for the view that loss of coolant in a breeder reactor would not result in a catastrophe, nor in the spread into the atmosphere of large amounts of radioactive fission products. In fact the reactor seemed to be 'the most stable and docile' of those used by the United Kingdom Atomic Energy Authority, according to Sir John Hill. Sir John contended that the result strengthened the case for constructing the first commercial breeder reactor.

It is reassuring to note, too, that as theoretical models of breeder behaviour are refined and improved as experimental results become available the severity of projected consequences of an accident have tended to diminish. In the *Annual Review of Energy* (**1**, 1976, p. 663) J. M. Hendric states that 'there is a reasonable prospect that detailed modelling of individual phenomena . . . may show that present estimates of the degree of core disruption from these accidents are grossly exaggerated'. He notes that extreme-design basis accidents, as defined in the early studies of breeder safety, were based on the assumption that shut-down systems would not function at all. In practice two separate shut-down systems, almost independent of each other, are incorporated. Hendric believes further improvements in accident assessment would favour simplification of this very conservative design for safety. He also argues that a major release of radioactive material in the event of an accident with a breeder would have harmful effects comparable to that from a light water reactor but that the probability of such an accident occurring with a breeder is less. The small size of the breeder core is one of the substantial safeguards, simplifying containment of plutonium in core and blanket.

The temperature of the core of the breeder is the same as that of a conservatively designed thermal reactor. The operation of the breeder at close to atmospheric pressure makes the possibility of loss of coolant much less than operation at higher pressures, such as is the case in most thermal types.

The arguments of Hendric and others appear to refute assumptions that breeder reactors are in any way less safe than thermal ones and they could well prove to be more safe. Additionally, the fact that plutonium is consumed in breeder reactors is a strong point in favour of their introduction. If only thermal reactors are used there would be a steady increase in the total weight of plutonium available in the world. E. Zebroski and M. Levenson (*Annual Review of Energy*, **1**, 1976, p. 129) state that 'in 10 to 15 years spent fuel will represent a large and technically available energy mine . . . comparable in value to one of the world's major oil fields if the residual uranium and plutonium values are used for recycle fuel in LWRs. If the plutonium in this energy mine is used to fuel breeder reactor capacity, starting in the 1990s the energy reserve from just this small amount of plutonium is greater than all of the oil and gas so far discovered in the world'. This extraordinary amount of energy wealth can be created by just over a decade of operation of thermal reactors on the comparatively modest scale presently envisaged. The plutonium consumption in breeders is not only the wise thing to do, it is a most rewarding exercise in its own right—one which could enhance the world energy store on a scale never previously experienced in history.

As we learnt on p. 168, it is also very likely that a large amount of thorium can be used to provide U^{233} as fuel. So far as safety is concerned this could prove in due course a more attractive possibility than the uranium/plutonium systems.

Another point closely concerned with safety in the broad sense, if not with safety of the breeder machine as such, should be stressed. If breeder reactors are used in conjunction with thermal reactors, uranium mining can be greatly reduced in scale.

The development of breeder reactors will not herald in a new 'plutonium economy' for that hazard is already with us. E. Zebroski and M. Levenson (*Ann. Rev. of Energy*, **1**, 1976, p. 101) think there are more than 80,000 nuclear weapons in existence, and state that 'plutonium production of the order of 100–200 tons has occurred in the US and at least a comparable amount overseas, primarily in Russia but also in the UK, France and China'. The breeder might some day, if nuclear weapons are no longer stocked, consume this plutonium and remove one element of the nuclear war threat. Presumably some of this weapon plutonium is carried in warships and military aircraft, and ships that are nuclear powered must have plutonium as a byproduct in their reactor cores. It would seem that a sea-borne or air-borne reactor with plutonium within it is not more safe than a stationary breeder equipped with all the sophisticated safeguards that can be devised. Nevertheless we should not forget the hazard implicit in the operation of such reactors were there to be a large number of them in the future, or were a few with a huge energy output to be constructed. The present-day large commercial reactors are designed to produce around 1,000 MW of electrical power, the total thermal energy released being roughly three times as much. It has been suggested reactors might be constructed so as to produce perhaps ten or more times as much energy; even if they are placed in huge reactor parks or islands it is essential that all safety factors be thoroughly explored.

Fusion Reactors. There is a tendency to contrast the known hazards of fission reactors with the theoretically safer fusion reactor. However, there is no proof at the present time that fusion reactors will become practical machines, and even if they do so it is not at all obvious that they really will be very much safer than fission reactors. They will have a core of reacting gases such as deuterium and tritium, and they will be 'fast' in the sense that the interacting particles, whether ions or neutrons, will be of extremely high energy. The fast neutrons can be slowed down after escape from the core and it will be arranged that many produce deuterium or tritium. The second of these is a radioactive gas, always more difficult to contain, and therefore more dangerous, than a radioactive solid. In addition the vessels surrounding the reacting gas core will become extremely radioactive, more so than any type of fission reactor. Very approximately we guess that the hazards of radioactive waste from such a fusion machine would be about one tenth of those of a fission reactor of about equivalent power output; therefore the hazards of a fusion machine will not be completely negligible. Furthermore, the fusion machine will be an extremely intense neutron generator, and obviously plutonium can be made in large amounts by exposing uranium to the flux. In the wrong hands fusion can be used to provide material for nuclear weapons. In this respect there is not a sharp dividing line between fusion and fission.

C. Processing and Reprocessing

Processing means the winning of natural uranium or thorium in the form of their ores, their conversion into fuel, and their treatment after use in the reactor to remove the wastes and recover unburnt material. The waste management is in itself a specialist activity, discussed in the next section.

During irradiation both uranium and thorium breed new fissile material which can be recovered, converted to fuel, re-irradiated, and repeatedly recovered. This is known as reprocessing. Processing and reprocessing are not as distinct as the definitions suggest and they proceed together in an operating plant. They are essential in nuclear breeding systems.

An informative, succinct, account of processing and reprocessing is given by G. N. Walton in *Nuclear Reactors to breed or not to breed* (ed. J. Rotblat, Taylor & Francis Ltd., 1977, p. 67).

The processing cycles presently utilized in UK reactors are as follows:

(*a*) The Magnox reactor uses fuel of natural uranium metal which produces plutonium and depleted uranium.

(*b*) The AGR starts with slightly enriched uranium oxide. Plutonium and depleted uranium, which may be returned for enrichment in U^{235}, are produced.

(*c*) The fast reactor uses initially enriched uranium, and later plutonium together with depleted uranium.

The three cycles are shown schematically in fig. 99 where the paths are labelled to coincide with the cycles (*a*), (*b*) and (*c*).

There are no major differences in the reprocessing treatments of fuel for fast breeders and for thermal reactors. The main steps in the reprocessing plant for irradiated fuel for the AGR at Windscale are shown in fig. 100.

Fuel for treatment is stored to allow the short-lived radioactivity to decay. The cladding is next removed, a step giving rise to a requirement to store the radioactive metal oxide. The uranium is dissolved in nitric acid and passed through a continuous solvent extraction plant. The highly active fission products come out in the first of the five extraction columns; the plutonium comes in the second. The other columns serve to decontaminate the products.

Walton emphasises that reprocessing the fuel from natural uranium reactors means handling all the uranium in order to burn approximately half of the 0·72 per cent of uranium which is U^{235}—for at that stage criticality, the concentration above which the process is self-generating, has ceased and new fuel is necessary. If the new fuel is made up to 2 or 3 per cent U^{235} the reprocessing is reduced about fourfold. With fast reactor fuel of 20 per cent plutonium content the reprocessing is reduced in frequency by a factor of about 30 for equal power output and equal fractional burn-up. In other words, much increased nuclear energy production certainly does not require a proportionate increase in reprocessing.

Blanket processing is independent of fuel treatment. The composition of both for a representative breeder reactor is given by Walton and shown in table 26.1.

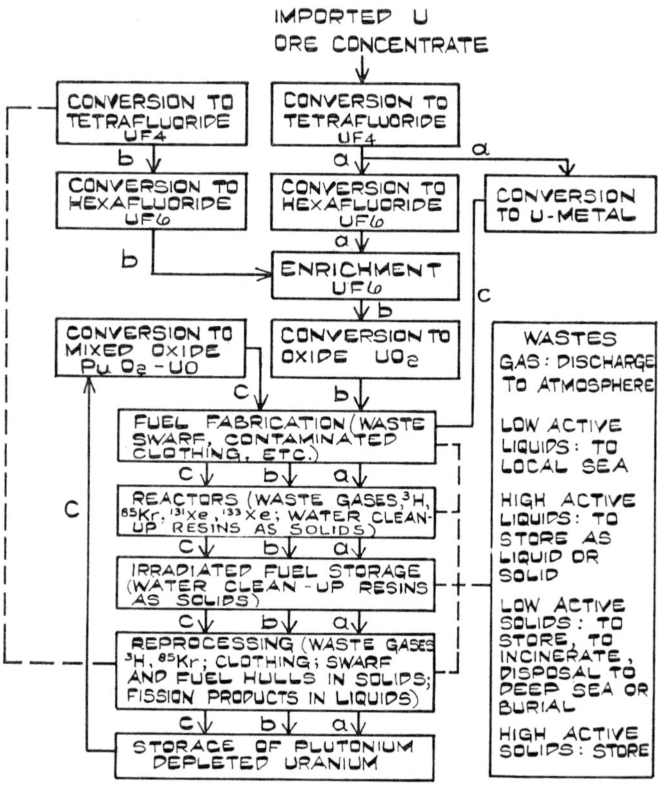

FIG. 99. Simplified version of the nuclear fuel cycles for (a) Magnox, (b) AGR, (c) fast breeder, showing main categories of waste. [From G. N. Walton in *Nuclear Reactors to breed or not to breed*, ed. J. Rotblat, Taylor & Francis, 1977, p. 68.] 'Swarf' is the industrial term for shavings of eroded metal.

TABLE 26.1

FAST BREEDER FUEL BURN-UP

(From G. N. Walton in *Nuclear Reactors to breed or not to breed*, ed. J. T. Roblat, 1977, p. 71)

	Fuel as loaded in reactor	Fuel on discharge from reactor
Core	25% PuO_2 75% UO_2 (as U^{238})	23% PuO_2 71% UO_2 6% fission products
Blanket	100% UO_2 (as U^{238})	95% UO_2 4% PuO_2 1% fission products

Regarding burn-up it is interesting that each megawatt of thermal power involves burning 1 g of fissile material per day, and this amount is nearly independent of the type of reactor or indeed the type of fissile material. For

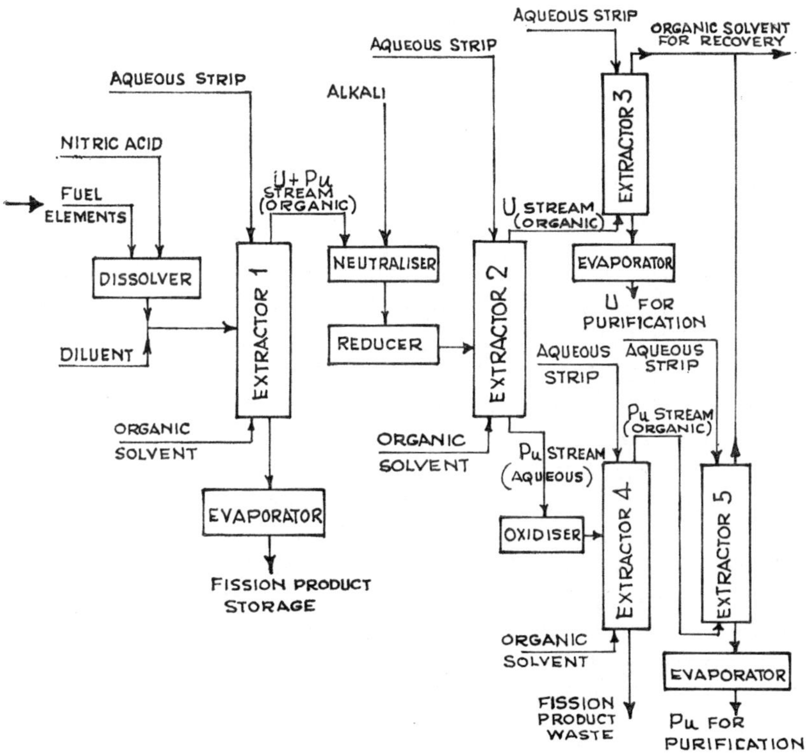

FIG. 100. The Windscale primary separation plant for the recovery of uranium and plutonium from irradiated fuel elements. The diagram shows in more detail the processes indicated by the box marked 'Reprocessing' in fig. 99. [From G. N. Walton in *Nuclear Reactors to breed or not to breed*, Taylor & Francis, 1977, p. 70]

a 1,000 MW electrical plant of around 33 per cent efficiency the thermal power is about 3,000 MW and the fissile material consumption about 3,000 g per day, that is 3 kg. In 1977 the total UK nuclear electrical output of some 10 GW meant a burn-up of some 30 kg per day. The Windscale design allows for the processing of 6,000 kg per day, well above the present load, encouraging the Government to look for contracts to process spent fuel from thermal reactors in other countries. While in principle a substantial breeder processing and reprocessing programme could be accommodated, there are differences in the nature of the irradiated fuel that make it desirable to construct a different kind of fuel treatment plant for breeder reactors. The higher rating of the breeder fuel (that is, less mass burning more rapidly) results in the material for reprocessing being more highly radioactive as compared with thermal material. Ruthenium (Ru^{106}), palladium (Pd^{107}), and rhodium (Rh^{101}) besides all being radioactive are chemically inert so they require special treatment since they cannot be manipulated into more convenient chemical forms. The result is that the reprocessing plant must be comparatively small with extensive shielding and considerable remote handling. The need for water and hydrogenous solvents introduces possibilities of parts of the chemical plant becoming critical; 'ever-safe' design principles can be used to overcome this complication.

Walton discusses other differences between highly irradiated fuel such as that in the typical fast breeder and the fuel in typical thermal reactors. Multiple neutron capture can occur much more readily in breeder arrangements and this can result in the formation of rather troublesome radio-elements which can act as 'poisons' in the fuel and which make handling more difficult in the reprocessing. The same difficulties attach to the thorium-U^{233} fuel cycle, perhaps more markedly. Consequently, though the breeder reprocessing is reduced so far as mass throughput is concerned, the material requires more sophisticated and careful chemical treatment.

Walton discusses concentration of fuel, particularly plutonium, and of waste, mainly fission products. He prefers high concentration of waste, a contrary view to that of some other experts, and the dilution of plutonium at all stages with UO_2. He advocates the adoption of a practice of reasonably long 'hold-up' time in the passage of fuel from reactor to the reprocessing. This increases costs but makes for improvements in safety and reliability. Finally he affirms his view that a breeder reactor construction programme should be followed and processing plants made. The plants should avoid all movement of separated plutonium, and should process completely high-level, medium-level, and low-level waste in all its forms.

Fears that reprocessing plants are particularly vulnerable to burglary by terrorists have been frequently expressed. This topic is dealt with in chapter 28.

D. Radioactive Waste

Many of those who are opposed to the expansion of nuclear power stress that there is continuing doubt about how to deal with the considerable amounts of waste radioactive substances. On the other hand a few scientists think far too much anxiety is expressed over this matter. Professor Frisch, for example, has written an amusing account of the problem of designing a coal reactor, arguing that if every possible disadvantage had been explored before constructing the first one possibly no power stations would ever have been built and no coal fires lit.

In the UK, ever since nuclear reactors first came into operation some twenty years ago, their waste, in common with all other radioactive material, has been handled according to instructions drawn up by government departments—the Department of the Environment, the Ministry of Agriculture and Fisheries, and the Welsh and Scottish Offices. Low-level waste has been discharged into the environment but the more active waste has been stored to enable it to decrease in its activity and to give time to find a disposal method acceptable to scientists, politicians, and the general public. Much the same situation obtains in other countries.

The nuclear process associated with different types of reactor all give rise to much the same wide variety of wastes, though some special difficulties arise with breeders (see pp. 278–80 above). Fig. 99 shows the most important wastes and disposal methods that occur in the three cycles that we mentioned in connection with processing.

In *The Breeder Reactor* (Scottish Academic Press, 1977, p. 6) S. C. Curran stresses that the total number of radioactive atoms produced in a fission power programme cannot possibly exceed the number of radioactive atoms of natural origin, generally uranium, by a factor of more than 3. Most of the man-made radioelements are more active than uranium, but as they also decay more quickly the fuel elements are stored to allow time for those left at the end of the run to decay. As a result most of the radioactivity has died away even before the spent fuel is withdrawn from the shielded reactor core. A few of the fission products, however, have awkwardly long half-lives, including the main product of neutron capture, Pu^{239}, and two other actinides—americum and curium. These three elements, like uranium itself, remain active for thousands of years and it is they that present the most serious waste problem. If the Pu^{239} is separated for use in breeders a significant part of the waste disposal problem disappears.

The highly active liquid waste, the result of extraction by nitric acid during reprocessing, contains more than 99·9 per cent of the residual radioactivity, including most of the americum and curium. After concentration by evaporation its volume is about ·25 cu.m for every tonne of spent fuel treated. At present it is stored in stainless steel tanks which have a cooling system to remove the heat of radioactive decay. Much research has been devoted to the problem of rendering this waste less mobile and less dangerous. It has been tentatively suggested by Marchetti that the containers should be buried directly at some selected sites, thus allowing the radioactive heat generated by their contents to heat the containers to the point where they would melt the rocks of the Earth's crust and drop down to bury themselves at great depths.

It is an attractive idea, for we must remember that there are, whether we like it or not, large amounts of uranium and thorium (with all their decay products such as radium) and of radioactive rubidium and potassium at comparatively shallow depths within the Earth's crust and in the oceans. These natural radioactive elements do not seem to constitute a substantial danger to health when they are left undisturbed in the Earth or the seas. Another idea is to put the radioactive waste at the North or South Poles and let it melt itself down into a trap of snow and ice; however, objections can be made to this procedure because it is conceivable that such a trap would split open in a relatively short period.

The present programme of work on radioactive waste disposal in the UKAEA is discussed in the Annual Report and Accounts of the UKAEA, 1976/77, p. 30. In the Harvest Process, which is being developed on a commercial scale by British Nuclear Fuels on the basis of experimental work done at Harwell, the concentrated waste is mixed with glass-forming chemicals so that, on cooling, they harden together into dense glass. Samples of this glassy material have withstood the equivalent of 8,000 years of alpha radiation dose. The principal advantages of the glass, other than its compactness which is convincingly demonstrated in plate 8, are that it will incorporate a wide variety of radioactive elements, and being highly insoluble, is not subject to leaching.

Highly active solid waste derives mostly from broken pieces of fuel element cladding, and appears in fig. 99 as 'swarf' and 'fuel hulls'. It is compressed and embedded in concrete. It is suggested that it should be stored near the surface of the Earth (engineered storage) for several decades to allow a large reduction of short-lived activity. The actinide concentration would be reduced by a factor of 10 by chemical processing or dilution, resulting in some tens of cubic metres of waste from each reactor per year. In this state the heat produced would be low enough to allow production of a stable insoluble matrix which would next be buried far underground in the rocks or under the sea.

Less active solid wastes, such as resins used to clean up the water coolant by ion exchange, and the residue filtered from gases, are stored, burnt, or dumped in the sea. Materials only slightly contaminated, such as clothing, are buried in approved areas.

Low-level liquid waste is discharged into the sea, and gases into the atmosphere.

A good account of waste management in the UK is N. J. Keen's article in *Geological Aspects of Uranium in the Environment* (Geological Society Miscellaneous Paper **7**, 1978, p. 46).

An interesting answer to the waste problem is found in the paper of Karl H. Puechl (*Nuclear Eng. Int.*, 1975, p. 950). His quantitative assessment of radioactivity levels leads him to conclude that if the curium and americum isotopes separated from the waste stream during reprocessing of the fuel elements were used or re-irradiated, instead of being treated as waste as at present, a dilution of the remaining fission product waste by a factor of 10^4 would achieve concentration levels of the material below 10^{-8} Ci/g for all time after a 600-year controlled storage. (A curie, the unit of radioactivity, equals $3{\cdot}7 \times 10^{10}$ disintegrations per second.) The re-irradiation of unwanted materials lying above the line of 10^4 dilution factor shown in fig. 101(a) would mean that the whole of the material diluted by 10^4 would present no appreciable health hazard. Puechl proposes dilution with sand, clay, or other aggregates after reprocessing rather than after the 600 years. He discusses in some detail the practical features of his scheme, and gives a flow sheet for a large reprocessing plant handling 5 tonnes per day which is shown schematically in fig. 101(b).

It is clear from our description of these various proposals that, though much research is being done, no perfect solution to the problem of radioactive waste has yet been established. Against our uncertainties about nuclear waste we have to weigh the known hazards of stack emissions from fossil fuels and of the increasing percentage of carbon dioxide in the atmosphere, both discussed in chapter 25. Beckmann, in *The Health Hazards of Not Going Nuclear* (*op. cit.*), stresses the fact that coal-fuelled plants also pass a good deal of radioactive material into the atmosphere via the stacks. He believes that even on radioactivity pollution grounds the nuclear path is preferable to the fossil-fuel route.

A recent flurry of activity suggests entirely novel solutions will be explored, probably successfully. *The Times* of 28th July 1978 reported the early stages

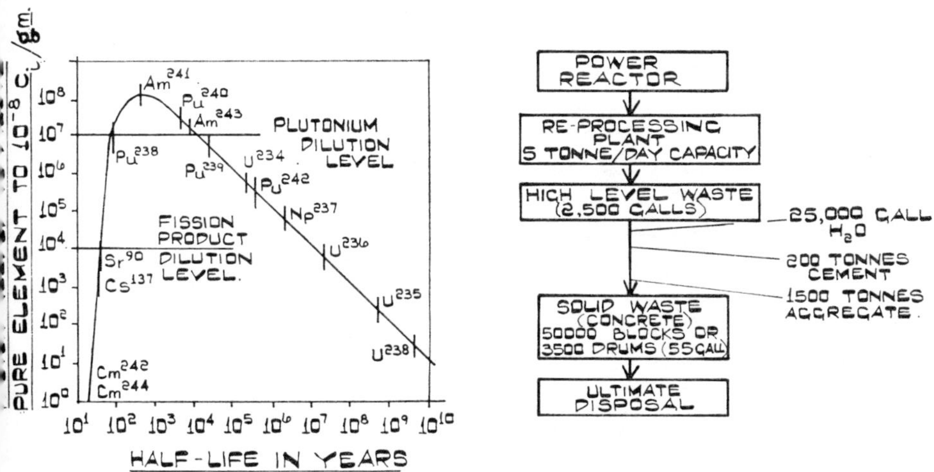

(a) (b)

FIG. 101. (a) Dilution factors needed to obtain safe concentrations ($< 10^{-8}$ C_i/gm) of pure isotopes after allowing for 600 years of controlled decay

(b) Flowsheet and materials balance for proposed waste treatment scheme (quantities are for a 5 tonne per day reprocessing plant)

[Both from K. H. Puechl, *Nuclear Eng. Int.*, 1975, p. 950]

of work by A. E. Ringwood at the Australian National University in Canberra. Ringwood observed that in nature small amounts of the elements that are present in radioactive form in nuclear waste are completely immobilised in the crystal lattices of certain minerals, and he has calculated that their imprisonment lasts for up to 2,000 million years, much longer than the time required for decay to safe levels. Among the effective minerals are perovskite, zirconia, hollandite, barium-felspar, and leucite. Ringwood's technique imitates nature. The minerals are mixed together in a melt at high temperature and to the melt are added radioactive elements in up to 10 per cent concentration. On cooling the radioactive atoms become locked in the minerals of the rock. Ringwood has named the synthetic substance Synroc, and contends that it will prove impervious to water. Supporting disposal procedures would include non-corrodible metal alloy containers, and burial in specially sealed, deep-drilled holes in granite rocks. Clearly much remains to be studied, but such an approach is a promising one (see Ringwood's own book, *Safe Disposal of High-Level Nuclear Wastes: A New Strategy*, Australian National University Press, 1978).

In conclusion, although the whole question of waste disposal is complex, it would not appear to be in any sense too difficult to resolve satisfactorily, and indeed the confidence of finding a solution when necessary might have contributed to the apparent delay in doing so.

E. Safety Legislation and Statistics

The assumption that all amounts of nuclear radiation however small may produce deleterious effects on the human body underpins the safety code that is followed in nuclear plants in the UK. The limit per annum for workers in atomic plants is set very conservatively, especially when it is remembered that all of us everywhere are exposed to penetrating radiation. The biological effect of a dose of ionizing radiation, measured in rads, is expressed in terms of the rem (roentgen equivalent man) as unit. This measure takes into account the nature of the radiation and adopts the biological effect of X-rays, generated at 250 kv, as standard. The rem is the product of the dose in rads and a quality factor (QF) dependent on the nature of the radiation. The QF value for alpha particles and neutrons is 10 while beta particles and gamma rays are regarded as equivalent to X-rays with QF equal to 1. Thus 1 rad of gamma radiation = 1 rem, and 1 rad of alpha particles = 10 rem. Protons can be equated with alpha particles and mesons with gamma rays. In cosmic radiation high energy protons, neutrons, mesons, and gamma rays are observed and these are sometimes absorbed in the body. In addition the potassium in the body, the isotope K^{40} in particular, is beta radioactive. The Earth itself and many materials around us are radioactive and all the background sources add up to a total average dose each year of one tenth rem, a rem being a dose of ionization equivalent to one which would produce ions carrying one electrostatic unit of electricity in dry air. The US law sets an upper limit of an additional 1/200 rem to the radiation dose that the public may be allowed to experience when living near a reactor plant. This shows the effectiveness of the reactor screening and the care with which all the processes are pursued. It must be remembered, too, that in certain parts of the world, for instance in areas where the rocks contain a higher than average amount of uranium and thorium, the background dose may be as high as 1 rem and even greater. Also at high altitudes the dose of cosmic radiation is always much greater than at sea level. So we are exposed to amounts varying between less than 1/10 rem and 1 rem according to where we live. The maximum amount that a worker in the US nuclear industry is allowed to absorb is 5 rem per year. This is approximately the dose received in an average X-ray examination, say, a TB test. It is clear that unless we take an extremely adverse attitude towards the use of X-rays in medicine (and none can deny the efficacy of X-ray diagnosis) the workers in the nuclear industry are well protected during normal operation of all plant.

The statistical data so far produced seem to establish that working in reactor plants is not an unhealthy occupation. In table 26.2 the mortality statistics of employees of the UKAEA and of British Nuclear Fuels Ltd. are shown and it has been assumed that the workers have an unusually good record, though it should be noted that special steps are taken to safeguard their health, including regular medical examination. Doubts have been expressed about the validity of some of the conclusions drawn from the data shown in the table but it does seem safe to conclude that there is at least no evidence of cause for concern to date.

TABLE 26.2

MORTALITY STATISTICS FOR MALE EMPLOYEES, 1962–74

Cause of death	Actual number	Actuarially anticipated for general public	Actual/anticipated ratio
All causes	2,730	3,652	0·75
All cancers	730	858	0·85
Circulatory	1,497	1,651	0·91
Respiratory	173	437	0·40
Digestive	65	90	0·72
Genito-urinary	26	47	0·55
Accidents and violence	153	197	0·78

The record of workers in the nuclear industry should be contrasted with the poor medical record of workers in many other occupations. Most of us realize coal-mining is dangerous to health and in fact the mortality rate for respiratory diseases in coal mining is thirty-five times as great as for the general public. Even company directors have a mortality rate almost eight times above the average for cancer, suicide, and accidents combined. In fact the most vulnerable people connected with the nuclear industry, the ones whose mortality is above average, are the uranium miners we have mentioned before. It is therefore reasonable to try to keep the amount of uranium we need to a minimum by making the best possible use of it; that is to consume it in breeder reactors or some near equivalent.

Two books stress strongly the excellent safety record of the nuclear industry. One is Petr Beckmann's *The Health Hazards of Not Going Nuclear* which we have often had occasion to quote before. In the second book, by Fred Hoyle, entitled *Energy or Extinction*, cited in connection with Hoyle's views on thorium (chapter 15) an outright attack is launched on those who oppose nuclear energy on health and other grounds. While much informative scientific material is lucidly presented in both books, there appear, as in the statements and arguments of some opponents of nuclear energy, some points that seem personal to the authors and speculative in nature.

Chapter 27

Political Aspects of the Civil Uses of Nuclear Power

People have become used to the vast amounts of spoil associated with coal and shale mining, and to some extent have learned to bear the spillage of oil from tankers which contaminates beaches and kills thousands of birds. There is little general awareness of the atmospheric pollution still caused by the burning of fossil fuels and the resulting huge tonnages of effluent which are spilt into the atmosphere (pp. 268–74). On the other hand, as we have seen, there is sharp controversy about nuclear power systens and particularly about the disposal of radioactive waste. Opposition to nuclear construction programmes has brought about a reduction in the scale of construction. For example, in the US in January 1975, President Ford announced that his administration would attempt to have at least 200 nuclear generating stations in the US by 1985, around a fourfold increase from the 55 then in operation. Soon a variety of factors led to the intended expansion being severely restricted, among them a hostile political environment in Washington, backed by a coalition, led by Ralph Nader, of anti-nuclear pressure groups which revealed their considerable strength and commitment in a protest march after the Harrisburg incident. Among the nuclear critics are the UCS (Union of Concerned Scientists) and the NRDC (Natural Resources Defence Council). Bills to curb the growth of the industry in various States resulted in considerable delay and frustration within the industry. In California the opposition brought about a referendum on the main safety questions but the voters came out strongly in favour of nuclear power expansion.

Colin Norman (*Nature* **257**, 1975, p. 346) has discussed political changes and delineated the main issues. These include enrichment policy; the recycling of uranium and plutonium for reactor fuel; the breeder reactor; and radioactive waste disposal. Although the opposition is seldom attacking nuclear energy as such, but its safety standards, no guarantee can be given for standards which are not infrequently impossibly high. The other area for attack is the idea that electricity consumption must continue to grow (for some years 6 per cent per year growth in the US) and nuclear fission energy be used to fill the gap left by dwindling oil supplies.

In the UK a number of important questions were the subject of the report *Nuclear Power and the Environment*, the Sixth Report of the Royal Commission on Environmental Pollution (Cmnd. 6618, HMSO, 1976) chaired by Sir Brian Flowers. The Commission concludes that thermal reactors of the current and proposed nuclear programme give no serious cause for disquiet except perhaps with respect to the long-term treatment of radioactive waste. They added that 'There should be no commitment to a large nuclear programme including fast reactors until the issues have been fully appreciated and weighed in the light of wide public understanding.'

The summary of the Report includes the statements:

(*a*) 'At the levels of radiation likely to be permitted in relation to possible somatic effects, the genetic effects should be of little concern.'

(*b*) 'No case has been made on the special carcinogenicity of "hot" particles of plutonium.'

(*c*) 'The risk of serious accidents in any single reactor is extremely small; the hazards posed by reactor accidents are not unique in scale nor of such a kind as to suggest that nuclear power should be abandoned for this reason alone.'

As regards radioactive waste the Commission view is expressed as follows:

(*d*) 'We are confident that an acceptable solution will be found and we attach great importance to the search; for we are agreed that it would be irresponsible and morally wrong to commit future generations to the consequences of fission power on a massive scale unless it has been demonstrated beyond reasonable doubt that at least one method exists for the safe isolation of these wastes for *the indefinite future*.'

A. R. Mackintosh has discussed the question of waste disposal and other topics in the light of the Flowers Report (*Physics Bull.*, **28**, no. 4, 1977, p. 165). He takes the view that (*a*), (*b*), and (*c*) are accepted by the great majority of informed people but argues that the last three words of (*d*) show the Commission failed to grasp a real understanding of the waste problem and confused *total* toxicity with *specific* toxicity (toxicity per unit mass).

It is not the total of radioactivity generated but the concentration of waste by the nuclear industry that introduces the true problem. The fission products, as Mackintosh stresses, must be kept out of the biosphere because they have an effective half-life of some thirty years and constitute a measurable amount of radioactivity, dwindling exponentially, over some three hundred years. He argues therefore that we must arrange for the wastes to be isolated for this period, but not for a time that could be described as 'the indefinite future'. He also points out that the actinides, including Pu^{239}, have half-lives of thousands or tens of thousands of years and their *total* toxicity, measured by their alpha activity, is not particularly large relative to the natural material from which they were formed (about 10^3 factor), and by no means large relative to the original amount of uranium and thorium involved. The total radioactive toxicity of nuclear waste after a few hundred years is in fact much lower than that of the radium in the core which was originally used to obtain reactor fuel. Mackintosh also stresses that the total relative increase in the amount of radioactivity in the biosphere due to the production of actinides in the generation of nuclear power is very small. The total amount of actinide activity can be reduced by consuming plutonium 239 in breeder reactors though Mackintosh does not himself consider this point. (He favours, however, an orderly development of the breeder on the grounds that it is the only new large-scale energy source that is known with reasonable certainty to be economically and technically viable.)

R. H. Mole, of the Medical Research Council Radiobiology Unit at Harwell, also discussed some of the weaknesses of the Flowers Report (*Nature*, **264**, 1976, p. 494). He notes the accent of the Commission on abnormal events rather than on the total effect of possible accidents, and he contrasts the exceptional radioactivity release of 10^7 curies of I^{131} or 10^6 curies of Cs^{137} giving 110–350 deaths from cancer with the same number of accidental deaths in UK coal production over a two to six year period at the current annual level. He says that such a radioactive release would occur once in 100,000 years if more than twice all the electricity generated in the UK in 1975 were of nuclear origin. On this basis nuclear power would seem to be more than 10,000 times safer than coal power. A substantial part of Mole's analysis is concerned with the need for more consistency in hazard analysis by bodies such as the Royal Commission, who refer to lack of research on atmospheric pollution by radioactivity yet appear not to know of the existence of more than one hundred scientific papers on the subject published from the Environmental Sciences and Medical Division of AERE alone. He also asks why no criteria are provided by which judgments of degrees of safety may be assessed.

In his own reflections on the Royal Commission Report, Sir Brian Flowers is quoted in *Nature* (**264**, 1976, p. 496) as wishing to see one specific site for waste disposal chosen with a consensus of expert opinion as soon as possible. He considers the site need not be the one eventually adopted but that it would be able to provide a fall-back position. He is apparently persuaded that 'the crucial long-term issues are on the one hand the competition between nuclear energy and coal and on the other hand between both and the renewable resources . . .'. We challenge his notion of this apparent conflict. No responsible person suggests that coal can fill all the energy needs when oil and gas become depleted. Moreover, there is plenty of opportunity to develop all forms of energy source and to arrive in due course at an optimum system in the light of actual experience.

A panel of US experts examined on behalf of the US Government issues similar to those of the Flowers Report. The twenty-one members were requested to provide a disinterested analysis and this appeared in *Nuclear Power: Issues and Choices* (Ballinger Publishing Co., 1977). They seemed to reach the conclusion that uranium is likely to be an important source of energy for at least several decades and that in some respects nuclear power could hold a slight edge over alternative energy sources. Colin Norman (*Nature*, **266**, 1977, p. 398) suggests the findings of the report, sponsored by the Ford Foundation and widely representative, could be reflected in the policy of President Carter. Among the findings are firstly that uranium may have an economic advantage over coal, and secondly although the use of uranium might introduce more health hazards than the use of coal there were uncertainties that could tip the disadvantages in the other direction. Norman suggests development of both coal and uranium will ensue in the wake of the report. The panel opposes reprocessing of spent reactor fuel and recycling of plutonium; this view has been challenged overseas as one only suitable for the US which has large reserves of coal and uranium and which, as the first nuclear weapon power, is also still conditioned to the concept of restricting the spread of nuclear weapon capability. In spite of these circumstances the report was not adverse to the 'insurance' provided by a breeder programme which the members thought should proceed with reduced priority. President Carter has endorsed this finding and charged ERDA to study the need for, and timing of, the breeder project. As regards waste the panel noted that 'spent fuel can be disposed of directly and probably at costs comparable to those for reprocessed wastes'. Conversely, Germany has legislated that reactor wastes must be reprocessed before disposal, and the deferment everywhere else on discussions about permanent disposal doubtless contributed to the belief such legislation was desirable.

It is clear that there is not a great divide in the recommendations of the UK and US expert panels, and that where differences do emerge they tend to be on grounds which can reasonably be attributed to differences in the national situations. All the panels of experts seem agreed that work on disposal methods should be given priority and, as we saw earlier, expanded programmes of study are under way. Much work on hazards has been pursued recently but the topic is a specialist one and rather too mathematical for inclusion in this book. Dr. B. A. J. Lister in his paper 'Nuclear Power—The Perspective of Risk' (*Atom*, **223**, 1975, p. 68) gives a well-balanced account of the situation at this time.

It is to be hoped that the administrators who decide our energy policies will relate hazard levels to the facts of everyday life and not impose impossibly onerous standards. The hazards of a nuclear energy programme must be compared to the likely repercussions of an energy crisis, and many occupations of everyday life carry a greater degree of risk. Most telling is the comparison with the risk of driving in a car. Since the mass-produced car came on to the market probably at least 5 million people have been killed on the roads throughout the world, but people do not therefore abandon cars and, in spite of the danger, as often drive for convenience as necessity. It is virtually inconceivable that nuclear energy production could cause such a loss of life, yet energy is at the heart of all our activities and lack of it could be a cause of discontent, hardship, and ill-health on a massive scale. An excellent survey of the factors influencing reactor safety and the assessment of reliability is given by F. R. Farmer, from whose work fig. 102 is taken. It vividly makes a case for nuclear power production.

On the other hand reasons other than scientific ones have been advanced recently by scientists themselves to explain the clash between nuclear and anti-nuclear lobbies. The editorial of *Nature*, **274**, 1978, p. 199, suggests that at the heart of the dispute lies 'the lack of any indication that the nuclear industry, with which the public identifies the scientist, is prepared to discuss values, those human judgements about what is best, by which most of us—including nuclear scientists—live our lives'. In the same issue of *Nature*, D. Pearce, Director of the Windscale Assessment and Review Project, writing after the conclusion of the Windscale Inquiry, argues that any future nuclear inquiry must deal effectively with the differing values adopted by the industry and its opponents, and that the industry must drop its haughty assumption that all opposition to it is irrational.

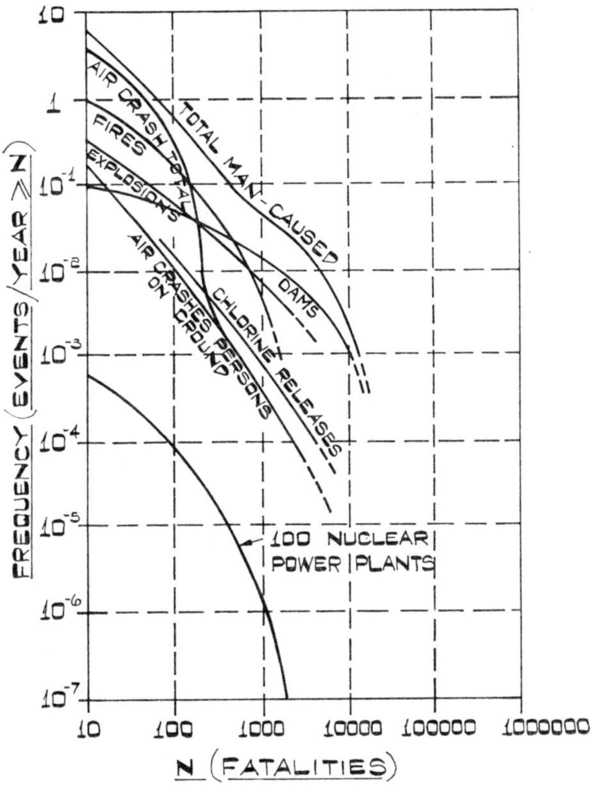

FIG. 102. Frequency of man-caused events with fatalities greater than N. (Those due to motor vehicle accidents not available.) [From F. R. Farmer, *Atom* **230**, 1975, p. 221]

Chapter 28

Proliferation and Terrorism

We have come at last to the most serious issue of all—the possible repercussions of civil nuclear programmes on the manufacture and use of nuclear weapons. At no time in the past have so many people been involved in public discussion and debate about the future of technological development. There have of course been many occasions when science has been at the centre of world-wide controversy, for example in the seventeenth century when Galileo substantiated Copernicus' helio-centric theory and in the last century when Darwin proposed his theory of evolution. In those situations it was the theological implications behind the clash of ideas that gave rise to serious concern, but today it is the question of the quality, or even the possibility, of life for all mankind that underlies the quarrel between those who would continue our present path and those who believe a very different route should be followed. This novel position has resulted partly from the sudden introduction in 1945 of the most destructive weapon ever devised by man. Without doubt the use of the atomic bombs at Hiroshima and Nagasaki made most thinking people aware of the awesome power contained in nuclear explosives and made them wonder if science and technology could imperil our future.

In the later stages of the second world war it seemed opposing views were held by scientists on the one hand and the military and politicians on the other. A large number of scientists, of whom one of the authors (S. C. Curran) was one, realized that their work was intended to result in the manufacture of atomic weapons. Most of them believed it was extremely important not to be threatened by the enemy with such a weapon while incapable of threatening a similar response. On the other hand there was no certainty until the first atomic weapon trial took place that a nuclear device would work. It was also believed that a nation with a reactor in operation might make an extremely radioactive mass of material and that this mass could be dispersed in a kind of block-buster device and so make large areas of enemy terrain uninhabitable. In fact this would have been possible even without the ability to design and manufacture a nuclear explosive weapon, provided of course a nuclear reactor could be designed and operated. A somewhat similar but more destructive device was discussed for some time after the world war under the heading of the cobalt bomb. This would be, in essence, a particularly vicious weapon with effects not unlike poison gas but capable apparently of more insidious and lethal effects by radiation.

It was not the scientists who decided to use nuclear weapons. It appears to have been entirely a decision of the politicians and the military. Here we have in stark form the dilemma that not infrequently arises with the advance of science. As our scientific capacities grow man often has the choice of using them for good or for evil. Until 1945 it was assumed that to defend one's country against aggression it was essential to be ahead, or at least at parity in strength, with potential enemies. Now we have all to decide whether we wish to be able to defend ourselves against an aggressor, especially an aggressor threatening nuclear attack.

In the aftermath of the second world war it was made clear by successive governments in the UK that an atomic bomb stockpile was essential to national defence, and later officially both Conservative and Labour governments were convinced of the necessity of having the hydrogen bomb. Today the US and USSR are believed to have a sufficient nuclear strength to render a major country, possibly even a major part of the world, virtually uninhabitable. The UK, China, France, and India are definitely known to have nuclear weapon capability and, presumably in the case of the first three at least, sufficient numbers of weapons to do a great deal of destruction and the Strategic Arms Limitation Treaty, even if observed, still allows each country to retain a massively destructive arsenal.

Nuclear bombs and nuclear warheads depend on the ability to produce plutonium or highly enriched U^{235} or both. Nuclear bombs of the fission type yield the equivalent of tens of thousands of tons of TNT explosive. The hydrogen bomb, which operates by fusion, yields in the order of one million tons of TNT equivalent. The fearful power of the hydrogen bomb seems to many people to offer the best safeguard against any use of nuclear devices in future wars. No hydrogen weapon has ever been used in warfare, and there has been no nuclear attack of any kind since 1945. Is it possible that it is the power of the hydrogen bomb that has given the world freedom from nuclear war for thirty-three years, and that mankind can hope that the devices will never be used? Yet in the meantime we continue to make ever more terrible weapons. The 'Enhanced Radiation Weapon' (ERW) is an atomic bomb tailored to provide a high proportion of short-lived radiation, which kills all living things, rather than blast, which destroys buildings. The afflicted territory could be inhabited (presumably by the conquerors) after about three months. The ERW is often referred to in the press as the 'neutron' bomb, and even, with supreme lack of understanding, as a 'clean bomb'.

Scientists are almost unanimously opposed to talk of 'clean' bombs or 'neutron' bombs. They do not wish people to pretend that nuclear devices are simply part of a continuous range of armaments. Indeed some scientists believe that 'the nuclear club' should outlaw the deliberate use of any device giving rise to radioactive products, another way of describing a nuclear weapon.

Proliferation Problems. It is in the light of these facts that all schemes for the peaceful uses of nuclear power should be regarded. As we learnt in chapter 15, in any reactor working for peaceful purposes, the potential explosive material, plutonium (Pu^{239}) is produced. Plutonium can be comparatively easily separated from the uranium from which it is derived and this separation is essential if nuclear bombs are to be made. It should be stressed that a reactor core can never be a bomb, exploding by nuclear means. Scientists have tried for more than thirty years to devise systems that would make transformation of plutonium into nuclear weapons impossible. The idea of 'poisoning' the plutonium was examined, and several other schemes were mooted, but in spite of many ingenious suggestions no scheme appears to offer entirely adequate safeguards. In principle any nation with reactors can make fissile weapons (though the making of fusion hydrogen weapons is more difficult). With good reason most people fear that the proliferation of nuclear weapons, even relatively simple devices of the fission type, greatly increases the chance of an atomic incident and, in consequence, the likelihood of a nuclear war. On these grounds there are many people who oppose the use of nuclear power for peaceful purposes and who wish to see the growth of the nuclear industry halted or even abandoned. They see in the breeder reactor, in particular, a means whereby a country could rapidly increase its amount of plutonium, and they believe this particularly increases the chance of nuclear war. Mostly on this account forty nations now participate in the International Nuclear Fuel Cycle Evaluation programme.

The authors are not convinced by the argument against the breeder. It is not in any truly significant sense more dangerous than any other power reactor in respect of weapon manufacture. The number of plutonium atoms produced in a breeder by fission is raised by between 10 and 50 per cent compared with the number produced in the various forms of thermal reactor operating at the same power level. It is this increase that makes the breeder what it is, but it does not make a major difference to the international political situation—all the hundreds of tons of plutonium in the weapon stockpiles of the major powers at the present time have been made in thermal reactors. So there cannot be a well-founded scientific argument that breeders increase the danger of a nuclear incident or nuclear war. It could be argued in fact that the reverse is true, because it is a form of reactor specifically designed to consume plutonium. Once a properly integrated system of thermal and breeder reactors is in operation the amount of plutonium available is constant, not indefinitely increasing as it would tend to do with only a thermal system in operation. It is true that to construct a typical large breeder more than two tons of plutonium have to be available in order to make the core, and there is therefore the temptation to set aside some of the plutonium for use in nuclear weapons. However, it is not ideal weapon plutonium (Pu^{239}) for it becomes 'poisoned' with Pu^{240} when it is not withdrawn specifically to make weapons. Probably the yield of a weapon made with it would be in tons, or tens of tons, rather than tens of thousands of tons of TNT equivalent.

Besides those people who oppose the civil use of nuclear power because they believe it will increase weapon manufacture, there are those who oppose civil use because of its own safety hazards while nevertheless believing it is necessary to retain nuclear weapons in order to protect their country. A third body of opinion holds that civil use is absolutely necessary to close the energy gap—for all the reasons enumerated throughout this book—and that nuclear weapons should also be retained. To this belief is sometimes added the hope that one day an international ban on the use of nuclear weapons may be devised.

As scientists the authors do not feel in any sense more qualified to put forward views on this difficult question than other people. It is nevertheless worth noting that history offers some reassurance as far as the deployment of new destructive weapons goes: poison gas was used for some time in the first world war but never again on a large scale; bacterial warfare has never been practised, though there have been rumours that the USSR may be preparing to equip its forces in Eastern Europe with biological war-heads (*The Times*, 31st Jan. 1978). Nuclear bombs have never been used since 1945 although the US has been involved in two wars since that time (in Korea and in Vietnam) and the USSR has been involved directly and indirectly in wars in the Far East and Africa. It is possible that nuclear, though not other forms of warfare, might be avoided by a forthright announcement from the two major nuclear military powers, the US and the USSR. If they declared that any individual, or group of individuals, using devices that released or produced radioactivity would be brought to trial as war criminals, it is difficult to see any person or group ignoring such a deliberate declaration and going ahead with the deployment of nuclear devices. In theory the United Nations Organization should take this step but it is unlikely ever to have vested in it the nuclear capability of either or both of the two major world powers. If China should come to be a third power of comparable strength it is likewise in that country's interest to ensure that no nuclear incidents occur. All these arguments having been advanced from the more optimistic viewpoints, there seems to be no good reason to believe that the world can avoid nuclear conflict of all kinds by less than a complete ban on all uses of such weapons.

Spread of Nuclear 'Know-how' for Civil Purposes. It is for reasons associated with weapon development that there is considerable anxiety about the spread of knowledge of the reprocessing of fuel elements. A good number of countries have purchased complete reactors or been involved as partners in the construction of such reactors. In 1977 most of these reactors were of the PWR type, sold overseas by US companies. The stage has been reached when the fuel elements in these reactors require reprocessing and suitable chemical treatment plants are usually available only in the countries with nuclear weapon capability. The Japanese, for example, have made arrangements to have their reactor fuel elements reprocessed both in Britain and in France. The Federal German government agreed to construct a reprocessing plant in Brazil and impart 'know-how' on uranium enrichment; this could mean that Brazil would have a ready-made nuclear industry with the capability in the near future to separate plutonium from the fuel elements. In this easy way a number of countries could come to the stage of having nuclear weapons and thus there is a lot of dispute about extending the 'know-how' of reprocessing. The reprocessing of fuel is certainly scientifically advanced and, when it is done with all the elaborate safety precautions scrupulously observed, it is complicated and expensive. At the same time it could doubtless be done much less elaborately if the chief aim is to obtain plutonium. Though it is apparently feasible for the powers with nuclear weapons to delay the acquisition of the techniques by other powers it is difficult to feel convinced that a country with a reasonable number of well-qualified scientists and engineers can be prevented from developing nuclear weapons without outside assistance if they are determined to do so. Possibly proliferation is not the real trouble but rather the will to use nuclear weapons. It may be that only the threat of retribution by the US and the USSR can really keep the peace.

Some experts think that in any case it is too late to stop proliferation. Since 1945 the number of countries with nuclear explosives has expanded from one to at least six, and possibly to seven if the view advanced by T. Bower, namely that Israel has an atomic weapon capability, is accepted (*Listener*, 29th June 1978, p. 822). A total of 104 nations have signed the Non-Proliferation Treaty of 1969 but 30 have not. Fishlock (*Nature*, **267**, 1977, p. 656) takes the view that no ban can prevent some of the 30 nations acquiring a nuclear weapon capability. A special research study, by the US Congress, indicated a nation could make 10 kg of plutonium per year with a simple reactor and the investment required would be between 13 and 26 million US dollars. It is estimated that around 20 nations might qualify for the nuclear club by 1987.

Fishlock has also pointed out that while the Federal German Republic-Brazil contract has been a matter of dispute, other contracts also could cause proliferation problems. In fact the US and Canada provided India with nuclear technology and know-how, and under the Columbo Aid programme Canada provided an NRX research reactor on very generous terms without international safeguards although India had always opposed international control as envisaged by the International Atomic Energy Authority (IAEA). This reactor has enabled India to produce plutonium for a nuclear weapon. Brazil has been refused help by the US government in enrichment and reprocessing because Brazil has not agreed to sign the Non-Proliferation Treaty of 1969.

France has engaged in negotiations to provide Pakistan and South Korea with reprocessing plants. In 1977 the US government met with representatives of the seven leading nuclear supplier nations in order to decide on policy for the future in the light of such possibilities. It was agreed that the reprocessing and enrichment of fuel were the most dangerous technologies to divulge, as these could be adapted to give fissile material for weapons and also heavy water for hydrogen bombs.

In September 1977 both the US and the USSR governments warned the government of South Africa that a nuclear weapon test would be regarded as a very serious matter and the South African government declared that it had no intention to test a weapon. It seemed both the major powers had reason to believe such a test was contemplated. It is known South Africa has experience with reactors, with enrichment techniques and with explosives.

In the light of all these signs of deterioration the US endorsed the IAEA idea of regional fuel cycle centres operating under international management. While the IAEA idea appeared to be practicable if some form of inspection could be agreed, the time lapse of about a decade before the international centres could be built and brought into operation presents a grave difficulty, and on 7th April 1977 President Carter declared a moratorium on reprocessing. He announced also that this was a US domestic policy for controlling proliferation but conceded that other nations without uranium deposits might not be willing to abandon the uranium and the plutonium locked up in the spent fuel due for reprocessing.

As Fishlock observes the US later amplified its policy on non-proliferation by stating it would expand its uranium enrichment capacity, under government control, and it could be inferred that the US would store spent fuel for customers and supply any required enriched fuel to customers from the US reserves—the fuel being enriched to a few per cent of U^{235}. Dr. Nye stated officially that the US would not give permission for fuel to be reprocessed elsewhere when the initial enriched fuel charge had been supplied from the US. This means that the countries with US-produced reactors—the vast majority of the commercial reactors in use outside the Warsaw Pact countries —are possibly obliged to conform to the new US policy. Any Japanese fuel of American origin which comes to Britain for reprocessing can only be transferred with the consent of the US government.

A number of countries are adversely affected by the new anti-proliferation measures of the US. Where there were plans for breeder technology the whole concept was threatened—and many countries had favoured the breeder idea because of their lack of indigenous uranium or capacity to enrich. The US with large stocks of uranium and enrichment facilities was seen to favour, in the interest of domestic expediency, one 'sensitive' technology, namely, enrichment, while renouncing the second, that is, reprocessing. The enrichment by the gas centrifuge was one that the US had previously considered to pose a serious threat to proliferation. The developing countries attacked what they regarded therefore as an attempt by a major power to debar them from a valuable technology, that of reprocessing with the breeding goal as the ultimate objective. The USSR condemned the proposals and seemed determined to proceed with its plans for breeder reactors.

The UK position has been stated clearly (1977) by the Foreign Secretary, Dr. Owen, who re-emphasised at the same time the UK commitment to non-proliferation. The British stockpile of uranium, in a reprocessing system involving a thermal plus breeder reactor arrangement, could yield energy exceeding that of the proven UK coal reserves. Dr. Owen said that the UK plans to offer reprocessing facilities to other countries, under IAEA safeguards, would be seriously hampered and possibly made useless unless re-export licences were granted to nations with reactors purchased from the US and provided with US enriched fuel. He pointed out that the US, by its reprocessing ban, had not managed to close other routes to nuclear explosives.

A knowledgeable attempt to resolve the problem of proliferation is Amory B. Lovins' book, *Soft Energy Paths* (Pelican Books, 1977). Lovins defines 'soft' technologies as having five characteristics: (*a*) reliance on renewable energy flows, such as sun, wind, and vegetation; (*b*) diversity, implying an aggregate of many modest contributions; (*c*) flexibility and the absence of esoteric skills; (*d*) scaling and location matching end-use; (*e*) matching energy quality to needs. He expects the soft technologies to replace the use of coal, oil, gas, and nuclear fission. Lovins points to the many inconsistencies in past US government efforts to restrict the spread of nuclear know-how outside the US. He notes that large amounts of fissile material, possibly of weapon enrichment, were supplied abroad and that a drive to sell reactors overseas led to a considerable spread of knowledge and capability. The problems of hindering theft or small-scale manufacture of nuclear explosive devices is explored, and the answer proposed is the complete abandonment of nuclear development programmes. He notes the fundamental flaws in the existing system of international safeguards (p. 195) and concludes that all that can be said about them is that they are better than nothing. He advocates that the US should, firstly, phase out the US nuclear power programme and US support of foreign nuclear power programmes; and, secondly, divert resources to implementing and promoting soft energy paths, freely and unconditionally helping other interested countries to do the same.

Lovins stresses the seriousness of the scientific difficulties that have beset nuclear power programmes in many countries including the US, and particularly the troubles of successful fuel reprocessing. There are indeed many hurdles to be overcome before nuclear power is the largest energy source available to man, and these should not be minimized. On the other hand Lovins overstates them. It is true, for example, that reprocessing is hampered by the extremely high standards of safety set for the personnel involved. The construction of reactors is also a costly business and very large capital investment is involved, but again a substantial part of this is due to the onerous imposition of safeguards ensuring safety of a very high order. On the other hand Lovins may well take too much for granted in his advocacy of the soft energy paths. He states that storage of energy is difficult on a large industrial scale but done easily and, with due care, safely, on a scale which is matched to needs of the consumer as an individual. It is not proven that storage for the soft energy economy is easier than for the hard one (exemplified in this case by the nuclear). If the same risks were taken in compressing air on a large scale as would be taken on the soft energy scale then the proposition that soft is easier than hard might not stand up. However, readers would be advised to consider the Lovins argument in detail as expounded in the book. His optimism is highlighted by his assumptions (p. 38) that US primary energy use will fall by the year 2025 to about 80 per cent of its 1975 value, and that all of this use will be met by soft technologies. The authors cannot share his conclusions.

Quite apart from Lovins' overoptimism about soft energy, the abandonment of nuclear energy in order to stop the spread of nuclear weapon capability seems to be a most questionable route to a lasting peace. The major powers would certainly retain their nuclear armaments—it is difficult to envisage the total destruction of nuclear weapon stockpiles being countenanced by the owner countries. Perhaps the simpler threat of treatment as war criminals is the most effective deterrent at the end of the day. Those who have power have the duty to use it to prevent abuse in such a fearsome area of human knowledge.

Terrorism. Closely linked to the threat of overt proliferation are the fears that plutonium may be stolen by governments for their national defence, and that a plutonium bomb of some kind might be used deliberately by a party of terrorists. While weapon plutonium must be assumed to be kept by governments beyond the grasp of terrorists and hi-jackers, plutonium in civil use is perhaps more vulnerable. The Flowers Report points out that the danger of theft multiplies as the number of sites where plutonium is stocked increases. It is argued that terrorists might steal sufficient plutonium to construct a simple bomb or explosive device of some kind. Most people assume, without any attempt at justifying the assumption, that constructing a bomb is relatively easy, but if such were the case it would scarcely be worthwhile trying to keep nuclear weapon technology highly secret. Moreover, much of the effect of a nuclear device with poor explosion performance would be long delayed and the fission products would present less hazard than the plutonium itself. It is probable, in fact, that such types of devices do not suit terrorists, who believe in quick results. It could be equally claimed that terrorists might also use poison gas or bacterial methods of terrorising. The similarity of poison gas and radioactive materials in the air or on the ground is more than superficial.

Nevertheless, in reprocessing plants there is possibly some opportunity for terrorist interference and everyone accepts that great care must be paid to guarding the plutonium. In a balanced reactor system some two-thirds of the total stock of plutonium is in reactors themselves, while one third is being reprocessed. How are the potentially explosive materials, Pu^{239} and U^{235}, handled and what steps are taken to ensure the handling is done responsibly? A good many experts have tried to explain the position and readers can find considerable information on the subject. T. N. Marsham's paper (*Atom* **253,** 1977, p. 297) is more pertinent than most in relating the question of terrorism to the breeder reactor and fuel fabrication and reprocessing.

The breeder reactor contains fuel rich in plutonium. It is sealed into fuel pins in the core which is submerged in a sodium pool at a temperature above 400 °C. In this state it is considered to be immune to attack. When not in the core, and except in a few places in the total process of handling, the plutonium exists as a mixed uranium-plutonium oxide ceramic welded into stainless steel pins, which are themselves held in hexagonal stainless steel sub-assemblies 12 ft long, 7 in in diameter and weighing about $\frac{1}{4}$ tonne. Theft could conceivably be attempted when:

1. Unirradiated new fuel sub-assemblies are in store in the fuel fabrication plant, or in transit, or in store before loading into the core.
2. Irradiated sub-assemblies are in store at the power plant, or in transit, or in store at a reprocessing plant.
3. Separated plutonium or plutonium-uranium is in the reprocessing plant or in a fabrication plant.

A terrorist would prefer to secure metallic plutonium, failing which plutonium oxide would be his second choice, and, failing either, plutonium-uranium oxide.

If the reprocessing and refabrication is on the fast reactor site no transport is required except within the plant. Even within the plant the forms in which plutonium is transported and handled make it extremely difficult to steal and to use in any terrorist activity. Since the steps are enumerated by Marsham they will not be repeated here. If, however, the terrorists secured fuel which was not highly radioactive it would be in the form of a mixture of plutonium and uranium oxides, and this material would also contain Pu^{240} which tends to inhibit the explosive action of the Pu^{239}. In these circumstances it would be unlikely to be useful in constructing a proper atomic weapon, though obviously it would be of high value as a potential threat. Even this theft, however, is improbable. The radioactive fuel that comes for reprocessing is necessarily handled with great skill to reduce the radiation hazards to the minimum with techniques that are well understood and carefully followed by an experienced work force. It is therefore difficult to understand the argument that reprocessing plants are particularly vulnerable.

Chapter 29

Summary and Some Thoughts for the Future

We have tried to show readers the great importance of energy resources for human well-being, and we have examined the rich diversity of proven and potential resources with particular regard to their future capabilities. This examination showed that a very serious shortfall in supply could arise in the next twenty to thirty years, especially if the industrialized nations continue to increase their demand and if the less advanced countries do likewise.

The 'gap' may be closed in different ways. Possibly by the end of the twentieth century the total contribution from new methods such as the conversion of solar, geothermal, wave, or tidal energy, until now a matter of research or at best marginally exploited, might amount to about 10 per cent of the total world consumption, given intensive research and development. At the same time we will be striving to exploit fossil fuels more efficiently. The only two large sources of extra supplies, however, are coal and nuclear power. Since the most optimistic estimates of an expanded coal industry are insufficient to fill the gap, the limiting factors being shortage of labour and time, nuclear power appears to be the only part of the existing energy industry that could increase its output sufficiently quickly to avoid a world shortfall.

Given that we go ahead with a nuclear programme the most difficult problem is to decide in what manner and at what rate to expand the industry. A large thermal reactor programme would use so much uranium that it is doubtful whether supplies would be adequate in the longer term. This leads many experts to propose that breeder reactors, which have worked well in prototype, should be constructed on a commercial scale, and some consider that the time to launch such a programme of construction is running very short. Other experts are against a breeder programme on grounds of safety; the reprocessing of fuel, the use of plutonium, and the storage of radioactive waste are regarded as presenting substantial difficulties. It is also necessary to ask the question whether enough reactors of this type could be made effective within the required timescale. All research, development, and prototype testing takes much time and effort, not least in the nuclear field. Decisions for investment in major energy programmes, especially those in the nuclear sector, can only be taken responsibly when all the possible snags have been thoroughly explored. Attractive and ingenious proposals have to be subjected to deep and searching analysis and all the results of a large experimental and development programme scrutinised. Only at the end of such exhaustive testing might a new method in the nuclear field be considered practicable on the industrial scale.

Trends in Nuclear Research. Given all the reservations about breeders it is encouraging to find scientists considering a new means of avoiding some of their problems and the associated use of plutonium as a fuel. In *Nature* **270,** 1977, p. 376, John Davies has given an excellent account of a scheme that brings together ideas from several branches of physics, including fusion and high energy physics. He shows how they may be used to achieve the principal goal of the breeder reactor programme by an alternative means. The concepts involved are illustrated in fig. 103. Here the effective breeding threshold is set at the value of 2·2 neutrons per fission. It is considered that an average of 0·2 is required to allow for escape and absorption in unhelpful processes. At a level of 2·2 there remains, after this loss of 0·2, one to continue the chain reaction and one to use for breeding, that is, making as much core fuel as is burned in the operating reactor. If we consider figs. 103(a) and (b) together we see that Pu^{239} gives a yield above 2·2 with fission neutrons and a comparable yield with fast reactor neutrons. The Pu^{239} yield exceeds that of U^{235}

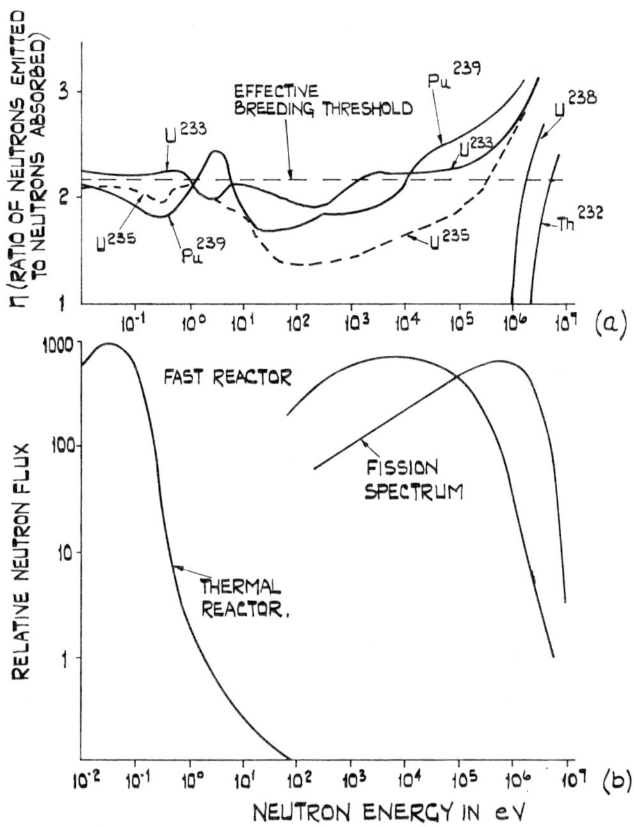

FIG. 103. (a) The ratio of neutrons emitted to neutrons absorbed plotted against neutron energy

(b) The neutron energy spectrum for fission neutrons and for those in a thermal reactor and a fast breeder reactor with both types of neutrons. The isotopes U^{238} and Th^{232} do not undergo fission as readily with either of these types of fast neutrons, while the isotope U^{233} does undergo fission readily, indeed comparably with Pu^{239} at all values of neutron energy.

For neutrons with the energy spectrum obtaining in a thermal reactor, the isotopes U^{233}, U^{235}, and Pu^{239} are all readily fissile and each can be used in thermal reactors to obtain a good yield of additional fissile material in a blanket of uranium or thorium. However, only U^{233} gives a value above 2·2, making breeding in thermal operation possible. A deuterium moderated reactor such as CANDU is found to be economical in neutron usage and, with the core consisting of natural uranium, a breeding gain as high as 0·92 is obtainable. Thus if natural thorium is the blanket material and the spare neutrons are allowed to be absorbed in it, a good yield of U^{233} is achieved in the reactions:

$$Th^{232}_{90}+n\rightarrow Th^{233}_{90}\rightarrow Pa^{233}_{91}+e_{-}\rightarrow U^{233}_{92}+e_{-}.$$

As a result 92 per cent of the fissile core material consumed, essentially U^{235}, is returned as fissile U^{233}. If 8 per cent of Th^{232} used in blankets could be converted by some other means to U^{233} the whole operation of a deuterium moderated thermal reactor would become self-sustaining and natural thorium, Th^{232}, could be burned in reactors. We would be using a thorium-U^{233} economy rather than a U^{238}-Pu^{239} one.

Davies emphasises some of the advantages of this alternative to the uranium-plutonium breeder route and outlines the possible means, based on findings in high energy physics, of making it a reality. By a process known as 'spallation', single protons with extremely high energies can be used to produce many neutrons. At an energy of 1 GeV a proton falling on a target of uranium can release about 60 neutrons. These neutrons can be absorbed in Th^{232} to give U^{233}. Each such U^{233} atom used in a reactor is able on average to return 200 MeV of fission energy. Thus by employing a proton with 1 GeV of energy a yield of 60×200 MeV can be obtained. If this spallation technique is applied to convert as much as 8 per cent of the thorium to be used later as the blanket for a thermal reactor, the thermal reactor with its blanket of thorium enriched with U^{233} constitutes a breeding system. The final return of energy is increased in this fashion by a factor of 100/8 so the accelerator has enabled a return of $12\times(100/8)$ GeV to be achieved, that is 150 GeV for each 1 GeV expended.

We appreciate that accelerated particles cannot normally give a net energy gain because in most circumstances there is a waste of kinetic energy in ionization and other processes. However, spallation is a somewhat remarkable process depending on the extremely high energy of the bombarding particles. The idea proposed in fusion of raising the temperature of the plasma by injecting very energetic neutral atoms is rather similar in principle. From fusion, however, a second possibility, more directly related to the thorium-U^{233} cycle, emerges. The large machines such as Tokamak could make very intense fluxes of 14 MeV neutrons, even if not self-sustaining, and these neutrons could readily convert atoms of fertile Th^{232} to fissile U^{233}. In principle, therefore, a very high-voltage accelerator or an intense fusion apparatus could become the tool enabling the gap of about 8 per cent in the conversion of Th^{232} to U^{233} to be closed. Such a machine coupled to a thermal reactor would provide a means of consuming natural Th^{232}. Being provided with a thorium blanket to be enriched in U^{233} a thermal uranium reactor like CANDU would be able to breed. Once sufficient U^{233} was available it could be used in a thermal breeder reactor clad with a natural thorium blanket. If this proved difficult such a blanket could be enriched as outlined above. In any event all natural thorium would become fissile fuel, just as in the uranium-plutonium system the aim is to be able to burn all available natural uranium instead of having to waste most of it.

Work is starting on high energy accelerators which could operate at very large beam currents, more than 100 mA at 1 GeV, or at smaller currents and very much higher voltage, say, 1,000 GeV. At this voltage a single proton can yield about 60,000 spallation neutrons. With this sort of result a practical reactor and accelerator system might be feasible.

This example of new thinking is unusually interesting but it is not without flaws. Davies states that accelerators are inherently cheaper to build than reactors and stresses their efficiency. It is difficult to accept that a 1 GeV machine operating at a few hundred milliamps of current is inexpensive. Such machines as are operating at the Centre Européen pour Recherche Nucléaire (CERN) at Geneva are supported by the contributions of several countries and their running costs are high. The price of a unit of electricity is the most important factor of many that are taken into account in reaching decisions on energy policy. If a brilliant concept adds appreciably to the cost to the consumer it will inevitably be discarded. As we have already stressed, reactor construction costs are high because their design incorporates so many safety features, but even though this adds materially to the cost of the output electricity produced by nuclear means is still more than competitive. Compared with accelerators of advanced design reactors are inherently simple machines; they are also already proven to be reliable and able to operate steadily for twenty or more years. All of these factors have to be assessed when comparisons are made. In addition, an immense neutron source like the 1 GeV accelerator or a large fusion device such as Tokamak could be used clandestinely to manufacture some tons of plutonium. Also if U^{233} is so comparable with plutonium in its interactions with neutrons, can it be guaranteed that it could not be used in atomic weapon threats by terrorists? When it is more or less pure after chemical separation from the Th^{232} from which it is derived, it could be diverted for illegal purposes. In short, if accelerators can be used to manufacture cheaply large amounts of U^{233} and Pu^{239} (the same spallation techniques could apply to the $U^{238} \rightarrow Pu^{239}$ transformation as apply to $Th^{232} \rightarrow U^{233}$ change) then they represent a cheaper way for a nation to make nuclear weapons.

In spite of the disadvantages linked to this particular scheme such fresh ideas show that exciting possibilities continue to emerge from research, and especially from the interplay of ideas between apparently unrelated research efforts. We should never assume that all the envisaged sources of available energy exhaust the possibilities. At the same time we have to remember that research schemes, even if well founded, are but the first step along a difficult and arduous road towards systems which may contribute only a comparatively small percentage of the huge world demand for energy. The investment involved, measured in terms of both financial and manpower resources, is inevitably huge. It would be foolhardy to delay progress with well-established systems in the expectation that new ideas could replace them in a short time. In the UK particularly, this point is most important. Rather too frequently the development and construction effort required in order to bring new ideas to fruition is underestimated. Of the total effort needed to make a new scheme economically viable more than $\frac{9}{10}$ths may be consumed in expanding and developing the prototype into a machine on a commercial scale. Inventive skill can prove a delusion in certain circumstances.

Energy and Longevity. R. H. Socolow has studied the energy consumption per capita in various countries and found, as might be expected, that it varied widely. His findings, which are mostly concerned with industrialized nations, are given in part in table 29.1. The figure for Mexico, however, gives some idea of the disadvantageous position of underdeveloped countries.

TABLE 29.1

ENERGY USE IN KW PER CAPITA FOR VARIOUS COUNTRIES
(From R. H. Socolow, *A. Rev. Energy* **2**, 1977, p. 249)

Country	Industry and agriculture	Transportation	Commercial and public facilities	Residences	Total
US	4·7	3·0	1·3	3·0	12·0
Canada	3·8	2·4	2·1	2·7	11·0
Norway	4·5	1·3	1·2	2·0	9·0
Sweden	3·4	1·3	1·3	2·0	8·0
Netherlands	2·8	2·4	0·6	1·7	7·5
German Federal Republic	2·7	1·0	0·6	1·4	5·7
Finland	2·7	0·9	0·6	1·4	5·6
UK	2·8	0·9	0·3	1·5	5·5
Denmark	1·7	0·9	0·9	1·8	5·3
Japan	3·0	0·5	0·4	0·4	4·3
France	1·4	1·3	0·5	0·9	4·1
Italy	1·7	0·7	0·2	0·6	3·2
Mexico	0·6	0·4	0·1	0·1	1·2

In an excellent review of energy systems W. Häfele and W. Sassin (*A. Rev. Energy* **2**, 1977, p. 1) discuss energy in a global context, its influence on population growth, on economic development and urbanisation and on national capacity to deploy modern technology. As an illustration of the influence of energy availability on international issues they show that human longevity and energy consumption are related. Life expectation rises rapidly with energy consumption and at a value around 1 kW per capita the rate of increase becomes very marked indeed. The longevity assumes an almost steady value exceeding 70 years for a per capita level of 5 kW and upwards. In 1977 the global average per capita value of consumption was 2 kW. This would appear to yield a life expectancy everywhere of almost 70 years but distribution among nations is so very uneven that over 70 per cent of the human population consume less than 2 kW per capita. The curve of fig. 104 is most striking and, while it would be easy to draw rash conclusions, it is obvious there are compelling reasons for seeking energy in abundance for all.

Proposals for expanding the Nuclear Power Programme in the UK. We believe this book has demonstrated that only nuclear energy can save our way of life from being dangerously disrupted when oil and natural gas begin to run out at the turn of the century. What rate of expansion will best secure its advantages while minimizing its risks?

Successful though fission reactors are, our practical experience on a commercial scale is limited to two decades and there is much to be said for regulating the extent of our dependence on fission energy until more substantial long-term experience has been acquired. We suggest it would be sensible to increase fission energy by 2000 A.D. to the point where it supplied half or more of the total electrical power needs, amounting to between 20 per cent and 30 per cent of the total energy consumption in the advanced countries. Dependence on coal or imported oil would then, for electricity generation, be much less critical and we could begin to conserve them for certain forms of transport and as feedstock for industries such as the petrochemical. At the same time we would be using nuclear energy to buy time so that some of the less proven energy sources can be more fully explored. In short there are excellent reasons for a flexible approach to energy programming.

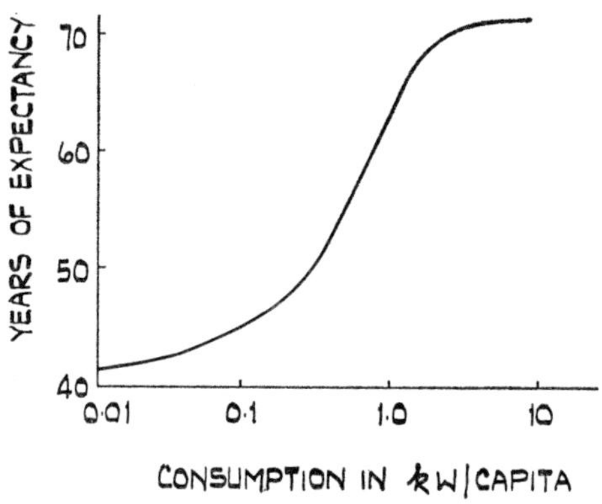

FIG. 104

Life expectancy and per capita energy consumption. [Based on Ififa and Sagan, *A. Rev. Energy*, **2**, p. 15]

The proportion of the energy supplied as electricity in the UK market has been discussed by D. Clark (*Phil. Trans. R. Soc.*, **276**, 1974, p. 559). His own estimates of the probable future market shares of the major energy sources are shown in fig. 105. The curves are for useful energy delivered to the customer and extrapolation is made to the year 2000. The coal, oil, and gas areas do not include their use for electricity generation and it is assumed that:

(1) total end energy grows at 4 per cent per annum;
(2) coal use, apart from conversion at power stations, falls to 20×10^6 tonnes per annum;
(3) gas levels off at 20×10^{17} joules per annum;
(4) electricity grows at about 5 per cent per annum.

With these assumptions the electricity share of the UK market would be 30 per cent by the year 2000 and nuclear generation would be expected to provide 80 per cent of the total electrical energy, almost 24 per cent of the total energy supply. Clark also considered a faster expansion of nuclear power in electricity generation (the appropriate amount being indicated in the figure as 'Additional Electricity') which implied boosting electricity generation by 7 per cent per annum from the early 1980s, and consequently replacing oil with electricity in about 20 per cent of the oil market. The authors feel that this swifter expansion of the electrical sector is the right rate of change and, although it would involve rapid development of a large thermal and perhaps breeder reactor programme, it would still necessitate the use of oil to meet both 43 per cent of the market needs and part of the electricity production which, at say 2 per cent, would mean 45 per cent dependence on oil, indigenous and imported. This is surely as large a sector as a single fossil fuel should occupy. Moreover, the increased generation of electricity by nuclear

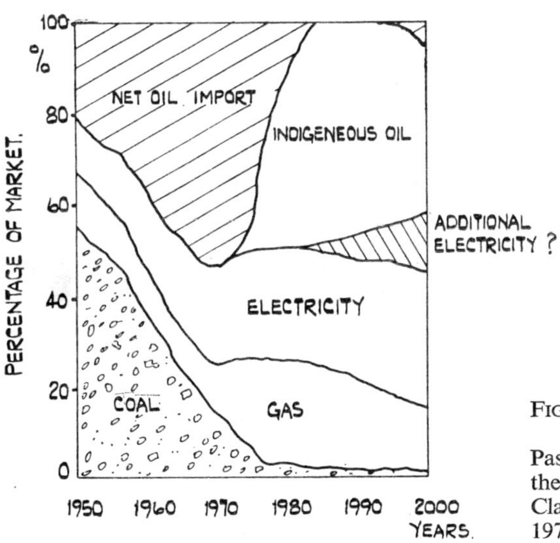

FIG. 105

Past and estimated future shares of the UK energy market. [From D. Clark, *Phil. Trans. R. Soc.*, **A276**, 1974, p. 568] means would set free our indigenous coal for other and possibly expanded uses such as we discussed elsewhere. While Clark makes no allowance for entirely new sources of energy, their total contribution is likely to be less than the variation in the electrical component. Nevertheless this should not be taken to imply that we should lessen our attempts to develop these sources.

We stress that a decision to expand the nuclear power output at 7 per cent per annum until 2000 would mean a major and swift programme of construction. The long construction time of plants limits the rate at which new stations can be brought into operation. Delay in deciding on the start of the programme could easily result in failure to meet our foreseeable needs.

Comparison of the Energy Gap in the US and UK. The UK energy position has been highlighted in this chapter, but much of the discussion applies in principle elsewhere and certainly in the US. Though there are notable differences in the size, energy situations, and prospects of the two countries, the problems facing both between now and the year 2000 are remarkably similar. Each has abundant potential supply of coal, but neither has confidence that the supply can be used to close the energy gap by more than a comparatively small fraction, say, between 20 and 30 per cent. The US is likely to achieve the upper end of the range as its coal beds are more readily worked. Both countries have oil and gas; but while the US is rapidly becoming

TABLE 29.2

UK NUCLEAR GENERATING CAPACITY, 1977

Station location	Start date	Reactor type number	Electrical output MW	Total MW
1. Calder Hall	1956	Magnox 4	50×4	200
2. Chapelcross	1959	Magnox 4	50×4	200
3. Berkeley	1962	Magnox 2	138×2	276
4. Bradwell	1962	Magnox 2	150×2	300
5. Hunterston	1964	Magnox 2	160×2	320
6. Trawsfynydd	1965	Magnox 2	250×2	500
7. Hinkley Point A	1965	Magnox 2	250×2	500
8. Dungeness A	1965	Magnox 2	275×2	550
9. Sizewell A	1966	Magnox 2	290×2	580
10. Oldbury	1967	Magnox 2	300×2	600
11. Wylfa	1971	Magnox 2	590×2	1,180
		Magnox total		5,206
12. Hinkley B	1977	AGR 2	625×2	1,250
13. Hunterston B	1977	AGR 2	625×2	1,250
		AGR total		2,500

Overall nuclear capacity in 1977=7·7 GW

Dungeness B, Hartlepool, and Heysham at advanced stage of construction

TABLE 29.3

WORLD NUCLEAR POWER PROGRAMME

(From N. J. Franklin, 'Irradiated Fuel Cycles', British Nuclear Fuels Ltd., 1975, p. 11)

	1975	1980	1985	1990	1995	2000
1. Installed nuclear electrical generating capacity (GW)	50	300	500	1,200	2,000	2,500
2. Cumulative nuclear electrical energy sent out (GW-y)	120	650	1,850	4,350	9,000	16,000
3. Spent fuel (1) reprocessing load (t/y)	700	3,000	8,000	14,000	31,000	46,000
4. Cumulative fuel reprocessed (t)	2,000	9,000	36,000	90,000	200,000	400,000
5. Cumulative electrical energy associated with reprocessed fuel:						
(GW-y)	60	270	1,100	2,700	6,100	12,000
($kW\text{-}h \times 10^{12}$)	0·5	2·5	10	25	50	100
6. Plutonium throughput (t/y)	7	30	80	140	400	900

NOTES: (1) All fuel quantities are based on the equivalent of high burn-up LWR fuel.
(2) t=tonne; y=year

very dependent on imported supplies of these fuels, in the shorter run the UK is less dependent on imported supplies, though by the year 2000 the UK supplies of oil and gas from the North Sea will be decreasing. It is a reasonable deduction that the US will require an expansion of nuclear electricity generation such as Clark proposed for the UK (fig. 105). This expansion amounts to a 90 per cent increase, by the year 2000, on the total electricity supply of 1977. In 1977 the first of the AGR nuclear stations, at Hunterston and Hinkley Point, became operational, and each added 1,250 MW capacity to the UK grid system (each station consisting of two reactors of 625 MW electrical output per reactor). This AGR contribution of 2,500 MW brought the total nuclear contribution to UK electricity supplies to approximately 7·7 GW, of which roughly two-thirds was due to Magnox reactors as shown in Table 29.2.

Taking 65 GW as the UK total generating capacity in 1977, we conclude that a 90 per cent additional capacity corresponds to a 58 GW increase between 1977 and the year 2000. If a typical station of AGR type delivers 1·25 GW, it follows that forty-six such stations have to be built and commissioned by the end of this century; that is two stations per year from 1977 should come into operation. An absolute minimum of one station per year would seem to be required; such a rate would add about 28 GW to the total available by the year 2000 and leave a substantial amount of the need to be met with coal. The US situation could prove more serious. In the US the total per capita consumption of energy exceeds that in the UK by more than a factor of 2 (table 29.1) and this is true of the electrical component also. In the US, between 1977 and the year 2000, almost eight stations, each of about 1 GW output, should be commissioned every year. In 1978 neither country showed much evidence of approaching the high level of expansion of the nuclear industry indicated by these figures.

The World Scene. The information available for most of the advanced countries suggests that similar difficulties will be encountered unless a marked change in the consumption of energy comes to be accepted during the next three decades. Such a change would of course necessitate a substantial alteration in lifestyle. A speculative world nuclear power programme, shown in Table 29.3, has been provided by N. L. Franklin. The world level of installed nuclear capacity of 50 GW in 1975 is shown to rise to 2,500 GW by the year 2000. Light water reactors are assumed to predominate until 1990, and fast reactors meet an increasing part of the load between 1990 and 2000 when they would constitute 20 per cent of the total. The associated reprocessing load and plutonium throughput are entered, and these, according to Franklin, mean that thirty reprocessing plants, each capable of handling 1,500 tonnes per year, are required by the year 2000. By that time the throughput of plutonium reaches 900 t/year.

From Franklin's estimates we are able to conclude that even the effort to close the energy gap by nuclear means will prove a daunting task. Unhappily no other source of energy provides an established credible route as an alternative.

Further Reading

Books and journals particularly apposite to our arguments are cited in the text. The following list is for the convenience of readers who wish to explore subjects in more detail. The material listed is not included in the Index.

CHAPTER NUMBER	REFERENCE

1. Many modern books on physics and chemistry provide the scientific background relevant to full appreciation of energy studies. Among them are:
   - W. Crouse, *Understanding Science*, McGraw-Hill (New York, 1973)
   - F. W. Sears & M. W. Zemansky, *University Physics*, Addison-Wesley (London, 1971)
   - D. Halliday & R. Resnick, *Physics*, pt. 1, 2nd ed. (1977); pt. 2, 3rd ed. (1978); Wiley (New York)
   - E. Thomas, *From Quarks to Quasars, An outline of Modern Physics*, Athlone Press (London, 1977)

2. M. Kenward, *Potential Energy*, Cambridge University Press (London, 1976)
   - G. M. Crawley, *Energy*, Macmillan (London, 1975)
   - E. J. Hoffman, *The Concept of Energy*, Ann Arbor Science (Ann Arbor, 1977)
   - McGraw-Hill Book Company, *Encyclopaedia of Energy* (New York, 1976)
   - Aspen Systems Corporation, *The Energy Sourcebook* (Pittsburgh, 1977)
   - L. C. Ruedisili & M. W. Firebaugh, *Perspectives on Energy*, Oxford University Press (London, 1975)

3. B. Booty, *Guide to Nuclear Physics*, Peter Peregrinus (Hitchin, Herts., 1977)
   - R. F. Pocock, *Nuclear Power—Its Development in the United Kingdom*, Unwin Brothers and Institute of Nuclear Engineers (Old Woking, Surrey, 1977)

   The following are historically interesting:
   - H. D. Smyth, *Atomic Energy for Military Purposes*, US War Department and Maple Press (York, Penn., 1945)
   - L. Bertin, *Atom Harvest*, Secker and Warburg (London, 1955)

CHAPTER NUMBER	REFERENCE

4. R. C. Tolman, *Relativity, Thermodynamics & Cosmology*, Oxford University Press (London, 1958)
   H. Alfvén, *On The Origin of The Solar System*, Greenwood Press (Westport, Connecticut, 1973)

5. I. G. Gass, P. J. Smith & R. C. L. Wilson, eds., *Understanding the Earth*, Open University Press (Milton Keynes, 1973)
   A. L. Hammond, 'Geothermal Energy; An Emerging Major Resource', *Science* **177**, no. 4053, 1972, p. 978
   J. D. Garnish, 'Geothermal Energy and the UK', *Atom* **263**, 1978, p. 242
   H. C. H. Armstead, 'Geothermal Energy Development', *Progress in Energy and Combustion Science*, **2**, 1977, p. 181
   E. Barbier & M. Fanelli, 'Non-electrical Uses of Geothermal Energy', *Progress in Energy and Combustion Science*, **3**, 1977, p. 73
   E. M. Wilson, 'Energy from the Tides', *Science Journal* **1**, pt. 5, July 1965, p. 50

6. L. Radaux and G. de Vaucouleurs, Larousse Encyclopedia of Astronomy, Paul Hamlyn, 1959

7. J. C. McVeigh, *Sun Power*, Pergamon Press (Oxford, 1977)
   H. Messel & S. T. Butler, eds., *Solar Energy*, Pergamon Press (Oxford, 1975)
   J. A. Duffie & W. A. Beckman, *Solar Energy Thermal Processes*, Wiley (New York, 1974)
   A. B. Lovins, *Soft Energy Paths*, Penguin Books (London, 1977)
   B. Brinkworth, *Solar Energy for Man*, Compton Press (Salisbury, Wilts., 1972)
   A. and M. Meinel, *Applied Solar Energy*, Addison-Wesley (London, 1977)
   W. Palz, *Solar Electricity—an Economic Approach to Solar Energy*, Committee of the European Communities (Brussels, 1978)
   J. Lenihan & W. W. Fletcher, eds., *Energy Resources and The Environment*, Blackie (Glasgow, 1975)
   J. R. Williams, *Solar Energy; Technology and Applications*, Wiley (New York, 1977)
   'Power from the Sun', *The Unesco Courier*, January 1974, p. 16
   J. D. Denton *et al.*, 'The Potential of Natural Energy Sources', *CEGB Research*, no. 2, 1975, p. 28 (London)
   E. Michelson, 'Prospects for Solar energy', *Nature* **277**, 91, 1979
   J. S. C. McKee, 'Power from the Sun', *Phys. Bull.* **29**, 565, 1978

CHAPTER NUMBER	REFERENCE

C. Marchetti, 'Genetic Engineering and the Energy System: How to make ends meet', *IIASA Res. memo*, 78–62, Dec. 1978

8. S. H. Salter, 'Wave Power', *Nature* **249**, no. 5459, 1974, p. 720
D. T. Swift-Hook, 'Wave Power', *CEGB Report*, R/MIN 790 (London, 1975)
Sandia Laboratories Energy Report, *Wind Energy—a revitalized pursuit*, Sandia Laboratories (Albuquerque, New Mexico, March 1975)
C. F. Clement, 'Windpower for the UK', *Nature* **268**, no. 5619, 1977, p. 396; also *The Times*, Oct. 19, 1977, p. 17

9. F. Acker, 'Prospects for hydro power', *Electrical Review* **201**, 1977, p. 51
F. P. Jenkin, 'Pumped storage is the cheapest way of meeting peak demands', *Electrical Review* **195**, 1974, p. 196

10. W. H. Freeman, *Materials* (book form of *Scientific American* issue) (San Francisco, 1967)
J. S. Blakemore, *Solid State Physics*, 2nd ed., Saunders (London, 1974)
A. I. Rosenblatt, 'Energy Crisis spurs development of photovoltaic power sources', *Electronics* **47**, no. 7, 1974, p. 991
NSF/NASA Solar Energy Panel, *An Assessment of Solar Energy as a National Energy Resource*, 1972
L. M. Magid, 'The Current Status of the US Photovoltaic Conversion Program', *Colloqué International sur l'Electricité Solaire* (Toulouse, March 1976)
M. Slesser & I. Hounam, 'Solar Energy Breeders', *Nature* **262**, no. 5566, 1976, p. 244
S. R. Elliott, 'Amorphous silicon: where now?' *Nature* **277**, 85, 1979

11. G. Porter & M. D. Archer, 'In Vitro Photosynthesis', *Interdisciplinary Science Reviews*, **1**, no. 2, 1976
D. O. Hall & K. K. Rao, *Photosynthesis*, Edward Arnold (London, 1972)
M. B. Wilkins, ed., *Physiology of plant growth and development*, McGraw-Hill (New York, 1969)
J. Phillipson, *Ecological Energetics*, Edward Arnold (London, 1966)
D. Pimentel *et al.*, 'Food Production and the Energy Crisis', *Science* **182**, no. 4111, 1973, p. 443
M. Calvin, 'Solar Energy by Photosynthesis', *Science*, energy issue, **184**, no. 4134, 1974, p. 375
A. Mitsui, S. Miyachi, A. San Pietro & S. Tamura, eds., *Biological Solar Energy Conversion*, Academic Press (New York, 1977)

CHAPTER NUMBER	REFERENCE

The Chemical Society, *Conservation of Resources*, special publication no. 27 of Symposium at Glasgow, April 1976

M. Slesser, 'The Energy Basis of Development', *Energy World* **9**, p. 3

M. Slesser, 'Accounting for Energy', *Nature* **254**, no. 5497, 1975, p. 170

R. A. Karam & K. Z. Morgan, *Energy and the Environment; Cost-Benefit Analysis*, 2 v., Pergamon Press (Oxford, 1976)

J. M. Hollander, M. K. Simmons & D. O. Wood, eds., *Annual Review of Energy*, v. 2, Annual Reviews (Palo Alto, 1977)

J. L. Monteith, *Principles of Environmental Physics*, Arnold (London, 1973)

C. Marchetti, 'On Energy and Agriculture', *IIASA Working Paper*, Jan. 1979

12. C. L. Wilson, ed., *Energy: Global Prospects 1985–2000* (Report of the Workshop on Alternative Energy Strategies), McGraw-Hill (New York, 1976)

A. F. Fox, *The World of Oil*, Pergamon Press (Oxford, 1964)

Teh Fu Yen & G. U. Chilingarian, eds., *Oil Shale*, Elsevier (New York, 1976)

C. E. B. Conybeare, *Geomorphology of Oil and Gas Fields in Sandstone Bodies*, Elsevier (Amsterdam, 1976)

Shell International Petroleum Co. Ltd., *The Petroleum Handbook*, 4th ed. (London, 1959)

A. J. Moll in *Proceedings of Second Annual Underground Coal Gasification Symposium*, ed. L. Z. Schuck, Energy Research and Development Admin., Morgantown, W. Va., US (1976)

Sir Peter Kent in *Phil. Trans. R. Soc.* **A276,** 1974, p. 405

'Assessment of Fuels for Power Generation by Electric Utility Fuel Cells', EPRI EM-695, Project 1042, Final Report, v. 1, March 1978 (report prepared by Arthur D. Little, Inc., Cambridge, Mass., for EPRI, Palo Alto, California 94304)

13. D. C. Ion, *Availability of World Energy Resources*, Graham and Trotman (London, 1975)

Wilson Clark, *Energy for Survival*, Anchor Press/Doubleday (Garden City, New York, 1974)

World Energy Conference (Istanbul) 1977: Executive Summaries of Conservation Commission Reports on Energy Conservation 1985–2020, Coal Resources, Unconventional Energy Resources, World Energy Demand, Oil Resources, Nuclear Resources; see also The Future of World Natural Gas Supply, 1985–2020

CHAPTER NUMBER	REFERENCE

S. Dix, *Energy: A Critical Decision*, Energy Education Publishers (Michigan, 1977)

L. G. Brookes, 'The Plain Man's Case for Nuclear Energy', *Atom* **234**, 1976, p. 95

J. C. Fisher, *Energy Crises in Perspective*, Wiley (New York, 1974)

'Geologists warn of Uranium Shortage' (unsigned), *Nature* **274**, no. 5671, 1978, p. 524

World Energy: looking ahead to 2020, I.P.C. Science and Technology Press, 1978.

T. Price, 'The Uranium Market—Economic and Political Factors', *Atom* **265**, 290, 1978; 'The Energy Tangle', Special Report, *Newsweek* **93**, April 16, 1979, p. 10

14. S. A. Durrani, 'Natural fission reactors—Oklo style', *Nature* **271**, no. 5643, 1978, p. 306

Physics in Technology: 'Perspectives on Energy', special issue, 8, July, 1977

H. A. C. McKay, 'The Discovery of Fission', *Atom* **265**, 301, 1978

15. S. E. Hunt, *Fission, Fusion and the Energy Crisis*, Pergamon Press (Oxford, 1974)

International Co-operation on Breeder Reactors, May 1978; Report of International Energy Associates Ltd (IEAL) to Rockefeller Foundation (Policy Study Series), P.O. Box 1265, Radio City Station, New York 10019

16. F. H. Schmidt & D. Bodansky, *The Energy Controversy—the fight over nuclear power*, Albion Publishing Company (San Francisco, 1976)

B. Commoner, *The Poverty of Power: Energy and the Economic Crisis*, Knopf (New York, 1976)

H. Hunt & G. Betteridge, 'The Economics of Nuclear Power', *Atom* **266**, 326, 1978

17. T. S. Green, *Thermonuclear Power*, George Newnes (London, 1963)

S. S. Penner, ed., *Lithium Needs and Resources*, Symposium in Corning, New York, Pergamon Press (1978)

L. M. Lidsky, 'The Quest for Fusion Power', *Technology Review*, Jan. 1972

D. J. Rose, 'Controlled Nuclear Fusion: status and outlook', *Science* **172**, no. 3985, 1971, p. 797

M. O. Hagler and M. Kristiansen, *An Introduction to Controlled Thermonuclear Fusion* (Farnborough, Lexington, 1977)

CHAPTER NUMBER	REFERENCE

18. F. Dollin, 'Some Design Problems Arising in the Development of Very Large High-speed Turbines', *Proc. Inst. Mech. Engrs.* **117**, 1963, p. 221

C. Marchetti, 'Transport and Storage of energy', *IIASA Research Report*, RR-75-38 (Laxenburg, Austria, Nov. 1975)

19. No material

20. T. N. Veziroglu, ed., *Introduction to Hydrogen Energy*, Pergamon Press (Oxford, 1975)

C. A. McAuliffe, *Hydrogen Energy*, Macmillan (London, 1976)

P. J. Hampson *et al.*, 'Can Hydrogen Transmission Replace Electricity', *CEGB Research* no. 2, 1975 (London)

J. Gretz, 'Some Aspects of the Potential of Solar Energy Conversion into Hydrogen and/or Other Fuels', Internal Working Paper, Euratom, Ispra, 1976

W. Häfele *et al.*, 'Second Status Report of the IIASA Project on Energy Systems, 1975', *Research Report* RR-76-1

'Hydrogen Production Using Electrolysis Versus a Hybrid Cycle', EPRI EM-789, Project 1086-2, Final Report, June 1978 (report prepared by Westinghouse Electric Corporation, South Large, Penn. 15236 for EPRI, Palo Alto, California 94304)

'A Technoeconomic Analysis of Large Scale Thermochemical Production of Hydrogen', EPRI EM-287, Project 467, Final Report, Dec. 1976 (report prepared by University of Kentucky, Lexington, Kentucky 40506 for EPRI, Palo Alto, California 94304)

21. E. G. Kovach, ed., *Thermal Energy Storage* (Conference of NATO Science Committee), Pergamon Press (Oxford, 1976)

H. A. Liebhafsky & E. J. Cairns, *Fuel Cells and Fuel Batteries: a guide to their research and development*, Wiley (New York, 1968)

I. Glendinning, 'Storing the Energy of Compressed Air', *CEGB Research*, no. 2, 1975, p. 21 (London)

'An Assessment of Energy Storage Systems Suitable for Use by Electric Utilities', EPRI EM-264, Project 225 or ERDA E (11-1)-2501, Final Report, v. 1, 2 and 3, July 1976 (report prepared by Public Service Electric and Gas Co., Newark, New Jersey 07101 for EPRI, Palo Alto, California 94304)

A. C. Gardner *et al.*, 'Storing Electrical Energy on a Large Scale', *CEGB Research*, no. 2, 1975, p. 12 (London)

'Economic Assessment of Utilization of Fuel Cells in Electric Utility Systems', EPRI EM-336, Project 729-1, Final Report, Nov. 1976 (report prepared by Public Service Electric and Gas Co., Newark,

CHAPTER NUMBER	REFERENCE

New Jersey 07101 for EPRI, Palo Alto, California 94304)

S. R. Shacket, *Electric Vehicles, past, present and future*, Quality Books (Warwick, 1977)

A. F. Anderson & A. Maeland, eds., *Hydrides for Energy Storage*, Pergamon Press (Oxford, 1978)

R. W. Cahn, 'Solar energy storage and conversion by hydrogen cycle', *Nature* **276**, 665, 1978

National Fuel Cell Seminar (Abstracts), July 11–13, San Francisco, 1978, coordinated by Courtesy Associates Inc.

B. C. Tofield, R. M. Dell and J. Jensen, 'Advanced Batteries', review article, *Nature* **276**, 1978, p. 217

B. C. Tofield, R. M. Dell and J. Jensen, 'Advanced batteries', *Nature* **276**, 217, 1978

22. C. B. Smith, ed., *Efficient Electricity Use*, Pergamon Press (Oxford, 1976)

23. British Plastics Federation, 'Effects of Increased Energy Costs on the Competitiveness of Plastics', Symposium Proceedings Publication no. 116/1, London, March 1975

The Watt Committee on Energy: Institution of Mechanical Engineers, *Deployment of national resources in the provision of energy in the UK, 1975–2025*. Report no. 2, August 1977

J. M. Hollander & M. K. Simmons, *Annual Review of Energy* v. 1, Annual Reviews (Palo Alto, 1976)

M. Slesser, 'The energy basis of development', *Energy World*, no. 9, Oct. 1974, p. 3

T. A. Markus, *Building Conversion and Rehabilitation*, Newnes-Butterworths (London, 1979)

24. E. G. Kovach, ed., *Technology of Efficient Energy Utilization* (Report of NATO Scientific Affairs Division Symposium, Brussels, 1973)

Energy Policy: A Consultative Document, Cmnd 7101, HMSO (London, 1978)

P. L. Cook & A. J. Surrey, *Energy Policy—Strategies for Uncertainty*, Martin Robertson (London, 1977)

M. Slesser, *Energy in the Economy*, Macmillan (London, 1978)

S. Baron, 'Solar Energy—will it conserve non-renewable resources?' *Public Utilities Fortnightly* (U.S.A.) **102**, 31, 1978

25. P. M. S. Jones, 'The Risks of Energy Production', *Atom* **262**, 1978, p. 223

The Royal Society, *Pollution in the Atmosphere: A Study Group Report* (London, 1978)

Lord Rothschild, 'Risk', Dimbleby Lecture, *The Listener* **100**, 1978, p. 715

CHAPTER NUMBER	REFERENCE

26. US National Research Council, *Reactor Safety Study: An Assessment of Accident Risks in US Commercial Nuclear Power Plants*, Wash-1400 (Washington, DC, 1975)
K. H. Puechl, 'The nuclear waste problem in perspective', Nuclear Eng. Int., Nov. 1975, p. 950
Royal Institution, *Nuclear Power and the Energy Future:* Forum, Oct. 1977, Royal Institution (London, 1977)
L. E. J. Roberts, *Radioactive Waste Disposal—Policy and Perspectives*, British Nuclear Eng. Soc., Nov. 1978, also *Atom* **267**, 9, 1979
R. F. Griffiths, 'Reactor Accidents and the Environment', *Atom* **266**, 314, 1978

27. W. B. Lewis, *Radioactive Waste Management in the Long Term*, Atomic Energy of Canada Ltd., DM-123, Ottawa, July 1971
J. B. Lewis, 'Treatment and Disposal of Liquid Waste in the Nuclear Power Industry', *Atom* **255**, 1978, p. 8
Windscale: Summary of The Evidence and The Argument, Guardian Newspapers (Manchester, 1977)
J. A. Reissland & G. W. Dolphin and also L. A. Sagan, 'Low Level Radiation Effects: The Mancuso Study', *Atom* **262**, 1978, p. 206
J. B. Lewis, 'Underground Waste Disposal', *Atom* **262**, 1978, p. 218
R. Greenberg, 'The growing concern over lung cancer and plutonium', *The Times*, May 5, 1977, p. 18
R. F. Griffiths, 'Reactor Accidents and the Environment', *Atom* **266**, 1978, p. 314
D. Burn, *Nuclear Power and the Energy Crisis*, Macmillan (London 1978)
'Annual Limits of Intake of Radionuclides for Workers', N.R.P.B. Report (based on ICRP recommendations), HMSO, 1978.

28. W. Marshall, 'Proliferation and the Recycling of Plutonium', *Atom* **263**, 1978, p. 234
J. S. Nye, 'Non-Proliferation: A Long-Term Strategy', *American Quarterly Review on Foreign Affairs*, Apl. 1978
V. Gilinsky, 'Nuclear Energy and Nuclear Proliferation', *Chemical and Engineering News*, Nov. 28, 1977

29. A. T. G. Ferguson & I. M. Blair, 'Conserving Uranium', *Nature* **271**, no. 5643, 1978, p. 300
W. Häfele *et al.*, 'Second Status Report of the IIASA Project on Energy Systems', RR-76-1 (Laxenburg, Austria, 1976)
P. Grand, 'The use of high energy accelerators in the nuclear fuel cycle', *Nature* **278**, 693, 1979

Subject and Author Index